CRC Series in Chromatography

Editors-in-Chief

Gunter Zweig, Ph.D. and Joseph Sherma, Ph.D.

General Data and Principles
Gunter Zweig, Ph.D. and
Joseph Sherma, Ph.D.

Lipids
Helmut K. Mangold, Dr. rer. nat.

Hydrocarbons
Walter L. Zielinski, Jr., Ph.D.

Carbohydrates
Shirley C. Churms, Ph.D.

Inorganics
M. Qureshi, Ph.D.

Drugs
Ram Gupta, Ph.D.

Phenols and Organic Acids
Toshihiko Hanai, Ph.D.

Terpenoids
Carmine J. Coscia, Ph.D.

Amino Acids and Amines
S. Blackburn, Ph.D.

Steroids
Joseph C. Touchstone, Ph.D.

Polymers
Charles G. Smith,
Norman E. Skelly, Ph.D.,
Carl D. Chow, and Richard A. Solomon

**Pesticides and Related
Organic Chemicals**
Joseph Sherma, Ph.D. and
Joanne Follweiler, Ph.D.

Plant Pigments
Hans-Peter Köst, Ph.D.

**Nucleic Acids and
Related Compounds**
Ante M. Krstulovic, Ph.D.

CRC Handbook of Chromatography: Peptides

Volume I

Author

Stanley Blackburn, Ph.D., C. Chem., F.R.S.C.

Leeds, England

CRC Series in Chromatography
Editors-in-Chief

Gunter Zweig, Ph.D.
President
Zweig Associates
Washington, D.C.

Joseph Sherma, Ph.D.
Professor of Chemistry
Lafayette College
Easton, Pennsylvania

CRC Press, Inc.
Boca Raton, Florida

7349. 1445

CHEMISTRY

Library of Congress Cataloging-in-Publication Data

Blackburn, S. (Stanley)
    Peptides.

    (CRC series of chromatography)
    Bibliography: v. 1,
    Includes index.
    1. Peptides--Analysis.   2. Chromatographic analysis.
I. Title.   II. Series.
QP552.P4B55   1986   547.7'56   86-6133
ISBN-0-8493-3065-3

    This book represents information obtained from authentic and highly regarded sources. Reprinted material is quoted with permission, and sources are indicated. A wide variety of references are listed. Every reasonable effort has been made to give reliable data and information, but the author and the publisher cannot assume responsibility for the validity of all materials or for the consequences of their use.

    All rights reserved. This book, or any parts thereof, may not be reproduced in any form without written consent from the publisher.

    Direct all inquiries to CRC Press, Inc., 2000 Corporate Blvd., N.W., Boca Raton, Florida, 33431.

© 1986 by CRC Press, Inc.
International Standard Book Number 0-8493-3065-3 (v. 1)

Library of Congress Card Number 86-6133
Printed in the United States

SERIES PREFACE

The present volume of the Handbook of Chromatography series on the subject of peptides by Dr. Stanley Blackburn is a continuation and extension of Dr. Blackburn's earlier book on amino acids and amines (CRC, 1983). We agree with Dr. Blackburn that the topic of peptide separation and identification has grown exponentially during the past 10 years since Volumes I and II of this series first appeared. There has been an actual revolution in the thinking and practice of protein characterization and amino acid sequence determination. Peptide identification and resolution have played a major role in these endeavors. We, as series editors, therefore, are grateful to the immense task that Dr. Blackburn has undertaken single-handedly to produce a most valuable and important volume in this series.

Other volumes of the Handbook of Chromatography that have previously appeared cover topics such as phenols and organic acids, drugs, terpenoids, pesticides, and lipids and fatty acids. Volumes that will appear in the near future will cover plant pigments, hydrocarbons, nucleic acid and derivatives, and inorganic ions and compounds.

We welcome suggestions and ideas from our faithful readers for other topics that should be included in the Handbook of Chromatography and prospective authors. Specific suggestions and corrections pertaining to this volume should be directed to the volume editor, Dr. Stanley Blackburn.

Gunter Zweig
Joseph Sherma

QP552
P4 B55
1986
v. 1
CHEM

PREFACE

The past few years have seen a tremendous increase in the number of publications describing the chromatography of peptides. This is due in part to the overall increase in chromatographic literature, reflected in a greater number of special journals, but other factors have also played a part. One such factor is the discovery of the opioid peptides and enkephalins, together with the need for efficient methods for their separation and estimation. Another is the introduction of greatly improved methods of peptide fractionation, such as the use of alkyl-bonded stationary phases for liquid chromatography. A new volume of the Handbook of Chromatography dealing with peptides is therefore timely.

The present book represents a logical extension of the *Handbook of Chromatography: Amino Acids and Amines* (Blackburn, 1983) to the topic of peptides. It supplements data given in Volume I of the Handbook of Chromatography and covers the literature since 1970. In addition to the tables of chromatographic data, sections of the volume describe methods of sample preparation and derivatization and detection reagents. A number of articles describing techniques applicable to peptide separation, including chromatographic procedures for peptide separation, fluorescence methods of peptide detection, methods of peptide mapping, the prediction of retention times and gel permeation chromatography supplement the information on methods and techniques presented in Volume II of the Handbook of Chromatography. A book directory gives information on sources for further detailed reading, while chromatographic materials listed in another section will direct the reader to commercial sources of supply.

S. Blackburn

THE EDITORS-IN-CHIEF

Gunter Zweig, Ph.D., received his undergraduate training at the University of Maryland, College Park, where he was awarded the Ph.D. in biochemistry in 1952. Two years following his graduation, Dr. Zweig was affiliated with the late R. J. Block, pioneer in paper chromatography of amino acids. Zweig, Block, and Le Strange wrote one of the first books on paper chromatography which was published in 1952 by Academic Press and went into three editions, the last one authored by Gunter Zweig and Dr. Joe Sherma, the co-Editor-in-Chief of this series. *Paper Chromatography* (1952) was also translated into Russian.

From 1953 to 1957, Dr. Zweig was research biochemist at the C. F. Kettering Foundation, Antioch College, Yellow Springs, Ohio, where he pursued research on the path of carbon and sulfur in plants, using the then newly developed techniques of autoradiography and paper chromatography. From 1957 to 1965, Dr. Zweig served as lecturer and chemist, University of California, Davis and worked on analytical methods for pesticide residues, mainly by chromatographic techniques. In 1965, Dr. Zweig became Director of Life Sciences, Syracuse University Research Corporation, New York (research on environmental pollution), and in 1973 he became Chief, Environmental Fate Branch, Environmental Protection Agency (EPA) in Washington, D.C. From 1980 to 1984 Dr. Zweig was Visiting Research Chemist in the School of Public Health, University of California, Berkeley, where he was doing research on farmworker safety as related to pesticide exposure.

During his government career, Dr. Zweig continued his scientific writing and editing. Among his works are (many in collaboration with Dr. Sherma) the now 11-volume series on *Analytical Methods for Pesticides and Plant Growth Regulators* (published by Academic Press); the pesticide book series for CRC Press; co-editor of *Journal of Toxicology and Environmental Health;* co-author of basic review on paper and thin-layer chromatography for *Analytical Chemistry* from 1968 to 1980; co-author of applied chromatography review on pesticide analysis for *Analytical Chemistry,* beginning in 1981.

Among the scientific honors awarded to Dr. Zweig during his distinguished career are the Wiley Award in 1977, Rothschild Fellowship to the Weizmann Institute in 1963/64; the Bronze Medal by the EPA in 1980.

Dr. Zweig has authored or co-authored over 80 scientific papers on diverse subjects in chromatography and biochemistry, besides being the holder of three U.S. patents. In 1985, Dr. Zweig became president of Zweig Associates, Consultants in Arlington, Va.

Joseph Sherma, Ph.D., received a B.S. in Chemistry from Upsala College, East Orange, N.J., in 1955 and a Ph.D. in Analytical Chemistry from Rutgers University in 1958. His thesis research in ion exchange chromatography was under the direction of the late William Rieman, III. Dr. Sherma joined the faculty of Lafayette College in September 1958, and is presently Charles A. Dana Professor and Head of the Chemistry Department. He is in charge of two courses in analytical chemistry, quantitative analysis and instrumental analysis. At Lafayette he has continued research in chromatography and had additionally worked a total of 14 summers in the field with Harold Strain at the Argonne National Laboratory, James Fritz at Iowa State University, Gunter Zweig at Syracuse University Research Corporation, Joseph Touchstone at the Hospital of the University of Pennsylvania, Brian Bidlingmeyer at Waters Associates, and Thomas Beesley at Whatman, Inc. and Advanced Separation Technologies, Inc.

Dr. Sherma and Dr. Zweig co-authored or co-edited the original Volumes I and II of the *CRC Handbook of Chromatography,* a book on paper chromatography, seven volumes of the series *Analytical Methods for Pesticides and Plant Growth Regulators,* and the Handbooks of Chromatography of drugs, carbohydrates, polymers, phenols and organic acids, amino acids and amines, pesticides, terpenoids, and lipids. Other books in the pesticide series and further volumes of the *CRC Handbook of Chromatography* are being edited with Dr. Zweig, and Dr. Sherma has co-authored the handbook on pesticide chromatography. Books on quantitative TLC and advances in TLC were edited jointly with Dr. Touchstone. A general book on TLC was written with Dr. Bernard Fried, the second edition of which is now being written. Dr. Sherma has been co-author of nine biennial reviews of column and thin layer chromatography (1968—1984) and the 1981, 1983, and 1985 reviews of pesticide analysis for the ACS journal *Analytical Chemistry*.

Dr. Sherma has written major invited chapters and review papers on chromatography and pesticides in *Chromatographic Reviews* (analysis of fungicides), *Advances in Chromatography* (analysis of nonpesticide pollutants), Heftmann's *Chromatography* (chromatography of pesticides), Race's *Laboratory Medicine* (chromatography in clinical analysis), *Food Analysis: Principles and Techniques* (TLC for food analysis), *Treatise on Analytical Chemistry* (paper and thin layer chromatography), *CRC Critical Reviews in Analytical Chemistry* (pesticide residue analysis), *Comprehensive Biochemistry* (flat bed techniques), *Inorganic Chromatographic Analysis* (thin layer chromatography), *Journal of Liquid Chromatography* (advances in quantitative pesticide TLC), and *Preparative Liquid Chromatography* (strategy of preparative TLC). Dr. Sherma is editor for residues and elements of the *AOAC* and is scientific coordinator of the AOAC.

Dr. Sherma spent 6 months in 1972 on sabbatical leave at the EPA Perrine Primate Laboratory, Perrine, Fla., with Dr. T. M. Shafik, two summers (1975, 1976) at the USDA in Beltsville, Md. with Melvin Getz doing research on pesticide residue analysis methods development, and one summer (1984) in the food safety research laboratory of the Eastern Regional Research Center, Philadelphia, with Daniel Schwartz doing research on the isolation and analysis of mutagens in cooked meat. He spent three months in 1979 on sabbatical leave with Dr. Touchstone developing clinical analytical methods. A total of more than 270 papers, books, book chapters, and oral presentations concerned with column, paper, and thin layer chromatography of metal ions, plant pigments, and other organic and biological compounds; the chromatographic analysis of pesticides; and the history of chromatography have been authored by Dr. Sherma, many in collaboration with various co-workers and students. His major research area at Lafayette is currently quantitative TLC (densitometry), applied mainly to clinical analysis and pesticide residue and food additive determinations.

Dr. Sherma has written an analytical quality control manual for pesticide analysis under contract with the USEPA and has revised this and the EPA Pesticide Analytical Methods Manual under a 4-year contract jointly with Dr. M. Beroza of the AOAC. Dr. Sherma has also written an instrumental analysis quality assurance manual and other analytical reports for the U.S. Consumer Product Safety Commission, and a manual on the analysis of food additives for the FDA, both of these projects as technical editor for the AOAC. He is preparing three additional FDA manuals on animal drug and food additives analysis, and analytical field operations.

Dr. Sherma taught the first prototype short course on pesticide analysis, with Henry Enos of the EPA, for the Center for Professional Advancement. He was editor of the Kontes TLC quarterly newsletter for 6 years and also has taught short courses on TLC for Knotes and the Center for Professional Advancement. He is a consultant for numerous industrial companies and federal agencies on chemical analysis and chromatography and regularly referees papers for analytical journals and research proposals for government agencies. At Lafayette, Dr. Sherma, in addition to analytical chemistry, teaches general chemistry and a course in thin layer chromatography.

Dr. Sherma has received two awards for superior teaching at Lafayette College and the 1979 Distinguished Alumnus Award from Upsala College for outstanding achievements as an educator, researcher, author and editor. He is a member of the ACS, Sigma Xi, Phi Lambda Upsilon, SAS, AIC, and AOAC.

THE AUTHOR

Dr. Stanley Blackburn gained both his Honours B.Sc. degree in Chemistry and his Ph.D. degree in Organic Chemistry at the University of Leeds, England. He is a former research scientist at the Wool Industries Research Association, Leeds, where his research interests included the development of chromatographic techniques, the structure and amino acid sequence of the proteins of wool keratin, and the end group determination of peptides and proteins.

Dr. Blackburn is a Chartered Chemist, a Fellow of the Royal Society of Chemistry, and a member of the Biochemical Society and the American Chemical Society. His current work is centered on scientific writing and documentation. He has written more than 30 scientific papers dealing with protein analysis and structure and is the author of several texts, including *Amino Acid Determination: Methods and Techniques*, *Protein Sequence Determination: Methods and Techniques*, and *Enzyme Structure and Function*, all published by Marcel Dekker. He is also the author of the recently published *CRC Handbook of Chromatography: Amino Acids and Amines*.

TABLE OF CONTENTS

Section I
Tables

Section I.I.

GAS CHROMATOGRAPHY TABLES

Wherever possible, tables are arranged according to classes of chemical compounds. This was not always possible when different chemical compounds were chromatographed under the same experimental conditions. The reader is referred to the compound index for specific compounds that may appear in different tables.

Table GC 1
TRIMETHYLSILYL DERIVATIVES OF
PEPTIDES

Packing	P1
Temperature	T1
Gas	He
Column	
Length (m)	12
Diameter (mm, I.D.)	0.2
Form	Capillary
Material	Fused silica
Detector	FID

Sequence peak no.	Dipeptide
1	Gly-Gly
2	L-Val-Gly
3	Gly-L-Leu
	L-Leu-Gly
4	L-Ala-L-Ala
5	Gly-L-Ser
6	Gly-L-Thr
7	L-Ala-L-Val
8	L-Ala-L-Leu
9	L-Ala-L-Ile
	L-Leu-L-Ala
10	α-L-Asp-Gly
11	Gly-L-Met
	L-Met-Gly
12	Gly-L-Asp
13	Gly-L-Ile
14	L-Val-L-Val
	L-Ser-L-Ala
15	DL-Ala-DL-Ser
16	DL-Ala-DL-Ser
17	L-Ala-L-Thr
18	Gly-L-Ser
19	L-Ala-L-Asn
20	Gly-L-Thr
21	Gly-L-Phe
	L-Phe-Gly
22	Gly-L-Glu
23	Gly-L-Val
	L-Leu-L-Leu
24	Gly-L-Leu
25	L-Ala-L-Asp
26	Gly-L-Ile
	L-Ala-L-Asn
27	L-Leu-L-Ser
28	L-Ala-L-Met
29	L-Met-L-Ala
30	L-Ala-L-Glu
31	L-Phe-L-Ala
32	L-Ala-L-Phe
33	γ-L-Glu-L-Leu
34	Gly-L-Asp
	L-His-Gly
35	L-Ser-L-Met
36	Gly-L-Met

Table GC 1 (continued)
TRIMETHYLSILYL DERIVATIVES OF PEPTIDES

Sequence peak no.	Dipeptide
37	L-Phe-L-Val
38	L-Val-L-Phe
39	DL-Leu-DL-Phe
40	Gly-L-Glu
41	DL-Leu-DL-Phe
42	L-Ser-L-Phe
43	Gly-L-Phe
44	L-His-L-Ala
45	L-Ala-L-His
46	L-Met-L-Met
47	L-Ala-L-Tyr
48	L-Tyr-L-Ala
49	L-Tyr-Gly
50	L-Trp-Gly
51	L-Met-L-Phe
52	L-His-L-Ser
53	Gly-L-His
54	L-Phe-L-Phe
55	L-Trp-L-Ala

Packing: Pl = SP-2100 (methyl silicone liquid phase) coating of capillary column.

Temperature: T1 = Programed at 4°C/min from 100 to 160°C, 6°C/min from 160 to 230°C and maintained at 230°C for 15 min.

REFERENCE

1. **Dizdaroglu, M. and Simic, M. G.**, *Anal. Biochem.*, 108, 269, 1980.

Reproduced by permission of Academic Press, Inc.

Table GC 2
TRIMETHYLSILYL DERIVATIVES OF DIPEPTIDE DIASTEREOMERS

Packing	P1
Temperature	T1
Gas	He
Column	
Length (m)	12
Diameter (mm, I.D.)	0.2
Form	Capillary
Material	Fused silica
Detector	FID

Sequence peak no. (see Figure GC 1)	Dipeptide
1	L-Ala-L-Ala
	D-Ala-D-Ala
2	L-Ala-D-Ala

Table GC 2 (continued)
TRIMETHYLSILYL DERIVATIVES OF
DIPEPTIDE DIASTEREOMERS

Sequence peak no. (see Figure GC 1)	Dipeptide
	D-Ala-L-Ala
3	L-Ala-L-Val
	D-Ala-D-Val
4	L-Ala-D-Val[a]
	D-Ala-L-Val[a]
5	L-Ala-L-Leu
	D-Ala-D-Leu[a]
6	L-Ala-D-Leu[a]
	D-Ala-L-Leu[a]
7	L-Ala-L-Ser
	D-Ala-D-Ser[a]
8	L-Ala-D-Ser[a]
	D-Ala-L-Ser[a]
9	L-Leu-L-Val
	D-Leu-D-Val[a]
10	L-Leu-D-Val[a]
	D-Leu-L-Val[a]
11	L-Leu-L-Leu
	D-Leu-D-Leu
12	L-Leu-D-Leu
	D-Leu-L-Leu
13	L-Ala-L-Phe
	D-Ala-D-Phe
14	D-Ala-L-Phe
15	L-Leu-L-Phe
	D-Leu-D-Phe[a]
16	L-Leu-D-Phe[a]
	D-Leu-L-Phe[a]
17	L-Leu-L-Tyr
18	D-Leu-L-Tyr

[a] The assignment of the peaks corresponding to these compounds was based on the order of elution of the other dipeptides.

Packing: P1 = SE-54 (5% phenyl, 1% vinylmethylsilicone gum, siloxane deactivated) coating a capillary column.

Temperature: T1 = Programed at 2°C/min from 70 to 150°C, then at 3°C/min from 150 to 250°C.

REFERENCE

1. **Dizdaroglu, M. and Simic, M. G.,** *J. Chromatogr.,* 244, 293, 1982.

Reproduced by permission of Elsevier Science Publishers B.V.

Table GC 3
RETENTION INDEX INCREMENTS OF AMINO ACID RESIDUES IN OLIGOPEPTIDE-DERIVED *O*-TRIMETHYLSILYLATED PERFLUORO-DIDEUTEROALKYL POLYAMINO ALCOHOLS

The amino acid sequencing of peptides with up to 30 amino acid residues can be carried out by gas chromatography/mass spectrometry (GC/MS). Key steps in the procedure are a partial hydrolysis to small peptides followed by derivatization, separation of the derivatives by GC, and finally identification by MS. The retention indexes of particular peptide derivatives can be related to the amino acid composition of the original oligopeptide and thus can be used as a reliable parameter for identification. The retention indexes of derivatives can be approximately calculated by adding increments which are assigned to each amino acid residue. The retention index increments are derived from the retention indexes of known oligopeptides.

Amino acid residue	Retention index increment
Ala	335
Gly	340
Val	425
Leu	475
Ile	475
Pro	495
Pro*	610
< Glu	610
Hyp	650
Ser	540
Thr	535
Asn	640
Asp	640
Gln	740
Glu	740
Met	825
Arg	705
Arg*	805
Lys	805
Phe	960
AEtCys	995
CMCys	1020
BzlCys	1285
Tyr	1225
His	1340
Trp	1660

Abbreviations: Pro*, derivative containing a pyrrolidine ring; Arg*, ω-trideuteromethyl ornithine derivative; < Glu, 2-pyrrolidone-5-carboxylic acid; AEtCys, *S*-aminoethyl cysteine; CMCys, *S*-carboxymethyl cysteine; BzlCys, *S*-benzyl cysteine.

REFERENCE

1. **Nau, H. and Biemann, K.**, *Anal. Biochem.*, 73, 139, 1976.

Reproduced by permission of Academic Press, Inc.

Table GC 4
$N^{\alpha,\epsilon}$-TRIFLUOROACETYL-N,O-PERMETHYLATED PEPTIDES

Column packing	P1	P2
Temperature	T1	T1
Gas	He	He
Column		
Length (m)[a]	8	8
Diameter (mm, I.D.)	0.31—0.34	0.31—0.34
Detection	D1	D1

Peptide	Retention index	
Thr-Ser	1772	1813
Leu-Met	2005	2054
Asn-Thr	2131	2200
Val-Gln	2152	2214
Asp-Phe	2235	2294
Trp-Leu	2560	2630
Gly-Thr-Phe	2582	2665
Lys-Tyr	2813	2888
Met-Asn-Thr	2959	3033
Ser-Lys-Tyr	3380	3461
Trp-Leu-Met	3420	3519
Val-Gln-Trp	3632	3753
Val-Gln-Trp-Leu	4282	4363

[a]   A column length of 5 m was used in conjunction with a film of thickness 0.4 μm.

Column packing: P1 = 0.12-μm thick film of CPSil 5 (100% methyl silicone).

P2 = 0.12-μm thick film of SE-54 (95% methyl silicone, 4% phenyl silicone, 1% vinyl silicone).

Temperature: T1 = 80°C, 1 min isothermal, programed at 4°C/min to 320°C.

Detection: D1 = Flame ionization detector.

REFERENCE

1. **Rose, K., Bairoch, A., and Offord, R. E.,** *J. Chromatogr.*, 268, 197, 1983.

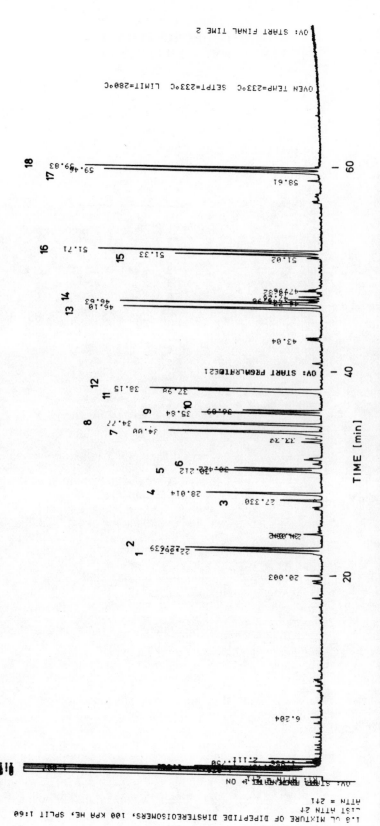

FIGURE GC1.   Separation of trimethylsilylated diastereomers of dipeptides by high resolution GC.   Column, fused silica SE-54, 12 m × 0.2 mm, I.D., programmed at 2°C/min from 70 to 150°C, then at 3°C/min from 150 to 250°C.   (See Table GC 2.)   Reference: Dizdaroglu, M. and Simic, M. G., *J. Chromatogr.*, 244, 293, 1982.   Reproduced by permission of Elsevier Science Publishers B.V.

Section I.II.

LIQUID CHROMATOGRAPHY TABLES

Wherever possible, tables are arranged according to classes of chemical compounds. This was not always possible when different chemical compounds were chromatographed under the same experimental conditions. The reader should refer to the compound index for specific compounds that may appear in different tables.

Table LC 1
ALANINE PEPTIDES

Packing	P1
Column	
Length (cm)	30
Diameter (cm)	0.436
Solvent	S1
Flow rate (mℓ/min)	0.50
Detection	D1

Peptide	k'
DL-Alanine	0.46
L-Alanyl-L-alanine	1.28
Tri-L-alanine	1.77
Tetra-L-alanine	1.99
Penta-L-alanine	2.47
Hexa-L-alanine	6.69

Packing: P1 =   Amberlite® XAD-4 resin, 45 to 65 μm.
Solvent: S1 =   Ethanol-water (10:90 v/v).
Detection: D1 =   UV at 254 or 208 nm.

REFERENCE

1. **Kroeff, E. P. and Pietrzyk, D. J.,** *Anal. Chem.,* 50, 502, 1978.

Reprinted with permission. Copyright 1978 American Chemical Society.

Table LC 2
HPLC OF ALANINE DIPEPTIDES AT DIFFERENT pH VALUES

Packing	P1	P1	P1	P1
Column				
Length (mm)	250	250	250	250
Diameter (mm)	3.2	3.2	3.2	3.2
Solvent	S1	S2	S3	S4
Flow rate (mℓ/min)	1.0	1.0	1.0	1.0
Detection	D1	D1	D1	D1

Dipeptide	k'			
L-Ala-L-Ala	0.28	0.17	0.00	0.08
D-Ala-D-Ala	0.29	0.18	0.00	0.07
L-Ala-D-Ala	1.13	0.65	0.16	0.21
D-Ala-L-Ala	1.11	0.64	0.17	0.20
Gly-Gly	0.00	0.00	0.00	0.00
L-Ala-Gly	0.14	0.07	0.00	0.04
D-Ala-Gly	0.14	0.08	0.00	0.04
Gly-L-Ala	0.26	0.15	0.00	0.03
Gly-D-Ala	0.26	0.15	0.00	0.03

Table LC 2 (continued)
HPLC OF ALANINE DIPEPTIDES AT
DIFFERENT pH VALUES

Packing: P1 = LiChrosorb® C₈, 10 μm.

Solvent: S1, S2, S3, and S4 are mixtures of ethanol-water (5:95 v/v), buffered with phosphate salts, 0.02 *M*, or HCl. Appropriate amounts of NaCl are added to give an ionic strength of 0.1. Their pH values are 2.10, 3.32, 5.75, and 7.89, respectively.

Detection: D1 = UV at 208 nm.

REFERENCE

1. **Kroeff, E. P. and Pietrzyk, D. J.,** *Anal. Chem.,* 50, 1353, 1978.

Reprinted with permission. Copyright 1978 American Chemical Society.

Table LC 3
HPLC OF TRIALANINE DIASTEREOMERS AT
DIFFERENT pH VALUES

Packing	P1	P1	P1	P1
Column				
Length (mm)	250	250	250	250
Diameter (mm)	3.2	3.2	3.2	3.2
Solvent	S1	S2	S3	S4
Flow rate (mℓ/min)	1.0	1.0	1.0	1.0
Detection	D1	D1	D1	D1
Tripeptide		*k'*		
L-Ala-L-Ala-L-Ala	0.68	0.62	0.09	0.26
D-Ala-D-Ala-D-Ala	0.69	0.63	0.09	0.27
L-Ala-L-Ala-D-Ala	1.59	1.14	0.22	0.59
L-Ala-D-Ala-L-Ala	3.32	2.58	0.69	1.18
Gly-Gly-Gly	0.05	0.00	0.00	0.00

Packing: P1 = LiChrosorb® C₈, 10 μm.

Solvent: S1, S2, S3, and S4 are mixtures of ethanol-water (5:95 v/v), buffered with phosphate salts, 0.02 *M*, or HCl. Appropriate amounts of NACl are added to give an ionic strength of 0.1. Their pH values are 2.10, 3.32, 5.75, and 7.83, respectively.

Detection: D1 = UV at 208 nm.

REFERENCE

1. **Kroeff, E. P. and Pietrzyk, D. J.,** *Anal. Chem.,* 50, 1353, 1978.

Reprinted with permission. Copyright 1978 American Chemical Society.

Table LC 4
POLYALANINE PEPTIDES

Packing	P1	P1	P1
Column			
Length (mm)	150	150	150
Diameter (mm)	4.1	4.1	4.1
Solvent	S1	S2	S3
Flow rate (mℓ/min)	1.0	1.0	1.0
Temperature (°C)	25	25	25
Detection	D1	D1	D1

Peptide		k′	
D,L-Ala	0.52	0.51	0.42
(L-Ala)$_2$	0.68	0.47	0.70
(L-Ala)$_3$	1.15	0.57	0.95
(L-Ala)$_4$	1.96	—	1.54
(L-Ala)$_5$	3.81	1.02	2.78
(L-Ala)$_6$	7.86	1.98	6.32

Packing: P1 = PRP-1, a 10-μm spherical microporous polystyrene-divinylbenzene copolymer.

Solvent: S1, S2, and S3 are 100% H_2O, 0.01 M phosphate buffer or HCl (pH 1.60) with NaCl added to give an ionic strength of 0.10 M. Their pH values are 1.75, 5.20, and 11.00, respectively.

Detection: D1 = UV at 208 or 254 nm.

REFERENCE

1. **Iskandarani, Z. and Pietrzyk, D. J.,** *Anal. Chem.,* 53, 489, 1981.

Reprinted with permission. Copyright 1981 American Chemical Society.

Table LC 5
DIPEPTIDES CONTAINING PHENYLALANINE

Packing	P1	P1	P1
Column			
Length (mm)	150	150	150
Diameter (mm)	4.1	4.1	4.1
Solvent	S1	S2	S3
Flow rate (mℓ/min)	1.0	1.0	1.0
Temperature (°C)	25	25	25
Detection	D1	D1	D1

Dipeptide		k′	
Phe-Gly	3.50	1.59	3.79
Phe-Ala	3.34	0.98	5.07
Phe-Val[a]	1.85	0.46	2.79
Phe-Leu[a]	5.79	1.65	7.17
Phe-Trp[a]	24.5	7.38	31.8
Gly-Phe	7.25	1.37	2.20
Ala-Phe	7.69	1.38	3.63

Table LC 5 (continued)
DIPEPTIDES CONTAINING PHENYLALANINE

Dipeptide	k'		
Val-Phe[a]	2.87	0.69	2.77
Leu-Phe[a]	7.19	2.13	6.66
Trp-Phe[a]	31.0	9.05	24.8
Phe-Ser	1.78	0.69	2.53
Phe-Thr	2.14	1.21	5.38
Ser-Phe	6.37	1.15	1.55
Thr-Phe	8.14	1.45	2.18
Phe-Arg	1.48	0.82	7.71
Arg-Phe	4.48	1.21	3.49
Phe-Asp	0.65	0.41	0.64
Phe-Tyr[a]	3.43	0.87	0.71
Asp-Phe	0.28	0.76	0.36
Tyr-Phe[a]	5.51	1.31	0.59

[a]   The acetonitrile-H_2O ratio was 13:87.

Packing: P1 = PRP-1, a 10-μm spherical microporous polystyrene-divinylbenzene copolymer.

Solvent: S1, S2, and S3 are mixtures of acetonitrile-H_2O (5:95 v/v), 0.010 *M* phosphate buffer or HCl (pH 1.60) with NaCl added to give an ionic strength of 0.10 *M*. Their pH values are 1.6, 5.00, and 11.00, respectively.

Detection: D1 = UV at 208 or 254 nm.

REFERENCE

1. **Iskandarani, Z. and Pietrzyk, D. J.,** *Anal. Chem.,* 53, 489, 1981.

Reprinted with permission. Copyright 1981 American Chemical Society.

Table LC 6
DIPEPTIDE DIASTEREOMERS

Packing	P1	P1
Column		
Length (mm)	250	250
Diameter (mm)	3.2	3.2
Solvent	S1	S2
Flow rate (mℓ/min)	1.0	1.0
Detection	D1	D1

Dipeptide	k'	
L(D)-Ala-L(D)-Val[a]	0.30	0.03
L(D)-Ala-D(L)-Val[a]	0.69	0.21
L-Ala-L-Leu	1.37	0.30
D-Ala-L-Leu	2.51	0.88
L-Ala-L-Phe	1.65	0.61
D-Ala-D-Phe	1.65	0.61
L-Ala-D-Phe	2.83	1.48
D-Ala-L-Phe	2.84	1.46
L-Leu-L-Tyr	1.74	0.62
D-Leu-L-Tyr	2.63	1.38
L-Val-L-Val	0.59	0.12
D-Val-D-Val	0.58	0.11

Table LC 6 (continued)
DIPEPTIDE DIASTEREOMERS

Dipeptide		k′
D-Val-L-Val	3.50	1.55
D-Leu-D-Leu	7.35	2.71
D-Leu-L-Leu	> 10	> 10

[a] Obtained as D,L-D,L mixture.

Packing: P1 = LiChrosorb® C_8, 10μm.
Solvent: S1 and S2 are mixtures of ethanol-water (10:90 v/v) buffered with phosphate salts, 0.02 *M*, or HCl. Appropriate amounts of NaCl are added to give an ionic strength of 0.1. Their pH values are 3.45 and 5.98, respectively.
Detection: D1 = UV at 208 nm.

REFERENCE

1. **Kroeff, E. P. and Pietrzyk, D. J.,** *Anal. Chem.*, 50, 1353, 1978.

Reprinted with permission. Copyright 1978 American Chemical Society.

Table LC 7
DI- AND TRIPEPTIDE
DIASTEREOMERS

Packing	P1
Column	
Length (cm)	250
Diameter (cm)	3.2
Solvent	S1
Flow rate (mℓ/min)	1.0
Detection	D1

Peptide	k′
L-Leu-L-Leu	0.92
D-Leu-D-Leu	0.92
L-Leu-D-Leu	4.78
D-Leu-L-Leu	4.79
L(D)-Leu-L(D)-Phe[a]	2.16
L(D)-Leu-D(L)-Phe[a]	6.94
L(D)-Leu-Gly-L(D)-Phe[a]	1.12
L(D)-Leu-Gly-D(L)-Phe[a]	1.58

[a] Obtained as D,L-D,L mixture.

Packing: P1 = LiChrosorb® C_8, 10 μm.
Solvent: S1 = ethanol-water (20:80 v/v), buffered to pH 6.4 with phosphate salts, 0.02 *M*. An appropriate amount of NaCl is added to give an ionic strength of 0.1.
Detection: D1 = UV at 208 nm.

Table LC 7 (continued)
DI- AND TRIPEPTIDE
DIASTEREOMERS

REFERENCE

1. **Kroeff, E. P. and Pietrzyk, D. J.**, *Anal. Chem.*, 50, 1353, 1978.

Reprinted with permission. Copyright 1978 American Chemical Society.

Table LC 8
TRIPEPTIDES

Packing	P1	P1	P1
Column			
Length (mm)	150	150	150
Diameter (mm)	4.1	4.1	4.1
Solvent	S1	S2	S3
Flow rate (mℓ/min)	1.0	1.0	1.0
Temperature (°C)	25	25	25
Detection	D1	D1	D1

Tripeptide		k′	
Gly-Gly-Gly	0.54	0.42	0.51
Gly-Gly-Ala	0.83	0.44	0.58
Gly-Gly-Val	5.05	0.90	1.50
Gly-Gly-Leu	24.4	3.20	6.00
Gly-Gly-Phe	20.0	10.6	20.2
Ala-Gly-Gly	0.62	0.46	0.68
Val-Gly-Gly	1.53	0.91	3.43
Leu-Gly-Gly	5.22	2.82	16.3
Phe-Gly-Gly	22.6	12.4	20.0

Packing: P1 = PRP-1, a 10-μm spherical microporous polystyrene-divinylbenzene copolymer.

Solvent: S1, S2, and S3 are 100% H_2O, 0.01 M phosphate buffer or HCl (pH 1.60) with NaCl added to give an ionic strength of 0.10 M. Their pH values are 1.56, 5.00, and 11.00, respectively.

Detection: D1 = UV at 208 or 254 nm.

REFERENCE

1. **Iskandarani, Z. and Pietrzyk, D. J.**, *Anal. Chem.*, 53, 489, 1981.

Reprinted with permission. Copyright 1981 American Chemical Society.

Table LC 9
DIASTEREOMERS

Packing	P1	P1	P1
Column			
Length (mm)	150	150	150
Diameter (mm)	4.1	4.1	4.1
Solvent	S1	S2	S3
Flow rate (mℓ/min)	1.0	1.0	1.0
Temperature (°C)	25	25	25
Detection	D1	D1	D1

Diastereomer		k'	
L-Ala-L-Ser, D-Ala-D-Ser	0.34	0.31	0.51
L-Ala-D-Ser, D-Ala-L-Ser	0.34	0.31	0.51
L-Ala-L-Met, D-Ala-D-Met	1.67	0.56	1.26
L-Ala-D-Met, D-Ala-L-Met	2.80	0.89	1.40
L-Ala-L-Leu, D-Ala-D-Leu	2.88	0.70	1.73
L-Ala-D-Leu, D-Ala-L-Leu	5.87	1.35	2.00
L-Ala-L-Phe, D-Ala-D-Phe	7.75	1.67	4.70
L-Ala-D-Phe, D-Ala-L-Phe	14.5	3.46	5.30
L-Leu-L-Ala, D-Leu-D-Ala[a]	0.62	0.60	0.69
L-Leu-D-Ala, D-Leu-L-Ala[a]	0.97	0.60	0.70
L-Leu-L-Val, D-Leu-D-Val[a]	1.11	0.67	1.19
L-Leu-D-Val, D-Leu-L-Val[a]	3.14	1.42	1.29
L-Leu-L-Leu, D-Leu-D-Leu[a]	2.51	1.11	2.21
L-Leu-D-Leu, D-Leu-L-Leu[a]	7.18	3.68	2.68
L-Leu-L-Phe, D-Leu-D-Phe[a]	5.29	2.25	5.01
L-Leu-D-Phe, D-Leu-L-Phe[a]	13.8	5.56	5.74
L-Ala-L-Ala-L-Ala[b]	1.15	0.57	0.95
L-Ala-L-Ala-D-Ala[b]	1.94	0.69	1.39
L-Ala-D-Ala-L-Ala[b]	3.64	1.18	1.74

[a] The acetonitrile-H_2O ratio was 15:85.
[b] The same as S1, S2, or S3 except that solvent was 100% H_2O.

Packing: P1 = PRP-1, a 10-μm spherical microporous polystyrene-divinylbenzene copolymer.
Solvent: S1, S2, and S3 are mixtures of acetonitrile-H_2O (5:95 v/v), 0.010 *M* phosphate buffer or HCl (pH 1.60) with NaCl added to give an ionic strength of 0.10 *M*.
Detection: D1 = UV at 208 or 254 nm.

REFERENCE

1. **Iskandarani, Z. and Pietrzyk, D. J.,** *Anal. Chem.,* 53, 489, 1981.

Reprinted with permission. Copyright 1981 American Chemical Society.

Table LC 10
HPLC OF PEPTIDES AT DIFFERENT pH VALUES

Packing	P1	P1	P1
Column			
Length (mm)	250	250	250
Diameter (mm)	3.2	3.2	3.2
Solvent	S1	S2	S3
Flow rate (mℓ/min)	1.0	1.0	1.0
Detection	D1	D1	D1
Peptide		k'	
Gly-Gly	0.00	0.00	0.00
Gly-L-Ala	0.01	0.00	0.00
Gly-L-Val	0.57	0.09	0.16
Gly-D-Val	0.58	0.09	0.16
Gly-L-Tyr	0.90	0.19	0.28
Gly-DL-Leu	2.45	0.56	0.82
Gly-DL-Phe	3.09	1.08	1.56
L-Ala-Gly	0.00	0.00	0.00
L-Ala-L-Val	0.55	0.09	0.29
L-Ala-L-Leu	2.52	0.49	1.22
L-Ala-L-Phe	3.57	1.12	2.67
L-Val-Gly	0.23	0.11	0.27
L-Val-L-Val	1.20	0.22	2.79
DL-Leu-Gly	0.99	0.60	1.10
L-Leu-L-Val	3.71	1.12	9.24
L-Leu-L-Tyr	5.17	1.53	10
L-Leu-L-Leu	> 10	6.45	> 10
L-Phe-Gly	2.21	1.30	2.69
L-Tyr-Gly	0.58	0.35	0.57
L-Phe-Gly-Gly	1.64	1.06	2.53
Gly-Gly-L-Phe	3.73	1.10	1.31

Packing: P1 = LiChrosorb® C$_8$, 10 μm.
Solvent: S1 = Ethanol-water (5:95 v/v), pH 3.38.
S2 = Ethanol-water (5:95 v/v), pH 5.82.
S3 = Ethanol-water (5:95 v/v), pH 7.91. The eluting ethanol-water mixtures were buffered with phosphate salts, 0.02 *M*, or HCl. Appropriate amounts of NaCl were added to give an ionic strength of 0.1.
Detection: D1 = UV at 208 nm.

REFERENCE

1. **Kroeff, E. P. and Pietrzyk, D. J.,** *Anal. Chem.*, 50, 1353, 1978.

Reprinted with permission. Copyright 1978 American Chemical Society.

Table LC 11
RETENTION OF SHORT CHAIN PEPTIDES ON
XAD-2 AS A FUNCTION OF pH

k′

Peptide	pH			
	3.68	5.70	7.40	10.35
Gly-DL-Phe	2.44	1.44	1.56	2.66
DL-Ala-DL-Phe	4.33	2.56	2.67	4.56
L-Val-L-Phe	5.89	3.11	7.78	21.2
DL-Leu-DL-Phe		10.67		
L-Ser-L-Phe	2.22	1.33	1.44	2.11
L-Pro-L-Phe	3.56	1.89	3.00	
L-Met-L-Phe	13.4	6.89	16.2	
L-Phe-Gly	1.89	1.56	2.67	4.11
L-Phe-L-Ala	1.78	1.22	3.22	5.89
L-Phe-L-Val	4.11	2.33	14.6	
L-Phe-Gly-Gly	1.56	1.33	2.56	3.78
Gly-Gly-L-Phe	3.00	1.44	1.56	2.00

Conditions: Column, 45 × 0.236 cm of Amberlite® XAD-2 resin, 45 to
              65 μm.
Flow rate: 0.50 mℓ/min. Phosphate buffers were at 0.1 *M* conc. in a solvent
              of ethanol-water (10:90 v/v).
Detection: UV at 254 or 208 nm.

REFERENCE

1. **Kroeff, E. P. and Pietrzyk, D. J.,** *Anal. Chem.,* 50, 502, 1978.

Reprinted with permission. Copyright 1978 American Chemical Society.

Table LC 12
RETENTION OF SHORT CHAIN PEPTIDES ON XAD-2 AS A FUNCTION OF ETHANOL-WATER CONCENTRATION

	k' (% ethanol)			
Peptide	10	15	20	30
Gly-DL-Phe	0.5	0.3	0.1	0.1
DL-Ala-DL-Phe	1.0	0.6		
L-Val-L-Phe	1.7	1.0		
DL-Leu-DL-Phe	7.3	5.1		
L-Ser-L-Phe	0.5	0.3		
L-Pro-L-Phe	2.2	1.4		
L-Met-L-Phe	4.6	3.1		
L-Phe-Gly	0.6	0.4	0.3	0.3
L-Phe-L-Ala	0.6	0.4	0.3	0.3
L-Phe-L-Val	1.4	0.8		
L-Phe-L-Leu	6.9	4.7	3.2	1.6
Gly-L-Tyr	0.1	0.1	0.1	0.1
Gly-DL-Phe	0.5	0.3	0.1	0.1
Gly-L-Trp	1.1	0.6	0.4	0.3
L-Phe-Gly-Gly	0.4	0.2	0.1	0
Gly-Gly-L-Phe	0.5	0.3	0.2	0.2
DL-Tyr	0.2	0.2	0.3	0.2
DL-Phe	0.6	0.6	0.5	0.3

Conditions: Column, 30 × 0.19 cm of Amberlite® XAD-2 resin, 45 to 65 μm.
Flow rate: 0.5 mℓ/min.
Detection: UV at 254 or 208 nm.

REFERENCE

1. **Kroeff, E. P. and Pietrzyk, D. J.,** *Anal. Chem.,* 50, 502, 1978.

Reprinted with permission. Copyright 1978 American Chemical Society.

Table LC 13
CHROMATOGRAPHY OF DIPEPTIDES ON A BONDED PEPTIDE STATIONARY PHASE

Packing	P1	P1	P1	P1	P1
Column					
Length (mm)	300	300	300	300	300
Diameter (mm, I.D.)	2.1	2.1	2.1	2.1	2.1
Solvent	S1	S2	S3	S4	S5
Detection	UV	UV	UV	UV	UV
Dipeptide			k'		
Gly-L-Trp	0.47	2.62	3.44	9.18	NE[a]
L-Trp-Gly	0.24	2.18	3.83	17.62	NE
Gly-L-Tyr	0.17	1.10	1.16	1.71	NE
L-Tyr-Gly	0.15	0.98	1.38	3.78	NE
Gly-L-Phe	0.22	1.28	1.35	1.83	NE
L-Phe-Gly	0.11	1.17	1.74	4.67	NE
L-Val-L-Phe	0.26	1.58	2.32	4.95	NE
L-Phe-L-Val	0.15	1.42	2.74	9.05	NE
L-Trp-L-Tyr	0.49	5.73	21.18	NE	NE
L-Trp-L-Trp	1.81	22.70	103.60	NE	NE
L-Tyr-L-Phe	0.36	2.54	8.03	NE	NE

[a]   NE, peaks did not elute in reasonable times.

Packing: P1 = The tripeptide L-Val-L-Ala-L-Pro was bonded to Partisil®-10 using the reagent Y-5918, 1-trimethoxysilyl-2-(4-chloromethylphenyl)-ethane. Elemental analysis showed 13.6% carbon.

Solvent: S1 = 1% citric acid/H_2O, pH 2.5, ionic strength 0.006 M.
　　　　　 S2 = 1% Na citrate/H_2O + HCl, pH 5.5, ionic strength 0.26 M.
　　　　　 S3 = 1% Na citrate/H_2O + citric acid, pH 7.4, ionic strength 0.20 M.
　　　　　 S4 = Distilled deionized water, pH about 5.5.
　　　　　 S5 = Methanol.

REFERENCE

1. **Fong, G. W.-K. and Grushka, E.,** *Anal. Chem.,* 50, 1154, 1978.

Reprinted with permission. Copyright 1978 American Chemical Society.

Table LC 14

CHROMATOGRAPHY OF DIPEPTIDES IN MOBILE PHASES HAVING VARIOUS pHs AND IONIC STRENGTHS

Packing	P1	P1	P1	P1	P1	P1	P1	P1	P1
Column									
Length (mm)	250	250	250	250	250	250	250	250	250
Diameter (mm, I.D.)	3	3	3	3	3	3	3	3	3
Solvent	S1	S2	S3	S4	S5	S6	S7	S8	S9
Detection	UV	UV	UV	UV	UV	UV	UV	UV	UV
Dipeptide					k'				
Gly-L-Trp	6.47	1.91	1.55	1.65	1.60	2.67	1.57	1.42	0.88
L-Trp-Gly	11.50	2.20	1.31	1.30	0.84	3.45	1.20	1.02	0.43
Gly-D-Phe	1.92	0.58	0.38	0.46	0.34	0.65	0.43	0.35	0.13
Gly-L-Phe	1.92	0.61	0.38	0.46	0.33	0.62	0.43	0.35	0.13
L-Phe-Gly	5.10	0.88	0.41	0.43	0.21	1.09	0.42	0.27	0.00
Gly-L-Tyr	1.87	0.46	0.32	0.36	0.26	0.55	0.30	0.26	0.13
L-Tyr-Gly	4.67	0.58	0.27	0.28	0.12	0.88	0.28	0.17	0.01
L-Phe-L-Val	9.42	1.43	0.73	0.72	0.48	2.38	0.60	0.50	0.12
L-Val-L-Phe	5.83	1.15	0.77	0.86	0.65	1.77	0.69	0.59	0.25
D-Leu-L-Tyr	2.50	1.09	0.82	0.87	0.72	1.11	0.72	0.63	0.34
L-Leu-L-Tyr	6.90	1.32	0.82	0.86	0.65	1.88	0.76	0.65	0.28
L-Tyr-L-Leu	13.5	1.61	0.80	0.90	0.67	2.56	0.70	0.57	0.26
L-Ile-L-Tyr	8.33	1.29	0.86	0.88	0.69	2.09	0.79	0.64	0.26
L-Tyr-L-Ile	12.9	1.45	0.73	0.80	0.60	2.59	0.70	0.54	0.24
L-Phe-L-Tyr		4.65	1.75	1.82	1.35	9.73	1.72	1.40	0.67
L-Tyr-L-Phe		4.66	1.89	1.92	1.56	10.1	1.83	1.31	0.76
L-Trp-L-Tyr		14.2	5.50	5.36	4.26		5.26	4.03	2.13
L-Trp-L-Trp		NE[a]	25.4	23.9				17.2	9.96

[a]   NE, solute did not elute in reasonable time.

Packing: P1 =   The tripeptide L-Val-L-Ala-L-Pro was bonded to Partisil®-10 using the reagent Y-5918, 1-trimethoxysilyl-2-(4-chloromethylphenyl)-ethane. Elemental analysis showed 12.3% carbon.

Table LC 14 (continued)

CHROMATOGRAPHY OF DIPEPTIDES IN MOBILE PHASES HAVING VARIOUS pHs AND IONIC STRENGTHS

Solvent: S1 = H$_2$O, pH 5.5.
S2 to S9 are citrate buffers with the following values of pH and ionic strength:

S2, pH 7.5, ionic strength 1.0 *M*.
S3, pH 5.0, ionic strength 1.0 *M*.
S4, pH 3.8, ionic strength 1.0 *M*.
S5, pH 2.1, ionic strength 1.0 *M*.
S6, pH 7.4, ionic strength 0.5 *M*.
S7, pH 5.2, ionic strength 0.5 *M*.
S8, pH 3.8, ionic strength 0.5 *M*.
S9, pH 2.2, ionic strength 0.5 *M*.

REFERENCE

1. **Fong, G. W.-K. and Grushka, E.**, *Anal. Chem.*, 50, 1154, 1978.

Reprinted with permission. Copyright 1978 American Chemical Society.

Table LC 15
DIPEPTIDES

Packing	P1	P1
Column		
Length (cm)	30	30
Diameter (mm, I.D.)	3	3
Material	SS	SS
Solvent	S1	S2
Temperature (°C)	21	20
Detection	D1	D1

Dipeptide	k′	
Trp-Phe	0.55	0.88
Val-Phe	1.90	1.34
Phe-Val	1.90	1.52
Trp-Tyr	0.39	0.62
Trp-Trp	0.55	0.86
Phe-Gly	0.92	1.32
Gly-Phe	0.88	0.96
Tyr-Gly	0.36	0.46
Gly-Tyr	0.40	0.44
Trp-Gly	0.38	0.80
Gly-Trp	0.56	0.70

Packing: P1 = BioSil A®.
Solvent: S1 = Sodium citrate-water (1:99 v/v), pH about 7.4.
S2 = Citric acid-water (1:99 v/v), pH about 2.5.
Detection: D1 = UV at 254 and 280 nm.

REFERENCE

1. **Kitka, E. J., Jr. and Grushka, E.,** *J. Chromatogr.,* 135, 367, 1977.

Reproduced by permission of Elsevier Science Publishers B.V.

Table LC 16
DIPEPTIDES ON A TRIPEPTIDE BONDED
STATIONARY PHASE

Packing	P1	P1	P1	P1
Column				
Length (cm)	30	30	30	30
Diameter (mm, I.D.)	3.8	3.8	3.8	3.8
Material	SS	SS	SS	SS
Solvent	S1	S2	S3	S4
Temperature (°C)	21	22	23	23
Detection	D1	D1	D1	D1
Dipeptide		k'		
Tyr-Phe	—[a]	0	2.18	3.89
Val-Phe	2.55	0.05	0.86	0.97
Phe-Val	4.05	0.05	0.82	1.20
Trp-Tyr	—[a]	0.25	7.59	10.7
Trp-Trp	—[a]	2.02	28.4	32.3
Phe-Gly	1.61	0.29		
Gly-Phe	1.01	0.43		
Tyr-Gly	1.48	0.24		
Gly-Tyr	1.00	0.33		
Trp-Gly	4.87	0.55		
Gly-Trp	3.93	0.79		

[a] Retention very long, peak could no longer be seen.

Packing: P1 = Tripeptide Gly-L-Val-L-Val bonded to BioSil A® through
reagent Y-5918 (1-trimethoxysilyl-2-chloromethylphenyl
ethane). Analysis showed 0.66% of nitrogen on the support.
Solvent: S1 = Sodium citrate-water (1:99 v/v), pH about 7.4.
S2 = Citric acid-water (1:99 v/v), pH about 2.5.
S3 = Citric acid-sodium citrate-water (0.5:0.5:99 v/v), pH about
4.2.
S4 = Methanol-water (1:3)-0.75% sodium citrate.
Detection: D1 = UV at 254 and 280 nm.

REFERENCE

1. **Kitka, E. J., Jr. and Grushka, E.**, *J. Chromatogr.*, 135, 367, 1977.

Reproduced by permission of Elsevier Science Publishers B.V.

Table LC 17

CHROMATOGRAPHY OF DIPEPTIDES ON A TRIPEPTIDE BONDED STATIONARY PHASE

Packing	P1	P1	P1
Column			
Length (mm)	250	250	250
Diameter (mm, I.D.)	2.1	2.1	2.1
Material	SS	SS	SS
Solvent	S1	S2	S3
Detection	D1	D1	D1

Dipeptide		k'	
Gly-Trp	0.60	11.4	24.1
Trp-Gly	0.31	14.6	NE[a]
Gly-Phe	0.25	3.62	5.78
Phe-Gly	0.08	4.62	22.5
Gly-Tyr	0.15	3.43	3.27
Tyr-Gly	0.13	4.38	10.8
Phe-Val	0.39	7.95	59.2
Val-Phe	0.43	7.90	28.0
Trp-Tyr	0.90	3.52	NE
Trp-Trp	3.4;	NE	NE
Tyr-Phe	0.38	NE	NE

[a] NE, solute did not elute in reasonable time.

Packing: P1 = The tripeptide L-Val-L-Phe-L-Val bonded to Partisil®-10 using the reagent, Y-5918, 1-trimethoxysilyl-2-chloromethylphenyl ethane.

Solvent: S1 = Citric acid in water (1:99 v/v), pH 2.5.
         S2 = Sodium citrate in water (1:99 v/v), pH 7.9.
         S3 = Distilled water, pH 5.5.

REFERENCE

1. **Fong, G. W.-K. and Grushka, E.,** *J. Chromatogr.*, 142, 299, 1977.

Reproduced by permission of Elsevier Science Publishers B.V.

Table LC 18
DIPEPTIDES ON A TRIPEPTIDE BONDED STATIONARY PHASE

Packing	P1	P1	P1	P1	P1	P1
Column						
Length (cm)	25	25	25	25	25	25
Diameter (mm, I.D.)	3.8	3.8	3.8	3.8	3.8	3.8
Material	SS	SS	SS	SS	SS	SS
Solvent	S1	S2	S3	S4	S5	S6
Temperature (°C)	21	22	18	19	17	21
Detection	D1	D1	D1	D1	D1	D1

Dipeptide			k'			
Tyr-Phe	6.13	3.42	1.96	0.38	0.34	3.22
Val-Phe	1.06	1.04	0.43	0.06	0.10	0.96
Phe-Val	1.06	1.04	0.43	0.06	0.03	0.86
Trp-Tyr	11.9	10.3	5.89	1.07	0.52	9.39
Trp-Trp	21.8	31.1	28.3	4.80	4.67	41.7
Phe-Gly	0.34	0.63	0.28	0.03	0.07	0.64
Gly-Phe	0.34	0.63	0.41	0.10	0.14	0.52
Tyr-Gly	0.25	0.46	0.23	0	0.02	0.67
Gly-Tyr	0.20	0.46	0.29	0.01	0.07	0.52
Trp-Gly	1.81	2.0	1.43	0.20	0.24	2.22
Gly-Trp	1.61	2.0	1.99	0.43	0.39	1.97

Packing: P1 = The tripeptide L-Val-L-Ala-L-Ser bonded onto silica gel CT through reagent Y-5918. Analysis showed 1.16% of nitrogen on the support.

Solvent: S1 = Distilled water.
           S2 = Sodium citrate-water (1:99 v/v), pH about 7.4.
           S3 = Citric acid-sodium citrate-water (0.5:0.5:99 v/v), pH about 4.2.
           S4 = Citric acid-water (1:99 v/v), pH about 2.5.
           S5 = Citric acid-water (2:98 v/v), pH about 2.0.
           S6 = KH_2PO_4-$(NH_4)_2CO_3$-water (0.25:0.25:99.5 v/v), pH about 7.6.

Detection: D1 = UV at 254 and 280 nm.

REFERENCE

1. **Kitka, E. J., Jr. and Grushka, E.,** *J. Chromatogr.,* 135, 367, 1977.

Reproduced by permission of Elsevier Science Publishers B.V.

Table LC 19
HPLC OF PEPTIDES ON PHENYL CORASIL

Packing	P1	P1	P1	P1	P1
Column					
Length (ft)	3	3	3	3	3
Diameter (in.)	$1/_8$	$1/_8$	$1/_8$	$1/_8$	$1/_8$
Material	SS	SS	SS	SS	SS
Solvent	S1	S2	S3	S4	S5
Flow rate (mℓ/min)	1.0	1.0	1.0	1.0	1.0
Temperature	rt	rt	rt	rt	rt
Detection	D1	D1	D1	D1	D1

Peptide			k'		
< Glu-His-OH	0.3			0	
< Glu-Ser-Gly-OH	0	0		0	0
< Glu-Ser-Asp-OH	0				
< Glu-His-Trp-OH	4.7	0.9	0.4	0.1	
< Glu-His-Pro-Gly-OH	1.6	0.2	0.2	0	
< Glu-His-Trp-Lys-Tyr-Pro-OH	15		15	3.4	1.3
< Glu-His-Tyr-Trp-Lys-Pro-OH	15		20	6.8	1.2
< Glu-Ser-Tyr-Gly-Leu-Arg-Pro-Gly-OH	10	2.8	0.3	0	0.2
<Glu-His-Trp-Ser-Tyr-Gly-Leu-Arg-Pro-Gly-NH$_2$(LHRH)			20	20	20

Packing: P1 = Bondapak® Phenyl-Corasil.
Solvent: S1 = Water.
S2 = Acetonitrile-water (10:90 v/v).
S3 = Acetonitrile-water (20:80 v/v).
S4 = Acetonitrile-water (40:60 v/v).
S5 = Acetonitrile-water (70:30 v/v).
Detection: D1 = UV at 254 nm or 220 nm.

REFERENCE

1. **Hansen, J. J., Greibrokk, T., Currie, B. L., Johansson, K. N.-G., and Folkers, K.,** *J. Chromatogr.,* 135, 155, 1977.

Reproduced by permission of Elsevier Science Publishers B.V.

Table LC 20
HPLC OF PEPTIDES ON PORAGEL® PN

Packing	P1	P1	P1	P1	P1
Column					
Length (ft)	3	3	3	3	3
Diameter (in.)	$^1/_8$	$^1/_8$	$^1/_8$	$^1/_8$	$^1/_8$
Material	SS	SS	SS	SS	SS
Solvent	S1	S2	S3	S4	S5
Flow rate (mℓ/min)	1.0	1.0	1.0	1.0	1.0
Temperature	rt	rt	rt	rt	rt
Detection	D1	D1	D1	D1	D1

Peptide			k'		
< Glu-His-OH	0.9			0	
< Glu-Ser-Gly-OH	0.5		0	0	0
< Glu-Ser-Asp-OH	0.3			0	
< Glu-His-Trp-OH	15	4.2	0.6	0	
< Glu-His-Pro-Gly-OH	1.7		0	0	
< Glu-His-Tyr-Trp-Lys-Pro-OH	20			0.3	0.5
< Glu-His-Trp-Ser-Tyr-Gly-Leu-Arg-Pro-Gly-NH$_2$(LHRH)				10	10

Packing: P1 = Poragel® PN, 35-75 μm.
Solvent: S1 = Water.
             S2 = Acetonitrile-water (10:90 v/v).
             S3 = Acetonitrile-water (20:80 v/v).
             S4 = Acetonitrile-water (40:60 v/v).
             S5 = Acetonitrile-water (70:30 v/v).
Detection: D1 = UV at 254 nm or 220 nm.

REFERENCE

1. **Hansen, J. J., Greibrokk, T., Currie, B. L., Johansson, K. N.-G., and Folkers, K.,** *J. Chromatogr.,* 135, 155, 1977.

Reproduced by permission of Elsevier Science Publishers, B.V.

Table LC 21
HPLC OF PEPTIDES ON PORAGEL® PS

Packing	P1	P1	P1	P1
Column				
Length (ft)	3	3	3	3
Diameter (in.)	$1/_8$	$1/_8$	$1/_8$	$1/_8$
Material	SS	SS	SS	SS
Solvent	S1	S2	S3	S4
Flow rate (mℓ/min)	1.0	1.0	1.0	1.0
Temperature	rt	rt	rt	rt
Detection	D1	D1	D1	D1
Peptide			k'	
< Glu-Ser-Gly-OH	20		10	15
< Gly-Ser-Asp-OH			20	15
< Glu-His-Pro-Gly-OH		7.1	1.2	4.4
< Glu-Ser-Tyr-Gly-Leu-Arg-Pro-Gly-OH		2.3	0.1	0.1
< Glu-His-Tyr-Ser-Trp-Gly-Leu-Arg-Pro-Gly-OH		20	0	0
< Glu-His-Trp-Ser-Tyr-Gly-Leu-Arg-Pro-Gly-NH$_2$(LHRH)	20		0.2	2.1

Packing: P1 = Poragel® PS, 35-75 μm.
Solvent: S1 = Water.
             S2 = Acetonitrile-water (20:80 v/v).
             S3 = Acetonitrile-water (40:60 v/v).
             S4 = Acetonitrile-water (70:30 v/v).
Detection: D1 = UV at 254 nm or 220 nm.

REFERENCE

1. **Hansen, J. J., Greibrokk, T., Currie, B. L., Johansson, K. N.-G., and Folkers, K.,** *J. Chromatogr.,* 135, 155, 1977.

Reproduced by permission of Elsevier Science Publishers B.V.

Table LC 22
HPLC OF PEPTIDES ON HYDROGEL IV

Packing	P1	P1
Column		
Length (ft)	3	3
Diameter (in.)	$^1/_8$	$^1/_8$
Material	SS	SS
Solvent	S1	S2
Flow rate (mℓ/min)	1.0	1.0
Temperature	rt	rt
Detection	D1	D1

Peptide	MW	k$'$	
< Glu-Ser-Gly-OH	273	0.3	1.3
< Glu-His-Trp-OH	452	10	1.7
< Glu-His-Pro-Gly-OH	420	0.9	1.3
< Glu-His-Tyr-Trp-Lys-Pro-OH	840	20	7.5

Packing: P1 = Hydrogel IV, 35—75 μm.
Solvent: S1 = Water.
          S2 = Acetonitrile-water (40:60 v/v).
Detection: D1 = UV at 254 nm or 220 nm.

REFERENCE

1. **Hansen, J. J., Greibrokk, T., Currie, B. L., Johansson, K. N.-G., and Folkers, K.,** *J. Chromatogr.,* 135, 155, 1977.

Reproduced by permission of Elsevier Science Publishers B.V.

Table LC 23
REVERSED-PHASE CHROMATOGRAPHY OF L,L-DIPEPTIDES

Packing	P1	P2	P3
Column			
Length (mm)	250	250	300
Diameter (mm)	2.6	4.6	4.0
Solvent	S1	S1	S1
Temperature	rt	rt	rt
Detection	D1	D1	D1

Dipeptide	k$'$		
Ala-Ala	1.00	0.40	0.54
Ala-Asn	0.94	0.40	0.42
Ala-Asp	0.38	0.29	7.28
Ala-Glu	0.44	0.29	5.48
Ala-Gly	0.94	0.40	0.54
Ala-His[a]	3.63	0.52	0.70
Ala-Lys[a]	3.25	0.31	
Ala-Met	1.25	0.63	0.68
Ala-Pro	1.44	0.54	
Ala-Ser	0.94	0.40	0.54
Ala-Thr		0.40	0.70

Table LC 23 (continued)
REVERSED-PHASE CHROMATOGRAPHY
OF L,L-DIPEPTIDES

Dipeptide		k′	
Ala-Tyr	0.88	0.71	0.90
Ala-Val	1.18	0.52	0.35
Glu-Ala	0.44	0.25	0.34
Gly-Ala	1.06	0.35	0.56
Gly-Gly	0.94	0.35	0.56
Gly-Gly-NH$_2$	2.75	0.52	0
His-Ala[a]	3.00	0.42	1.50
His-Gly[a]	2.38	0.50	1.68
His-Leu[a]	6.00	0.79	2.52
His-Lys[a]		0.77	0.62
His-Phe[a]	6.00	1.46	3.16
His-Ser[a]	3.50	0.50	2.28
His-Tyr[a]	3.00	0.98	3.20
Leu-Leu	2.06		1.00
Leu-Met	1.94	0.85	1.06
Leu-Phe	2.13	2.00	0.80
Leu-Ser	1.19	0.48	0.78
Leu-Trp	1.81	3.90	1.78
Leu-Tyr	1.19	1.13	1.02
Lys-Ala[a]	3.81	0.38	0.06
Met-Ala	1.19	0.48	1.04
Phe-Ala	1.38	0.81	1.38
Phe-Phe	1.88	4.56	2.44
Pro-Ala		0.40	1.00
Pro-Gly	1.69	0.43	1.08
Pro-Ile	2.50	0.65	1.38
Pro-Leu	2.63	0.71	2.46
Pro-Met	1.94	0.67	1.26
Pro-Trp	2.19	2.43	2.42
Ser-Ala	0.88	0.38	0.92
Trp-Ala	1.19	1.69	1.66
Tyr-Ala	0.88	0.63	1.20
Tyr-Tyr	0.88	1.58	2.70

Dose-dependent; data for approximately 20-μg doses.

Packing: P1 = Phenyl-Sil-X-I, 13-μm porous particles.
P2 = Nucleosil® 5 CN, 5-μm porous particles.
P3 = μBondapak® NH$_2$, 10-μm porous particles.
Solvent: S1 = 0.01 *M* ammonium acetate.
Detection: D1 = UV and/or RI.

REFERENCE

1. **Lundanes, E. and Greibrokk, T.,** *J. Chromatogr.*, 149, 241, 1978.

Reproduced by permission of Elsevier Science Publishers B.V.

Table LC 24
REVERSED-PHASE CHROMATOGRAPHY OF
L,L-DIPEPTIDES ON SPHERISORB® ODS

Packing	P1	P1	P1
Column			
Length (mm)	250	250	250
Diameter (mm)	4.6	4.6	4.6
Solvent	S1	S2	S3
Temperature	rt	rt	rt
Detection	D1	D1	D1
Dipeptide		k′	
Ala-Asp	0.36	0.36	0.36
Ala-Glu	0.36	0.36	0.36
Ala-Asn	0.79	0.88	0.94
Ala-Thr	0.73	0.94	0.97
Ala-Gly	0.73	0.79	0.94
Ala-Ser	0.73	0.76	0.94
Ala-Ala	0.79	0.82	0.91
Ala-Val	0.82	1.00	1.58
Ala-Pro	1.06	1.42	2.06
Ala-Tyr	0.94	1.73	3.15
Ala-Leu	1.33	1.76	3.30
Ala-Ile	1.12	1.58	
Ala-Met	1.18	2.09	4.88
Ala-NorVal	1.21	2.00	4.21
Ala-His[a]	2.12	3.24	
Ala-Lys[a]	2.61	3.39	
Ala-Phe	1.70	3.45	
Ala-Gly-NH$_2$[a]	1.91	3.45	
Gly-Gly-NH$_2$[a]	1.85	3.03	
Gly-Gly	0.73	0.79	0.97
Gly-His[a]	2.42	3.61	
His-Gly[a]	2.64	3.21	
His-Ala[a]	2.42	3.40	
His-Leu[a]	6.3	11.4	
His-Lys[a]	5.2	7.5	
His-Phe[a]	9.9	17.7	
His-Ser[a]	3.15	3.33	
His-Tyr[a]	5.72	11.4	
Leu-Ala	1.09	1.70	3.45
Leu-Leu	2.97		
Leu-Met	1.82		
Leu-Phe	5.67		
Leu-Ser	0.97	1.36	2.30
Leu-Trp	5.42		
Leu-Tyr	1.61		
Leu-Val	1.48	3.24	
Phe-Phe	7.36		
Pro-Gly	0.82	1.21	2.03
Pro-Leu	1.79	3.42	
Pro-Ile	1.61	2.88	
Pro-Met	1.27	2.24	
Pro-Trp	3.00	12.2	
Tyr-Tyr	1.27	4.52	
Glu-Ala	0.39	0.36	0.36
Gly-Ala	0.67	0.85	0.94
Ser-Ala	0.67	0.76	0.88
Pro-Ala	0.85	1.09	1.55

Table LC 24 (continued)
REVERSED-PHASE CHROMATOGRAPHY OF L,L-DIPEPTIDES ON SPHERISORB® ODS

Dipeptide	k'		
Met-Ala	0.94	1.21	1.94
Leu-Ala	1.09	1.76	3.45
Tyr-Ala	0.94	1.79	3.94
Phe-Ala	1.48	3.06	
Lys-Ala[a]	3.30	3.48	
Trp-Ala	2.03	5.85	

[a] Dose-dependent; data for approximately 20-μg amounts.

Packing: P1 = Spherisorb® S5W-ODS, 5-μm porous particles.
Solvent: S1 = 0.01 *M* ammonium acetate in 30% methanol.
S2 = 0.01 *M* ammonium acetate in 10% methanol.
S3 = 0.01 *M* ammonium acetate in water.
Detection: D1 = UV and/or RI.

REFERENCE

1. **Lundanes, E. and Greibrokk, T.,** *J. Chromatogr.*, 149, 241, 1978.

Reproduced by permission of Elsevier Science Publishers B.V.

Table LC 25
REVERSED-PHASE CHROMATOGRAPHY OF L,L-DIPEPTIDES ON ODS-HYPERSIL®

Packing	P1	P1	P1
Column			
Length (mm)	250	250	250
Diameter (mm)	4.6	4.6	4.6
Solvent	S1	S2	S3
Temperature	rt	rt	rt
Detection	D1	D1	D1

Dipeptide	k'		
Ala-Ala	0.44	0.58	0.69
Ala-Asn	0.53	0.64	0.67
Ala-Asp	0.36	0.36	0.39
Ala-Glu	0.39	0.36	0.36
Ala-Gly	0.47	0.58	0.61
Ala-Gly-NH₂	0.61	0.81	0.92
Ala-His	0.53	0.75	0.86
Ala-Ile	0.83	1.56	2.94
Ala-Leu	0.89	1.78	3.83
Ala-Lys	0.53	0.72	0.69
Ala-Met	0.83	2.14	7.03
Ala-NorVal	0.81	2.08	6.03
Ala-Phe	1.28	4.00	9.0
Ala-Ser	0.50	0.58	0.58

Table LC 25 (continued)
REVERSED-PHASE CHROMATOGRAPHY
OF L,L-DIPEPTIDES ON ODS-HYPERSIL®

Dipeptide	k′		
Ala-Pro	0.58	0.69	0.92
Ala-Thr	0.55	0.61	0.58
Ala-Trp	1.39	6.31	
Ala-Val	0.64	0.83	1.17
Ala-Tyr	0.69	1.42	3.19
Arg-Asp	0.44	0.58	0.64
Glu-Ala	0.38	0.36	0.36
p-Glu-His	0.47	0.81	1.83
Gly-Ala	0.50	0.58	0.61
Gly-Asn	0.53	0.58	0.58
Gly-Asp	0.38	0.36	0.36
Gly-Glu	0.36	0.36	0.36
Gly-Gly	0.47	0.58	0.58
Gly-Gly-NH$_2$	0.58	0.75	1.25
Gly-His	0.58	0.75	0.92
Gly-Ile	0.92	1.69	3.44
Gly-Leu	0.97	2.03	4.11
Gly-Leu-NH$_2$	1.64	4.94	16.2
Gly-Met	0.69	1.08	2.14
Gly-NorLeu	1.11	2.22	4.75
Gly-NorVal	0.69	1.00	1.61
Gly-Phe	1.31	3.81	11.6
Gly-Pro	0.58	0.69	1.06
Gly-Ser	0.47	0.56	0.67
Gly-Thr	0.47	0.58	0.72
Gly-Trp	1.39	6.14	
Gly-Tyr	0.64	1.42	3.69
Gly-Val	0.64	0.92	1.50
His-Ala	0.50	0.67	1.14
His-Gly	0.53	0.69	1.33
His-Leu	1.08	2.97	11.8
His-Lys	0.64	0.92	1.58
His-Phe	1.56	7.33	
His-Ser	0.50	0.67	0.92
His-Tyr	0.72	2.06	9.5
Leu-Ala	0.78	1.44	4.11
Leu-Gly	0.92	2.31	7.11
Leu-Leu	3.76	13.00	
Leu-Met	1.53	7.67	
Leu-Phe	6.97		
Leu-Ser	0.69	1.14	2.14
Leu-Trp	5.28		
Leu-Tyr	1.22	6.56	
Leu-Val	1.31	4.47	
Lys-Ala	0.64	0.78	0.83
Lys-Asp	0.39	0.56	0.58
Lys-Gly	0.56	0.72	0.81
Lys-Leu	1.11	1.86	4.78
Lys-Lys	0.72	0.78	1.00
Lys-Phe	1.42	4.11	14.7
Met-Ala	0.61	1.06	2.03
Met-Asn	0.61	0.83	1.58
Met-Glu	0.47	0.58	0.83
Met-Gly	0.69	1.19	3.06

Table LC 25 (continued)
REVERSED-PHASE CHROMATOGRAPHY
OF L,L-DIPEPTIDES ON ODS-HYPERSIL®

Dipeptide	k'		
Met-Leu	1.97	9.0	
Met-Met	1.11	4.03	22
Met-Phe	3.64	16	
Met-Ser	0.56	0.86	1.53
Met-Val	0.92	2.39	12.9
Phe-Ala	1.22	3.94	15.4
Phe-Gly	1.39	4.67	
Phe-Phe	12.7		
Pro-Ala	0.58	0.64	0.67
Pro-Gly	0.56	0.69	1.14
Pro-Ile	0.97	1.86	5.56
Pro-Leu	1.08	2.31	7.72
Pro-Met	0.75	1.28	4.00
Pro-Phe	1.58	5.33	
Pro-Phe-NH$_2$	2.92	5.67	
Pro-Trp	1.67	6.89	
Pro-Tyr	0.75	1.47	5.17
Pro-Val	0.72	1.06	1.89
Ser-Ala	0.47	0.58	0.67
Ser-Gly	0.53	0.56	0.64
Ser-Leu	0.97	1.61	3.28
Ser-Phe	1.22	3.14	10.2
Trp-Ala	1.47	5.67	
Trp-Gly	1.69	8.4	
Tyr-Ala	0.72	1.39	5.11
Tyr-Gly	0.81	1.89	8.3
Tyr-Tyr	1.00	5.1	
Val-Ala	0.64	0.78	1.39
Val-Gly	0.69	1.03	2.22

Abbreviation: *p*-Glu, pyrrolidone carboxylic acid.

Packing: P1   = ODS-Hypersil®, 5-μm porous particles.
Solvent: S1   = 0.01 *M* ammonium acetate in 30% meth-
                    anol.
           S2   = 0.01 *M* ammonium acetate in 10% meth-
                    anol.
           S3   = 0.01 *M* ammonium acetate in water.
Detection: D1   = UV and/or RI.

REFERENCE

1. **Lundanes, E. and Greibrokk, T.**, *J. Chromatogr.*, 149, 241, 1978.

Reproduced by permission of Elsevier Science Publishers B.V.

Table LC 26
DIASTEREOMERS OF DIPEPTIDES

Packing		P1	P1	P1	P1	P1	P2	P2	P2
Column									
Length (mm)		250	250	250	250	250	150	150	150
Diameter (mm)		4.6	4.6	4.6	4.6	4.6	4.6	4.6	4.6
Solvent		S1	S2	S3	S4	S5	S2	S4	S5
Temperature		rt	rt	rt	rt	rt	rt	rt	rt
Detection		D1	D1	D1	D1	D1	D1	D1	D1

Dipeptide	Configuration				k′				
Ala-Ala	L,L + D,D	0.70	0.70	0.73	0.82	0.91	0.42	0.58	0.67
	L,D + D,L					1.24	0.42	0.72	1.08
Ala-Leu	L,L + D,D	0.97	1.33	1.36	1.76	3.30	0.89	1.78	
	L,D + D,L	1.12	1.94	2.36	3.88	9.82	1.44	5.28	
Ala-Val	L,L + D,D	0.82	0.89	0.94	1.00	1.58	0.61	0.83	1.17
	L,D + D,L		1.09	1.39	1.88	3.94	0.83	1.86	
Leu-Ala	L,L + D,D	0.85	1.06	1.33	1.80	3.27	0.78	1.39	4.11
	L,D + D,L	1.06	1.79	2.70	4.36	12.64	1.47	5.67	
Leu-Leu	L,L + D,D	1.42	2.97				3.39	13	
	L,D + D,L	2.67	10.63				13.8		
Leu-Phe	L,L + D,D	1.85	5.67				6.97		
	L,D + D,L	3.06	15.00				13		

Packing: P1   =   Spherisorb® S5W-ODS, 5 μm porous particles.
        P2   =   ODS-Hypersil®, 5 μm porous particles.
Solvent: S1   =   0.01 *M* ammonium acetate in 50% methanol.
        S2   =   0.01 *M* ammonium acetate in 30% methanol.
        S3   =   0.01 *M* ammonium acetate in 20% methanol.
        S4   =   0.01 *M* ammonium acetate in 10% methanol.
        S5   =   0.01 *M* ammonium acetate in water.
Detection: D1   =   UV and/or RI.

REFERENCE

1. **Lundanes, E. and Greibokk, T.,** *J. Chromatogr.*, 149, 241, 1978.

Table LC 27
DIPEPTIDES

Packing	P1
Column	
Length (cm)	25
Diameter (mm, I.D.)	4.6
Solvent	S1
Flow rate (mℓ/min)	0.5
Detection	RI

Dipeptide	t_r(min)
Ala-Tyr	6.5
Leu-Ala	7.3
Ala-Leu	8.6
Ala-Phe	11.5
Trp-Ala	12.3

Packing: P1 = Partisil® 10 CCS/C₈.
Solvent: S1 = Methanol–water (25:75 v/v).

Table LC 27 (continued)
DIPEPTIDES

REFERENCE

1. Whatman Liquid Chromatography Product Guide, p. 27.

Table LC 28
HPLC OF SMALL PEPTIDES

Packing	P1
Column	
Length (mm)	200
Diameter (mm, I.D.)	4
Solvent	S1
Flow rate (mℓ/min)	3
Temperature (°C)	31
Detection	D1

Peptide	t_r(min)
Val-Gly	2.1
Gly-Val	3.0
Tyr-Gly	5.5
Leu-Gly	5.5
Gly-Tyr	6.8
Gly-Leu	6.8
Phe-Gly	8.0
Gly-Phe	9.8
Ala-Ala-Tyr-Ala-Ala	10.5
Gly-Trp	13.2
Phe-Tyr	14.6
Leu-Trp-Met-Arg	18.3
Met-Glu-His-Phe-Arg-Trp-Gly	22.5
Leu-Leu-Val-Tyr	24.0
Renin inhibitor (an octapeptide)	33.6

Packing: P1 = ODS particles, 5-μm (Nucleosil® 5 C-18).
Solvent: S1 = A linear gradient was applied, the initial eluent being 0.05 M KH_2PO_4 adjusted to pH 2 with H_3PO_4, the final eluent being methanol. The gradient was terminated after 60 min.
Detection: D1 = UV at 230 nm.

REFERENCE

1. **Mönch, W. and Dehnen, W.**, *J. Chromatogr.*, 140, 260, 1977.

Reproduced by permission of Elsevier Science Publishers B.V.

Table LC 29
ELUTION CONDITIONS FOR DIASTEREOMERIC DIPEPTIDES ON AN AMINO ACID ANALYZER

Dipeptide	pH	Column length (cm)	t_r (min)
L-Leu-D-Tyr	4.25[a]	58	140
L-Leu-L-Tyr	4.25[a]	58	217
L-Leu-D-Phe	5.26	58	104
L-Leu-L-Phe	5.26	58	86
$N^{\alpha,\epsilon}$-(di-L-Leu)-D-Lys	7.00	58	240
$N^{\alpha,\epsilon}$-(di-L-Leu)-L-Lys	7.00	58	128
L-Leu-D-His	4.66	10.5	148
L-Leu-L-His	4.66	10.5	198
L-Leu-D-Arg	5.26	10.5	200
L-Leu-L-Arg	5.26	10.5	161

[a]   Containing 1% benzyl alcohol and 2% 1-propanol.

Conditions: Ion-exchange chromatography was performed with a Beckman Model 120 B amino acid analyzer. A long column, 0.9 × 58 cm, containing Beckman AA-15 sulfonated polystyrene or a short column, 0.9 × 10.5 cm, containing Beckman PA-35 sulfonated polystyrene, was used. The pH 4.25 buffer (0.2 N in sodium) and the pH 5.26 (0.35 N in sodium) citrate buffers were prepared from Beckman concentrates. The pH 4.66 buffer was prepared by mixing 3 parts of pH 4.25 buffer with 1 part of pH 5.26 buffer. The pH 7.00 buffer was prepared by titration of the pH 5.26 buffer with 50% NaOH. The columns were operated at a flow rate of 61 mℓ/hr and 57°C.

REFERENCE

1.   **Mitchell, A. R., Kent, S. B. H., Chu, I. C., and Merrifield, R. B.,** *Anal. Chem.*, 50, 637, 1978.

Reprinted with permission. Copyright 1978 American Chemical Society.

Table LC 30
ION-EXCHANGE CHROMATOGRAPHY OF SMALL PEPTIDES

Packing	P1
Solvent	S1
Flow rate (mℓ/min)	30

Peptide	R_{AH} × 100	Color yield (area/μmol)	
		440 nm	570 nm
Gly-Ser	22.2	2.9	27.1
Ala-Asp	25.4	5.0	33.6
Gly-Asp	28.8	5.6	56.0
Ala-Ser	28.9	3.3	24.9
Gly-Gly	30.6	5.1	46.5
Ser-Gly	35.3	6.4	58.0

Table LC 30 (continued)
ION-EXCHANGE CHROMATOGRAPHY OF SMALL PEPTIDES

Peptide	$R_{AH} \times 100$	Color yield (area/µmol)	
		440 nm	570 nm
Glu-Ala	37.1	22.0	77.0
Ala-Gly-Gly	37.7	3.0	18.8
Ala-Glu	40.7	5.2	33.0
Ala-Gly	44.0	7.4	50.4
Val-Ser	45.5	0.4	2.4
Ala-Ala	47.9	5.4	44.6
Gly-Gly-Gly	48.2	5.2	36.8
Gly-Ala	48.5	5.8	74.0
Val-Gly	54.1	1.2	5.4
Val-Ala	55.3	0.2	1.4
Val-Gly-Gly	57.4	0.2	1.2
Gly-Val	58.9	7.8	66.5
Gly-Pro	62.3	8.9	75.0
Pro-Gly[a]	—	—	—
Ala-Pro[a]	—	—	—
Val-Pro	65.8	8.0	43.4
Leu-Val	66.5	5.0	16.0
Val-Val	67.0	0.6	2.6
Leu-Gly	70.2	8.1	59.0
Val-Ala	70.5	0.8	4.8
Gly-Ile	71.6	9.6	69.5
Gly-Leu	72.4	9.2	78.0
Val-Met	72.4	1.0	4.0
Val-Leu	73.8	1.2	6.5
Gly-Tyr	84.1	5.6	48.8
Val-Tyr-Val	84.7	1.0	6.5
Val-Tyr	86.0	0.6	2.8
Ala-Phe	87.0	8.8	63.0
Val-Phe	90.0	0.8	4.2
Val-His	112.5	0.8	4.8
Gly-Lys	114.0	8.2	69.2
Val-Trp	121.0	0.4	2.2
Ile-Ala	67.8	0.7	3.1
Ile-Gly	70.7	1.0	4.0
Ile-Glu	61.0	0.3	3.3
Ile-Leu	80.5	0.8	3.0
Ile-Lys	125.0	1.5	8.3
Ile-Met	74.5	0.6	3.2
Ile-Phe	98.0	0.4	2.3
Ile-Pro	69.3	5.9	40.5
Ile-Ser	44.4	0.6	2.9
Ile-Trp	135.0	0.3	2.6
Ile-Val	70.0	0.4	4.0
Leu-Leu	82.7	5.4	35.0
Leu-Met	76.0	2.7	17.5
Leu-Phe	98.9	3.2	19.8
Leu-Ser	54.8	10.4	78.2
Leu-Trp	137.0	2.5	43.8
Leu-Tyr	91.7	11.5	81.0
Pro-Hyp	28.1	0.5	0.4
Pro-Val	56.2	0.8	0.5
Pro-Met	65.5	2.4	1.4

Table LC 30 (continued)
ION-EXCHANGE CHROMATOGRAPHY OF SMALL PEPTIDES

Peptide	$R_{AH} \times 100$	Color yield (area/μmol)	
		440 nm	570 nm
Pro-Ile	73.2	0.7	0.1
Pro-Leu	74.9	1.0	0.1
Pro-Tyr	85.1	0.6	0.4
Pro-Phe	88.3	0.5	0.3
His-Ser	100.0	7.0	23.2
His-Gly	108.1	7.8	31.6
Lys-Ala	108.2	15.5	55.0
His-Ala	109.5	5.4	22.9
Lys-Gly	110.0	18.0	63.6
Lys-Val	112.5	18.0	66.5
His-Leu	124.5	6.6	27.0
Lys-Leu	126.5	20.7	76.0
Pro-Trp	131.0	0.5	0.4
His-Tyr	146.0	6.0	25.0
Glu-Glu	0	3.9	30.0
Glu-Gly[b]	0		
Glu-Gly[b]	9.2		
Asp-Gly[b]	12.2		
Asp-Gly[b]	14.6		
Glutathione (oxidized)	22.8	19.2	88.0
Met-Ala-Ser	39.3	8.5	46.0
Glu-Val	46.2	13.2	52.8
Met-Ser	51.7	12.1	66.4
Met-Glu	60.0	7.2	83.4
Met-Gly	64.3	9.6	56.0
Met-Ala	67.2	8.8	56.0
Ser-Leu	70.0	11.2	68.0
Tyr-Glu	72.9	10.8	54.8
Met-Pro	73.5	4.9	19.2
Met-Met	77.0	Coincides with ammonia peak	
Tyr-Gly	81.5	11.1	60.3
Tyr-Ala	83.6	10.8	51.8
Met-Leu	85.6	12.4	74.4
Phe-Gly	86.0	9.7	56.6
Phe-Ala	89.2	9.5	52.5
Tyr-Val	90.5	11.6	60.3
Phe-Pro	91.7	7.3	14.6
Phe-Val	94.8	11.7	84.3
Met-Phe	97.8	9.0	56.4
Tyr-Leu	100.0	10.2	64.2
Phe-Leu	106.6	10.4	62.5
Tyr-Tyr	109.1	16.0	81.6
Phe-Tyr	115.9	11.1	63.2
Tyr-Phe	116.5	14.0	60.6
Phe-Phe	116.9	8.8	50.3
Phe-Trp[c]	—		

[a] Not observed at a column loading of 2 μmol.
[b] The peptide gives double peaks.
[c] Not eluted from the column even after 25 hr (that is an $R_{AH} \times 100$ value of about 150).

Table LC 30 (continued)
ION-EXCHANGE CHROMATOGRAPHY OF SMALL PEPTIDES

$$R_{AH} = \frac{\text{elution time (mins) of peak}}{\text{elution time (mins) between standards}}$$

The time interval between the elution peaks of aspartic acid (zero time) and histidine is designated as unity. A change in R_{AH} of 0.03 corresponds to a time interval of about 20 min.

Packing: P1 = Chromobeads (8% cross-linked ion-exchange resin. Type B, Technicon). A Technicon Autoanalyzer was used.
Solvent: S1 = Sodium citrate buffers of pH 2.875, 3.80, and 5.00 as described in the Technicon manual (Hamilton P.B., *Techniques in Amino Acid Analysis*, Technicon International Division S.A., Geneva, 1966, 17). The total elution time was about 19 hr.

REFERENCES

1. **Heathcote, J. G., Washington, R. J., Keogh, B. J., and Glanville, R. W.,** *J. Chromatogr.*, 65, 397, 1972.
2. **Heathcote, J. G., Keogh, B. J., and Washington, R. J.,** *J. Chromatogr.*, 79, 187, 1973.
3. **Heathcote, J. G., Washington, R. J., and Keogh, B. J.,** *J. Chromatogr.*, 92, 355, 1974.
4. **Heathcote, J. G., Washington, R. J., and Keogh, B. J.,** *J. Chromatogr.*, 104, 141, 1975.

Reproduced by permission of Elsevier Science Publishers B.V.

Table LC 31
HPLC OF PEPTIDES USING METAL ION-MODIFIED MOBILE PHASES

Packing	P1	P1
Solvent	S1	S2
Flow rate (mℓ/min)	1.5	1.5
Temperature	rt	rt
Detection	D1	D1

Peptide	t_r(min)	
Gly-Gly	11.1	11.0
Gly-Leu	9.0	12.3
Gly-Thr	4.0	10.2
Ala-Gly	5.4	9.3
Ala-Ala	5.2	10.0
Ala-Val	4.1	8.4
Ala-Leu	4.5	10.2
Val-Val	3.2	8.9
Gly-Gly-Gly	13.7	13.6
Gly-Gly-Gly-Gly	15.2	14.2
Gly-Gly-Gly-Gly-Gly	15.8	

Packing: P1 = 600-NH amino column (Alltech). A precolumn of silica gel (Universal Scientific) was used to improve the life and stability of the column.

Table LC 31 (continued)
HPLC OF PEPTIDES USING METAL
ION-MODIFIED MOBILE PHASES

Solvent: S1 = 89 mℓ 5 × 10^{-4} M CdSO$_4$ + 10
mℓ acetonitrile + 1 mℓ
trifluoroethanol.

S2 = 10^{-3} M zinc acetate.

Detection: D1 = UV at 202 nm.

REFERENCE

1. **Dua, V. K. and Bush, C. A.,** *J. Chromatogr.,*
244, 128, 1982.

Table LC 32
NONAPEPTIDES

Packing	P1	P1	P2	P2	P3
Column					
Length (cm)	25	25	15	15	25
Diameter (cm, I.D.)	0.3	0.3	0.4	0.4	0.3
Solvent	S1	S2	S1	S3	S1
Flow rate (mℓ/min)	3.0	3.0	2.0	2.0	3.0
Temperature	rt	rt	rt	rt	rt
Detection	D1	D1	D1	D1	D1
Compound			**k$'$**		
Ornipressine	1.4	2.8	2.7	1.1	2.3
Lypressine	1.4	2.8	2.7	1.1	2.3
Oxytocine	3.9	9.0	7.0	2.2	7.3
Felypressine	5.6	11.7	9.5	2.8	10.0
Demoxytocine	10.8	29.0	17.5	4.3	21.1

Packing: P1 = RP8, 10 μm.

P2 = Nucleosil® C$_8$, 5 μm.

P3 = Nucleosil® C$_{18}$, 5 μm.

Solvent: S1 = Phosphate buffer, pH 7-acetonitrile, 4:1.

S2 = Phosphate buffer, pH 7-acetonitrile, 33:7.

S3 = Phosphate buffer, pH 7-acetonitrile, 3:1.

Detection: D1 = UV at 210—220 nm.

REFERENCE

1. **Krummen, K. and Frei, R. W.,** *J. Chromatogr.,* 132, 27, 1977.

Table LC 33
CYCLIC HEXAPEPTIDES CONTAINING GLUTAMINE

Packing	P1
Solvent	S1
Flow rate (mℓ/min)	1.0

Peptide	t_r(min)
Cyclo-(Gly-Pro-D-Gln)$_2$	3.3
Cyclo-(Ala-Pro-D-Gln)$_2$	3.3
Cyclo-(Val-Pro-D-Gln)$_2$	4.8
Cyclo-(Leu-Pro-D-Gln)$_2$	12.6
Cyclo-(Phe-Pro-D-Gln)$_2$	16.7

Packing: P1 = Whatman Partisil® 10/25 ODS-3 column.
Solvent: S1 = 25% CH_3CN in water containing 0.1%
                TFA (trifluoroacetic acid).

REFERENCE

1. **Kopple, K. D. and Parameswaran, K. N.,** *Int. J. Peptide Protein Res.,* 21, 269, 1983.

Table LC 34
REVERSED-PHASE HPLC OF PEPTIDES AND PROTEINS WITH ION-PAIRING REAGENTS

Compound	Column	Solvent	t_r(min)
Acyl carrier protein	P1	S1	1.8
Acyl carrier protein	P1	S2	4.2
ACTH 1-24 pentaacetate	P2	S3	2.1
Glucagon	P2	S3	4.6
Insulin (porcine)	P3	S4	6.0
Linear antamanid	P1	S5	100
	P1	S6	2.4
Leu-Trp-Met-Arg-Phe	P1	S7	40.6
Leu-Trp-Met-Arg-Phe	P1	S8	7.3
Leu-Trp-Met-Arg-Phe	P1	S9	13.6

Column: P1 = μBondapak®-Fatty acid analysis column, 30 cm × 4 mm, I.D.
        P2 = Bondapak®-C_{18}-Corasil®.
        P3 = μBondapak® C_{18}, 10-μm particles, 30 cm × 4 mm, I.D.
Solvent[a]: S1 = 5% CH_3CN, 0.1% H_3PO_4.
        S2 = 30% CH_3CN, 5 mM sodium hexane sulfonate, pH 6.5.
        S3 = 40% CH_3OH, 0.1% H_3PO_4.
        S4 = 60% CH_3OH, 0.1% H_3PO_4.
        S5 = 60% CH_3OH.
        S6 = 60% CH_3OH, 0.1% H_3PO_4 + 0.1 M KH_2PO_4.
        S7 = 50% CH_3OH.
        S8 = 50% CH_3OH + 0.1% H_3PO_4.
        S9 = 50% CH_3OH + 5mM sodium hexane sulfonate, pH 6.5.

[a] Expressed as a percentage of the organic component; in all cases the other solvent was water.

Table LC 34 (continued)
REVERSED-PHASE HPLC OF PEPTIDES AND PROTEINS WITH ION-PAIRING REAGENTS

REFERENCE

1. Hancock, W. S., Bishop, C. A., Prestidge, R. L., and Harding, D. R. K., *Science*, 200, 1168, 1978.

Table LC 35
EFFECT OF PHOSPHORIC AND ACETIC ACIDS ON THE RETENTION OF PEPTIDES

Packing	P1	P1
Column		
Length (cm)	30	30
Diameter (mm, I.D.)	4	4
Solvent	S1	S2[a]
Flow rate (mℓ/min)	1.5	1.5
Temperature (°C)	22	22
Detection	S1	S1

Peptide	t_r(min)	
Gly-Phe	2.4	3.5
Gly-Gly-Tyr	1.9	3.0
Gly-Leu-Tyr	2.4	3.6
Arg-Phe-Ala	2.05	2.2
Met-Arg-Phe	2.5	3.8
Leu-Trp-Met-Arg	2.3	4.1
Leu-Gly-Met-Arg-Phe	5.1	8.3
Phe-Ser-Lys-Leu-Gly-Asp-Gly	2.4	3.8

[a]   Broad peaks unsuitable for analytical separations were observed with the solvent.

Packing: P1 = μBondapak® alkylphenyl, 10 μm.
Solvent: S1 = 50% Methanol containing 0.1% H_3PO_4, pH 2.5 (apparent pH value measured with a glass electrode).
           S2 = 50% Methanol containing 0.1% acetic acid, pH 4.
Detection: D1 = UV at 205—225, 254, or 280 nm, depending on the nature of the sample and the mobile phase.

REFERENCE

1. Hancock, W. S., Bishop, C. A., Prestidge, R. L., Harding, D. R. K., and Hearn, M. T. W., *J. Chromatogr.*, 153, 391, 1978.

Table LC 36
RETENTION TIMES OF THE DECAPEPTIDE LINEAR ANTAMANID[a]

Mobile phase	t_r(min)
Methanol-water, 50:50	∞
Methanol-water, 55:45	∞
Methanol-water, 60:40	100
Methanol-water, 50:50[b]	6.01
Methanol-water, 55:45[b]	2.74
Methanol-water, 60:40[b]	2.37
Methanol-water, 55:45, pH 2.5[c]	2.45
Methanol-water, 55:45, pH 3.0[c]	2.70
Methanol-water, 55:45, pH 4.0[c]	3.30
Methanol-water, 55:45, pH 5.1[c]	4.95
Methanol-water, 55:45, pH 7.0[c]	4.07

[a]   Amino acid sequence Val-Pro-Pro-Ala-Phe-Phe-Pro-Pro-Phe-Phe.
[b]   0.1% H_3PO_4 added.
[c]   0.1% H_3PO_4 + 0.1 M KH_2PO_4 + K_2HPO_4.

Conditions: μBondapak® C_{18}, 10 μm, column, 30 cm × 4 mm I.D., operated at 22°C and a flow rate of 1.5 mℓ/min. Detection was in the range 205—225, 254, or 280 nm depending on the nature of the sample and the mobile phase.

REFERENCE

1. Hancock, W. S., Bishop, C. A., Prestidge, R. L., Harding, D. R. K., and Hearn, M. T. W., *J. Chromatogr.*, 153, 391, 1978.

Reproduced by permission of Elsevier Science Publishers B.V.

Table LC 37
EFFECT OF HYDROPHOBIC ION PAIRING ON THE RETENTION OF PEPTIDES

Packing	P1	P1	P1	P1
Column				
Length (cm)	30	30	30	30
Diameter (mm)	4	4	4	4
Solvent	S1	S2	S3	S4
Flow rate (mℓ/min)	1.5	1.5	1.5	1.5
Temperature (°C)	22	22	22	22
Detection	UV	UV	UV	UV
Peptide		t_r(min)		
Gly-Phe	2.3	2.3	2.3	2.5
Gly-Gly-Tyr	64.5	1.9	2.4	6.2
Gly-Leu-Tyr	2.5	2.4	2.4	2.7
Arg-Phe-Ala	48	2.05	3.0	33.2
Met-Arg-Phe	32.5	2.4	3.6	58
Leu-Trp-Met-Arg	112	2.3	4.0	16.2
Leu-Trp-Met-Arg-Phe	120	5.1	10.2	40.5

Table LC 37 (continued)
EFFECT OF HYDROPHOBIC ION PAIRING ON THE RETENTION OF PEPTIDES

Packing: P1 = μBondapak®-alkylphenyl, 10 μm.
Solvent: S1 = Methanol-water, 1:1 v/v.
        S2 = Methanol-water, 1:1 v/v, containing 5 mM H$_3$PO$_4$, pH 2.5.
        S3 = Methanol-water, 1:1 v/v, containing 5 mM sodium hexanesulfonate, pH 6.5.
        S4 = Methanol-water, 1:1 v/v, containing 5 mM sodium dodecylsulfate, pH 7.15.

REFERENCE

1. Hancock, W. S., Bishop, C. A., Meyer, L. J., Harding, D. R. K., and Hearn, M. T. W., *J. Chromatogr.*, 161, 291, 1978.

Table LC 38
EFFECT OF pH ON THE RETENTION OF PEPTIDES IN THE PRESENCE OF AN ION-PAIRING AGENT

Packing	P1	P1	P1	P1
Column				
Length (cm)	30	30	30	30
Diameter (mm)	4	4	4	4
Solvent	S1	S2	S3	S4
Flow rate (mℓ/min)	1.5	1.5	1.5	1.5
Temperature (°C)	22	22	22	22
Detection	UV	UV	UV	UV
Peptide		**t$_r$(min)**		
Gly-Phe	2.5	2.3(−)[a]	8.9	2.5(−)
Gly-Gly-Tyr	2.1	2.4(+)	5.5	6.2(+)
Gly-Leu-Tyr	2.5	2.3(−)	9.3	2.7(−)
Arg-Phe-Ala	2.5	3.0(+)		33.2
Met-Arg-Phe	3.1	3.6(+)		58
Leu-Trp-Met-Arg	3.2	3.0(−)	25	16.2(−)
Leu-Trp-Met-Arg-Phe	5.6	10.2(+)		40.5

[a] The + or − sign indicates whether an increased or decreased retention time was caused by an increase in pH.

Packing: P1 = μBondapak®-alkylphenyl, 10 μm.
Solvent: S1 = Methanol-water, 1:1 v/v, containing sodium hexanesulfonate, pH 2.1.
        S2 = Methanol-water, 1:1 v/v, containing sodium hexanesulfonate, pH 6.5.
        S3 = Methanol-water, 1:1 v/v, containing sodium dodecyl sulfate, pH 2.9.
        S4 = Methanol-water, 1:1 v/v, containing sodium dodecylsulfate, pH 7.15.

REFERENCE

1. Hancock, S. W., Bishop, C. A., Meyer, L. J., Harding, D. R. K., and Hearn, M. T. W., *J. Chromatogr.*, 161, 291, 1978.

Table LC 39
EFFECT OF CATIONIC REAGENTS ON THE RETENTION OF PEPTIDES

	S1	S2	S3	S4	S5	S6	S7	S8	S9	S10	S11	S12	S13	S14	S15	S16	S17	S18	S19	S20
Packing	Pl	Pl	Pl	Pl	Pl	Pl	Pl	Pl	Pl	Pl	Pl	Pl	Pl	Pl	Pl	Pl	Pl	Pl	Pl	Pl
Column Length (cm)	30	30	30	30	30	30	30	30	30	30	30	30	30	30	30	30	30	30	30	30
Diameter (mm, I.D.)	4	4	4	4	4	4	4	4	4	4	4	4	4	4	4	4	4	4	4	4
Solvent	S1	S2	S3	S4	S5	S6	S7	S8	S9	S10	S11	S12	S13	S14	S15	S16	S17	S18	S19	S20
Flow rate (ml/min)	1.5	1.5	1.5	1.5	1.5	1.5	1.5	1.5	1.5	1.5	1.5	1.5	1.5	1.5	1.5	1.5	1.5	1.5	1.5	1.5
Temperature (°C)	22	22	22	22	22	22	22	22	22	22	22	22	22	22	22	22	22	22	22	22
Detection	UV	UV	UV	UV	UV	UV	UV	UV	UV	UV	UV	UV	UV	UV	UV	UV	UV	UV	UV	UV
Peptide																				
											t_r(min)									
Leu-Trp-Met-Arg	2.8	3.0	3.05	3.5	2.7	2.7	3.05	3.1	2.9	2.4	1.6	2.9	2.9	3.0	3.4	2.8	3.2	16.2	1.6	2.3
Leu-Trp-Met-Arg-Phe	5.7	6.8	6.5	8.7	4.0	5.7	7.0	6.1	6.8	4.4	2.1	4.7	4.7	4.8	4.8	4.5	4.9	40.5	2.1	4.45
Gly-Phe	2.4	2.4	2.6	2.7	2.05	2.4	2.5	2.4	2.3	2.25	2.0	2.2	2.2	2.35	2.5	2.2	2.5	2.5	2.0	2.2
Gly-Gly-Tyr	2.0	2.0	2.2	1.8 2.05[a]	1.7	2.0	2.2	2.3	2.0	1.8	1.4	2.1	2.1	2.3	2.4	2.2	2.3	6.2	1.4	1.6
Met-Arg-Phe	2.6	2.7	2.8	3.0	1.7	2.6	2.75	2.7	2.5	2.2	1.45	2.6	2.6	2.7	2.8	2.6	3.0	58	1.45	2.1
Gly-Leu-Tyr	2.4	2.5	2.8	2.7	2.1	2.4	2.6	2.45	2.45	2.25	2.0	2.3	2.3	2.5	2.7	2.3	2.5	2.7	2.0	2.2
Arg-Phe-Ala	2.3	2.3	2.6	2.1 2.7[a]	1.8	2.3	2.3	2.45	2.1	2.0	1.3	2.2	2.2	2.45	2.8	2.3	2.6	33.2	1.3	1.9

[a] Two peaks were observed.

Packing: Pl = μBondapak®-alkylphenyl, 10 μm.

Solvent: Unless otherwise indicated, S1 to S20 consisted of methanol-water, 1:1 v/v, a 2 mM solution of the reagent, acetate as the anion, and the pH was adjusted to 4. The cationic reagents used were as follows: for S1 to S5 the ammonium salt R_4NOAc, where R = H for S1, CH_3 for S2, CH_2CH_3 for S3, $(CH_2)_2CH_3$ for S4, and $(CH_2)_3CH_3$ for S5; for S6 to S11 the amine salt R^+NH_3, where R = H for S6, CH_3 for S7, CH_2CH_2OH for S8, $(CH_2)_3CH_3$ for S9, $(CH_2)_5CH_3$ for S10 and $(CH_2)_1CH_3$ for S11; Li^+ for S12, Na^+ for S13, K^+ for S14, Cs^+ for S15, Mg^{2+} for S16, Ca^{2+} for S17, $CH_3(CH_2)_{11} SO_3^- Na^+$ (the pH being adjusted to 6.5) for S18; $CH_3(CH_2)_{11}NH_3^{+-}OAc$ for S19 and for S20 a mixture of the reagents used in S18 and 19.

REFERENCE

1. **Hancock, W. S., Bishop, C. A., Battersby, J. E., Harding, D. R. K., and Hearn, M. T. W.,** *J. Chromatogr.,* 168, 377, 1979.

Reproduced by permission of Elsevier Science Publishers, B. V.

Table LC 40
EFFECT OF ION-PAIRING REAGENTS ON RETENTION OF PROCTOLIN AND RELATED PENTAPEPTIDES

Packing	P1	P1	P1	P1	P1	P1	P1
Column							
Length (cm)	30	30	30	30	30	30	30
Diameter (mm, I.D)	3.9	3.9	3.9	3.9	3.9	3.9	3.9
Solvent	S1	S2	S3	S4	S5	S6	S7
Flow rate (mℓ/min)	1.5	1.5	1.5	1.5	1.5	1.5	1.5
Temperature	rt	rt	rt	rt	rt	rt	rt
Detection	D1	D1	D1	D1	D1	D1	D1

Peptide				t_r(min)			
Arg-Tyr-Leu-Pro-Thr	7.1	11.3	3.5	5.5	8.2	5.1	3.7
Arg-Tyr-Leu-Pro-Ala	7.2	12.6	3.7	6.1	9.4	5.7	3.9
Arg-Tyr-Leu-Ala-Thr	4.6	7.0	3.0	4.1	6.9	4.4	3.4
Arg-Tyr-Ala-Pro-Thr	3.1	4.3	2.4	3.3	4.5	3.4	2.7
Arg-Ala-Leu-Pro-Thr		5.2	2.6	3.6	5.2		
Ala-Tyr-Leu-Pro-Thr	12.5	16.4	4.0	5.2	5.4	4.4	2.4
Arg-Tyr-Leu-Pro-D-Thr	10.2	17.4	3.8	6.7	10.1	5.8	4.6
Arg-Tyr-Leu-D-Pro-Thr	9.7	16.1	4.1	7.2	11.2	5.7	4.6
Arg-Tyr-D-Leu-Pro-Thr	16.5		5.5	10.1	14.8	8.0	7.4
Arg-D-Tyr-Leu-Pro-Thr	21.7		5.7	10.5	15.9	8.7	5.7
D-Arg-Tyr-Leu-Pro-Thr	14.3	24.8	4.5	8.4	12.7	7.2	4.7
Arg-Tyr-Leu-Pro-Ser	5.9	9.3	3.0	4.8	7.2	4.5	3.4
Arg-Phe-Leu-Pro-Thr	15.0		5.3	10.1	16.1	8.5	6.0

Packing: P1 = μBondapak® fatty acid analysis column (Waters).
Solvent: S1 = 5 mM H_3PO_4 in acetonitrile-water, 1:9 v/v.
            S2 = 5 mM trifluoroacetic acid in acetonitrile-water, 1:9 v/v.
            S3 = 5 mM trifluoroacetic acid in acetonitrile-water, 1:4 v/v.
            S4 = 5 mM trichloroacetic acid in acetonitrile-water, 1:4 v/v.
            S5 = 5 mM heptafluorobutyric acid in acetonitrile-water, 1:4 v/v.
            S6 = 5 mM PIC reagent B-6 (5 mM hexane sulfonic acid containing sufficient acetic acid to yield a pH of about 3.5) in acetonitrile-water, 1:4 v/v.
            S7 = 5 mM sodium hexane sulfonate, pH 6.4, in acetonitrile-water, 1:4 v/v.
Detection: D1 = UV at 225 nm.

REFERENCE

1. **Starratt, A. N. and Stevens, M. E.,** *J. Chromatogr.,* 194, 421, 1980.

Reproduced by permission of Elsevier Science Publishers B.V.

Table LC 41
EFFECT OF TRIFLUOROACETIC ACID (TFA) CONCENTRATION ON RETENTION OF PEPTIDES

	k'		
TFA (%)	Secretin	des-His¹-Secretin	Calcitonin
0.50	6.0	6.5	3.6
0.25	4.7	5.2	3.3
0.10	3.6	4.2	3.2
0.05	3.7	4.6	4.2
0.02	15.4		
0.01	10	10	10
0	10	10	10

Table LC 41 (continued)
EFFECT OF TRIFLUOROACETIC ACID (TFA)
CONCENTRATION ON RETENTION OF PEPTIDES

Column: LiChrosorb® RP-18.
Solvent: Methanol-water, 70:30 v/v.
Flow rate: 1 mℓ/min.
Detection: UV.

REFERENCE

1. **Voskamp, D., Olieman, C., and Beyerman, H. C.,** *Rec. Trav. Chim. Pays-Bas,* 99, 105, 1980.

Reproduced by permission of the Koninklijke Nederlandse Chemische Vereniging.

Table LC 42
CHROMATOGRAPHY OF PORCINE
SECRETIN ON DIFFERENT STATIONARY
PHASES

Column	Solvent	k'
Nucleosil® C18	S1	4.8
	S2	5.4
Polygosil CN	S3	0.7
Polygosil phenyl	S4	4.1

Solvent: S1 = Methanol-water-trifluoroacetic acid, 65:35:1 v/v.

S2 = Acetonitrile-water-trifluoroacetic acid, 32:68:1 v/v.

S3 = Methanol-water-trifluoroacetic acid, 50:50:1 v/v.

S4 = Methanol-water-trifluoroacetic acid, 65:35:1 v/v.

Flow rate: 1 mℓ/min.
Detection: UV.

REFERENCE

1. **Voskamp, D., Olieman, C., and Beyerman, H. C.,** *Rec. Trav. Chim. Pays-Bas,* 99, 105, 1980.

Reproduced by permission of the Koninklijke Nederlandse Chemische Vereniging.

Table LC 43
THE EFFECT OF HYDROPHILIC ION-PAIRING
REAGENTS ON THE RETENTION OF PEPTIDES

	k'		
Solvent	Secretin	des-His¹-Secretin	Calcitonin
S1[a]	2.1		
S2[a]	0.1		
S3	6.8		3.6
S4	11.0	11.1	5.8

[a] Very broad peaks were observed with this solvent.

Column: LiChrosorb® RP18.
Solvent: S1 = Methanol-water, 60:40 v/v containing 0.01 M H_3PO_4.
         S2 = Methanol-water, 70:30 v/v containing 0.01 M H_3PO_4.
         S3 = Methanol-water, 67:33 v/v containing 0.02 M KH_2PO_4.
         S4 = Methanol-water, 67:33 v/v containing 0.01 M KH_2PO_4.
Flow rate: 1 mℓ/min.
Detection: UV.

REFERENCE

1. **Voskamp, D., Olieman, C., and Beyerman, H. C.,** *Rec. Trav. Chim. Pays-Bas,* 99, 105, 1980.

Reproduced by permission of the Koninklijke Nederlandse Chemische Vereniging.

Table LC 44
PEPTIDES IN METHANOL-WATER-
TRIFLUOROACETIC ACID MIXTURES

	Peptide	Solvent	k'
1	Calcitonin	S1	3.2
	Calcitonin sulfoxide	S1	0.6
2	Secretin (S)	S1	3.7
3	Motilin	S2	3.7
	Motilin sulfoxide	S2	2.5
	S(1-12)	S3	3.1
4a	S(13-18)	S4	2.6
4b	S(13-18)	S4	1.9
	S(13-14)	S5	4.4
	S(15-18)	S5	1.0
	S(19-21)	S5	7.9
	S(22-27).NH₂	S6	1.5
5a	S(1-6)NH₂	S7	6.1
5b	S(1-6)NH₂	S7	5.4
6	Substance P	S6	7.1
7	VIP(1-13)	S8	8.5
8	Enkephalin	S6	9.7

Table LC 44 (continued)
PEPTIDES IN METHANOL-WATER-
TRIFLUOROACETIC ACID MIXTURES

Note: The synthetic segments of porcine secretin are named S(n-m), in which n amd m are the sequence numbers of the corresponding amino acid residues in secretin, if the histidine residue occupies position 1. VIP(1-13) is the N-terminal 13-peptide of porcine vasoactive intestinal peptide. The formulae of the peptides are (1) Cys-Gly-Asn-Leu-Ser-Thr-Cys-Met-Leu-Gly-Thr-Tyr-Thr-Gln-Asp-Phe-Asn-Lys-Phe-His-Thr-Phe-Pro-Gln-Thr-Ala-Ile-Gly-Val-Gly-Ala-Pro-NH$_2$. (2) His-Ser-Asp-Gly-Thr-The-Thr-Ser-Glu-Leu-Ser-Arg-Leu-Arg-Asp-Ser-Ala-Arg-Leu-Gln-Arg-Leu-Leu-Gln-Gly-Leu-Val-NH$_2$. (3) Phe-Val-Pro-Ile-Phe-Thr-Tyr-Gly-Glu-Leu-Gln-Arg-Met-Gln-Glu-Lys-Glu-Arg-Asn-Lys-Gly-Gln. (4a) 15α-Asp-Ser. (4b) 15β-Asp-Ser. (5a) 3α-Asp-Gly. (5b) 3β-Asp-Gly. (6) Arg-Pro-Lys-Pro-Gln-Gln-Phe-Phe-Gly-Leu-Met-NH$_2$. (7) His-Ser-Asp-Ala-Val-Phe-Thr-Asp-Asn-Tyr-Thr-Arg-Leu. (8) Tyr-Gly-Gly-Phe-Met.

Column: Nucleosil® C18.
Solvents: S1, S2, S3, S4, S5, S6, S7, and S8 are methanol-water-trifluoroacetic acid mixtures of the compositions 65:35:1, 56:44:0.5, 25:65:0.1, 13:87:0.1, 4:96:0.1, 50:50:0.1, 15:85:0.1, and 45:55:0.1, respectively.
Flow rate: 1 mℓ/min.
Detection: UV.

REFERENCE

1. **Voskamp, D., Olieman, C., and Beyerman, H. C.,** *Rec. Trav. Chim. Pays-Bas*, 99, 105, 1980.

Reproduced by permission of the Koninklijke Nederlandse Chemische Vereniging.

Table LC 45
PORCINE SECRETIN (S) AND ANALOGUES

Peptide	k′
Secretin-(S)	3.7
(3β-Asp)-S	3.5
(Aspartoyl^3)-S	3.5
(L-Ala^4)-S	3.5
(D-Ala^4)-S	3.2
(Sar^4)-S	3.3
(Tyr^{10})-S	2.2
(Ala^4,Tyr^{10})-S	2.3
(Glu^3,Tyr^{10})-S	2.3
(Gln^3,Tyr^{10})-S	2.1
(L-Pyr(3)Ala^1)-S	4.4
(D-Pyr(3)Ala^1)-S	4.3

Table LC 45 (continued)
PORCINE SECRETIN (S) AND ANALOGUES

Peptide	k'
(L-Thi[1])-S	6.0
(D-Thi[1])-S	5.8
(Phe[1])-S	6.6
(Tyr[1])-S	4.7
(Ala[1])-S	4.3
(desamino-His[1])-S	4.4
S(2-27).NH$_2$	4.5

Detection: UV

REFERENCE

1. **Voskamp, D., Olieman, C., and Beyerman, H. C.,** *Rec. Trav. Chim. Pays-Bas,* 99, 105, 1980.

Reproduced by permission of the Koninklijke Nederlandse Chemische Vereniging.

Table LC 46
SECRETIN AND ANALOGUES

Packing	P1	P1	P1	P1	P1	P1	P1	P1	P1	P1	P1	P1
Column												
Length (cm)	15	15	15	15	15	15	15	15	15	15	15	15
Diameter (cm, I.D.)	0.4	0.4	0.4	0.4	0.4	0.4	0.4	0.4	0.4	0.4	0.4	0.4
Solvent	S1	S2	S3	S4	S5	S6	S7	S8	S9	S10	S11	S12
Temperature	rt	rt	rt	rt	rt	rt	rt	rt	rt	rt	rt	rt
Detection	D1	D1	D1	D1	D1	D1	D1	D1	D1	D1	D1	D1
Peptide						k'						
Secretin (S)	1.3	2.2	2.7	5.7	8.5	19.0	1.9	5.4	5.8	15.6	4.8	12.5
(Aspartoyl[3])-S	1.2	2.2	2.8	5.7	9.2	19.9	2.2	6.4	6.9	18.6	5.9	15.1
(β-Asp[3])-S	1.3	2.2	2.7	5.2	7.9	16.5	1.7	4.7	5.0	13.4	4.0	10.7
(Ala[1])-S	1.6	2.9	2.7	5.8	7.7	15.7	1.2	3.0	3.0	6.9	2.0	5.0
(desamino-His[1])-S	1.6	2.8	2.9	6.0	8.0	15.6	1.2	3.0	3.0	7.1	2.0	4.2

Packing: P1 = Nucleosil® C$_{18}$, 7 μm.
Solvent: S1 = 5 mM trifluoroacetic acid (TFA) in methanol-water, 65:35.
S2 = 10 mM TFA in methanol-water, 65:35.
S3 = 5 mM pentafluoropropionic acid in methanol-water, 67:33.
S4 = 10 mM pentafluoropropionic acid in methanol-water, 67:33.
S5 = 5 mM heptafluorobutyric acid in methanol-water, 67:33.
S6 = 10 mM heptafluorobutyric acid in methanol-water, 67:33.
S7 = 5 mM perfluoroheptanoic acid in methanol-water, 79:21.
S8 = 10 mM perfluoroheptanoic acid in methanol-water, 79:21.
S9 = 5 mM perfluorooctanoic acid in methanol-water, 79:21.
S10 = 10 mM perfluorooctanoic acid in methanol-water, 79:21.
S11 = 5 mM perfluorodecanoic acid in methanol-water, 85:15.
S12 = 10 mM perfluorodecanoic acid in methanol-water, 85:15.
Detection: D1 = UV, 205—225 nm.

REFERENCE

1. **Schaaper, W. M. M., Voskamp, D., and Olieman, C.,** *J. Chromatogr.*, 195, 181, 1980.

Reproduced by permission of Elsevier Science Publishers B.V.

Table LC 47
HEXAPEPTIDES RELATED TO SECRETIN

Packing	P1	P1	P1	P1	P1	P1	P1	P1	P1	P1	P1	P1
Column Length (cm)	15	15	15	15	15	15	15	15	15	15	15	15
Diameter (cm, I.D.)	0.4	0.4	0.4	0.4	0.4	0.4	0.4	0.4	0.4	0.4	0.4	0.4
Solvent	S1	S2	S3	S4	S5	S6	S7	S8	S9	S10	S11	S12
Temperature	rt	rt	rt	rt	rt	rt	rt	rt	rt	rt	rt	rt
Detection	D1	D1	D1	D1	D1	D1	D1	D1	D1	D1	D1	D1
Peptide							k'					
S*(1-6)-NH$_2$	4.0	5.5	6.4	8.9	3.0	3.7	3.2	4.6	3.2	4.3	5.8	9.5
(Aspartoyl^3)-S(1-6)-NH$_2$	3.9	5.7	6.6	9.2	3.2	3.9	3.5	4.7	3.3	4.6	6.0	9.6
(β-Asp^3)-S(1-6)-NH$_2$	3.4	4.8	5.3	7.7	2.7	3.4	3.0	4.1	3.0	4.0	5.4	8.5
S(13-18)	1.3	2.2	3.9	7.2	3.6	5.0	8.4	14.5	9.0	15.3	25.7	32.2
(β-Asp^{15})-S(13-18)	0.9	1.5	2.9	5.5	3.1	4.5	7.9	13.9	8.5	14.9	24.6	32.1

Note: The segments of porcine secretin are designated S(n-m), in which n and m are the sequence numbers of the corresponding amino acid residues in secretin, if the histidine residue occupies position 1.

[a] S, secretin.

Packing: P1 = Nucleosil® C$_{18}$, 7 μm.
Solvent: S1 = 5 mM TFA in methanol-water, 13:87.
          S2 = 10 mM TFA in methanol-water, 13:87.
          S3 = 5 mM pentafluoropropionic acid in methanol-water, 20:80.
          S4 = 10 mM pentafluoropropionic acid in methanol-water, 20:80.
          S5 = 5 mM heptafluorobutyric acid in methanol-water, 40:60.
          S6 = 10 mM heptafluorobutyric acid in methanol-water, 40:60.
          S7 = 5 mM perfluoroheptanoic acid in methanol-water, 60:40.
          S8 = 10 mM perfluoroheptanoic acid in methanol-water, 60:40.
          S9 = 5 mM perfluorooctanoic acid in methanol-water, 65:35.
          S10 = 10 mM perfluorooctanoic acid in methanol-water, 65:35.

Table LC 47 (continued)
HEXAPEPTIDES RELATED TO SECRETIN

S11 = 5 mM perfluorodecanoic acid in methanol-water, 70:30.
S12 = 10 mM perfluorodecanoic acid in methanol-water, 70:30.
Detection: D1 = UV, 205—225 nm.

REFERENCE

1. **Schaaper, W. M. M., Voskamp, D., and Olieman, C.,** *J. Chromatogr.,* 195, 181, 1980.

Reproduced by permission of Elsevier Science Publishers B.V.

Table LC 48
THE SECRETIN-GLUCAGON
FAMILY OF PEPTIDES

Peptide	Solvent	t_r(min)
Vasoactive intestinal peptide	S1	4.5
Gastric inhibitory polypeptide	S2	8.8
Secretin	S3	6.5
Glucagon	S4	7.5
Pancreatic polypeptide	S5	5.6

Conditions: μBondapak® C-18 column, 0.39 × 30 cm, operated at room temperature.

Solvents: S1, S2, S3, S4, and S5 are triethylammonium phosphate (TEAP) buffer, 0.25 N, pH 3.5 containing 25, 28, 30, 30, and 32% of acetonitrile, respectively.

Flow rate: 2 mℓ/min.

Detection: UV at 210 nm.

Note: The natural peptides vasoactive intestinal polypeptide, gastric inhibitory polypeptide, and synthetic secretin were found to contain impurities.

REFERENCE

1. **Fourmy, D., Pradayrol, L., and Ribet, A.**, *J. Liq. Chromatogr.*, 5, 2123, 1982.

Table LC 49
PROTECTED PEPTIDES USED IN THE SEQUENTIAL
SYNTHESIS OF SECRETIN

Packing	P1	P1	P1	P1	P1	P1
Column						
Length (cm)	30	30	30	30	30	30
Diameter (mm, I.D.)	0.4	0.4	0.4	0.4	0.4	0.4
Solvent	S1	S2	S3	S4	S5	S6
Flow rate (mℓ/min)	1.2	1.2	1.2	1.2	1.2	1.2
Temperature	rt	rt	rt	rt	rt	rt
Detection	D1	D1	D1	D1	D1	D1

Sequence number[a]	Last coupled amino acid residue		k′	
27	Z-Val-NH$_2$	2.8		
26	Z-Leu	10.4	1.2	
25	Boc-Gly-	5.0	0.7	
24	Boc-Gln-	3.9		
23	Boc-Leu-		1.8	
22	Boc-Leu-		3.2	
21	Boc-Arg(NO$_2$)-		2.7	
20	Boc-Gln-		3.0	
19	Boc-Leu-		12.9	1.6
18	Boc-Arg(NO$_2$)-		8.8	1.1
17	Boc-Ala-			1.7

Table LC 49 (continued)
PROTECTED PEPTIDES USED IN THE SEQUENTIAL SYNTHESIS OF SECRETIN

Sequence number[a]	Last coupled amino acid residue	k'			
16	Boc-Ser(Bzl)-	5.9			
15	Boc-Asp(Bzl)-	8.6	2.1		
14	Boc-Arg(NO$_2$)-	7.8	1.8		
13	Boc-Leu-	5.3			
12	Boc-Arg(NO$_2$)-		1.6		
11	Boc-Ser(Bzl)-		3.6		
10	Boc-Leu-		2.2	0.5	
9	Boc-Glu(Bzl)-		4.1	0.8	
8	Boc-Ser(Bzl)-			1.6	
7	Boc-Thr(Bzl)-			2.3	
6	Boc-Phe-			3.5	
5	Boc-Thr(Bzl)-			7.6	
4	Boc-Gly-			5.4	
3	Boc-Asp(Bzl)-				6.4
2	Boc-Ser(Bzl)-				10.3
1	Z-His(Z)-				16.2

[a] The sequence number is the number of the last coupled amino acid residue, if the histidyl residue occupies position one.

Packing: P1 = LiChrosorb® RP18, mean particle size of 10 μm.
Solvents: S1, S2, S3, S4, and S5 are methanol-water mixtures with the compositions 50:50, 70:30, 80:20, 85:15 and 90:10 v/v, respectively; S6 has the composition methanol-water-acetic acid 90:10:1 v/v.
Detection: D1 = UV at 254 nm.

REFERENCE

1. **Bakkum, J. T. M., Beyerman, H. C., Hoogerhourt, P., Olieman, C., and Voskamp, D.,** *Rec. Trav. Chim. Pays-Bas,* 96, 301, 1977.

Reproduced by permission of the Koninklijke Nederlandse Chemische Vereniging.

Table LC 50
REVERSED-PHASE HPLC OF OXYTOCIN DIASTEREOMERS

Packing	P1	P1	P1	P1
Column				
Length (cm[a])	60	60	60	60
Diameter (cm)	0.39	0.39	0.39	0.39
Solvent	S1	S2	S3	S4
Flow rate (mℓ/min)	1.6	1.6	2.0	1.5
Detection	D1	D1	D1	D1

Oxytocin diastereomer	k'			
Oxytocin	7.73	16.9	7.31	7.70
4-D-Glutamine	9.12	20.3	8.33	
7-D-Proline	9.12	22.8	10.4	

Table LC 50 (continued)
REVERSED-PHASE HPLC OF OXYTOCIN
DIASTEREOMERS

6-hemi-D-Cystine	9.66	22.6	9.43	
5-D-Asparagine	11.1	20.4	9.28	
1-hemi-D-Cystine	13.1	31.6	14.6	
8-D-Leucine	13.3	29.1	12.0	13.7
2-D-Tyrosine	17.7	42.9	10.7	12.7

[a]   Two 30-cm columns connected in series using a 3-cm stainless steel connector.

Packing: P1 = μBondapak® C_{18} reversed phase.
Solvent: S1 = 10% Tetrahydrofuran, 90% 0.05 M ammonium acetate brought to pH 4.0 with acetic acid.
S2 = 10% Tetrahydrofuran, 90% 0.05 M ammonium acetate brought to pH 6.0 with acetic acid.
S3 = 18% Acetonitrile, 82% 0.01 M ammonium acetate brought to pH 4.0 with acetic acid.
S4 = 16% Dioxane, 84% 0.05 M ammonium acetate brought to pH 4.0 with acetic acid.
Detection: D1 = UV at 280 nm.

REFERENCE

1. **Larsen, B., Fox, B. L., Burke, M. F., and Hruby, V. J.,** *Int. J. Pept. Prot. Res.,* 13, 12, 1979.

Table LC 51
PARTITION CHROMATOGRAPHY OF OXYTOCIN AND ITS DIASTEREOMERS

Packing	P1
Solvent	S1

Compound	R_F
Oxytocin	0.24
(1-hemi-D-Cystine)oxytocin	0.33
	0.37
(2-D-Tyrosine)oxytocin	0.38
(4-D-Glutamine)oxytocin	0.24
(5-D-Asparagine)oxytocin	0.24
(6-hemi-D-Cystine)oxytocin	0.33
(7-D-Proline)oxytocin	0.37
(8-D-Leucine)oxytocin	0.36

Packing: P1 = Sephadex® G-25.
Solvent: S1 = 1-Butanol-3.5% aqueous acetic acid in 1.5% pyridine.

Note: Here, $R_F = V_H/V_E$, where V_H is the hold-up volume and V_E is the elution volume of the peak.

REFERENCE

1. **Larsen, B., Fox, B. L., Burke, M. F., and Hruby, V. J.,** *Int. J. Pept. Prot. Res.,* 13, 12, 1979.

Table LC 52
PEPTIDE HORMONE DIASTEREOMERS

Packing	P1
Column[a]	
Length (cm)	60
Diameter (cm)	0.39
Solvent	S1
Flow rate (mℓ/min)	2.0
Detection	D1

Table LC 52 (continued)
PEPTIDE HORMONE
DIASTEREOMERS

Peptide	t_r(min)
Oxytocin	27
(1-hemi-D-(α-^{2}H)cysteine) oxytocin	35
(2-D-tyrosine) oxytocin	38
(6-hemi-D-(α-^{2}H) cysteine) oxytocin	52
(8-D-(2-^{13}C) leucine) oxytocin	43
(3-L-(2-^{13}C) leucine) oxytocin	17
(3-D-(2-^{13}C) leucine) oxytocin	22

[a] Two 30-cm columns connected in series with a 3-cm stainless steel connector were used.

Packing: P1 = μBondapak® C_{18} reversed phase.

Solvent: S1 = Solution A, 0.1 M ammonium acetate brought to pH 4.0 with acetic acid. Solution B, acetonitrile. The solvent was composed of 82% solution A and 18% solution B.

Detection: D1 = UV at 254 and 280 nm simultaneously.

REFERENCE

1. **Larsen, B., Viswanatha, V., Chang, S. Y., and Hruby, V. J.**, *J. Chromatogr. Sci.*, 16, 207, 1978.

Reproduced from the Journal of Chromatographic Science by permission of Preston Publications, Inc.

Table LC 53
DIASTEREOMERS OF OXYTOCIN ANALOGUES

Compound[a]	R^1	R^2	R^3	R^4	R^5	k'	K_{OS}^b	Solvent
1	NH$_2$	H	H	S-S	L-Phe(Et)	2.09		S1
2	NH$_2$	H	H	S-S	D-Phe(Et)	3.65		S1
3	H	H	H	S-S	L-Phe(Et)	9.25		S2
4	H	H	H	S-S	D-Phe(Et)	14.87		S2
5	NH$_2$	CH$_3$	CH$_3$	S-S	L-Phe(Et)	3.05		S1
6	NH$_2$	CH$_3$	CH$_3$	S-S	D-Phe(Et)	5.89		S1
7	H	H	H	S-CH$_2$	L-Phe(Et)	5.18	4.32	S3
8	H	H	H	S-CH$_2$	D-Phe(Et)	7.48	6.04	S3
9	H	H	H	S-CH$_2$	L-Phe(Me)	4.75	3.84	S4
10	H	H	H	S-CH$_2$	D-Phe(Me)	6.60	5.18	S4
11	H	H	H	S-CH$_2$	L-Phe	3.90	3.12	S4
12	H	H	H	S-CH$_2$	D-Phe	5.51	4.30	S4
13	H	H	H	S-CH$_2$	L-Phe(Cl)	4.45	3.67	S2
14	H	H	H	S-CH$_2$	D-Phe(Cl)	6.18	5.22	S2
15	H	H	H	S-CH$_2$	L-Tyr(Et)	7.03	5.48	S5
16	H	H	H	S-CH$_2$	D-Tyr(Et)	9.14	6.28	S5

Table LC 53 (continued)
DIASTEREOMERS OF OXYTOCIN ANALOGUES

Compound[a]	R[1]	R[2]	R[3]	R[4]	R[5]	k'	K$_{os}^{b}$	Solvent
17	H	H	H	S-CH$_2$	L-Tyr	10.44	7.56	S6
18	H	H	H	S-CH$_2$	D-Tyr	9.92	4.77	S6
19	H	H	H	CH$_2$-S	L-Tyr	17.8	11.4	S6
20	H	H	H	CH$_2$-S	D-Tyr	17.4	11.6 and 8.25	S6

[a]   The oxytocin analogues have the structure

[b]   K value for the corresponding sulfoxide.

Conditions: Chromatography was performed on a 25 × 0.4 cm, I.D. column packed with Separon SI-C-18 (Laboratory Apparatus, Prague, Czechoslovakia). A UV detector was employed.

Solvents: S1 = Methanol-0.1 M ammonium acetate, pH 7, 7:3.
S2 = Methanol-0.05% TFA, 13:7.
S3 = Methanol-0.05% TFA, 7:3.
S4 = Methanol-0.05% TFA, 3:2.
S5 = Methanol-0.05% TFA, 3:2.
S6 = Methanol-0.05% TFA, 2:3.

REFERENCE

1. **Lebl, M.,** *J. Chromatogr.,* 264, 459, 1983.

Reproduced by permission of Elsevier Science Publishers B.V.

Table LC 54
ENKEPHALIN ANALOGUES

Packing	P1	P1	P1	P1	P1
Column					
Length (cm)	30	30	30	30	30
Diameter (mm, I.D.)	3.9	3.9	3.9	3.9	3.9
Solvent	S1	S2	S3	S4	S5
Flow rate (mℓ/min)	1.0	1.0	1.0	1.0	1.0
Temperature	rt	rt	rt	rt	rt
Detection	D1	D1	D1	D1	D1

Peptide			k'		
Tyr-D-Ala-Gly-Phe-Met-NH$_2$	2.54	1.94	1.77	2.07	1.45
Tyr-D-Ala-Gly-Phe-Met	1.37	1.26			1.25
Tyr-D-Ala-Gly-Phe-Leu-NH$_2$	3.15	2.38	2.00	1.64	1.33
Tyr-Gly-Gly-Phe-Leu-NH$_2$	2.75	1.95	1.77	2.00	1.70
Tyr-Gly-Gly-Phe-Leu	1.30	1.17			1.15
Tyr-D-Met-Gly-Phe-Pro-NH$_2$	3.40	2.32	2.00	2.21	
Tyr-2-Me-Ala-Gly-Phe-Met-NH$_2$	3.35	2.25	2.00	2.21	
Tyr-Gly-Gly-Phe-Met-NH$_2$	2.20	1.69			
Tyr-Ala-Gly-Phe-Met-NH$_2$	2.35	2.10	1.85	2.15	
Tyr-Gly-Gly-Phe-Met	1.21	1.16			

Table LC 54 (continued)
ENKEPHALIN ANALOGUES

Peptide	k′	
3,5-Br₂-Tyr-Gly-Gly-Phe-Leu	1.97	1.80
Ala-Ala-Ala-Tyr-Gly-Gly-Phe-Leu	1.47	1.40
Ala-Ala-Ala-Tyr-Gly-Gly-Phe-Met	1.16	1.11
3,5-Br₂-Tyr-Gly-Gly-Phe-Met	1.94	1.56

Packing: P1 = μBondapak®/Phenyl.

Solvent: S1, S2, S3, and S5 consist of aqueous acetonitrile containing 0.01 *M* ammonium acetate adjusted to pH 4.5 with acetic acid, the percentages of acetonitrile being 25, 30, 40, and 55, respectively. S4 is aqueous acetonitrile containing 0.005 *M* ammonium acetate adjusted to pH 4.5 with acetic acid.

Detection: D1 = UV at 254 and 280 nm.

REFERENCE

1. **Currie, B. L., Chang, J.-K., and Cooley, R.,** *J. Liq. Chromatogr.,* 3, 513, 1980.

Reprinted by courtesy of Marcel Dekker, Inc.

Table LCᵃ 55
β-ENDORPHIN ANALOGUES
MODIFIED IN POSITIONS 2 & 5

Packing or layer	P1	P2	L1
Column			
Length (cm)	48	20.4	
Diameter (cm)	1.76	1.06	
Solvent	S1	S1	S2
Detection	D1	D1	D1

Endorphin analogue	$R_F \times 100$		
(Sar²)-β$_c$-EPᵇ	26	14.8	48
(Ala²)-β$_c$-EP	29	16.7	46
(D-Leu²)-β$_c$-EP	51	34	50
(D-Lys²)-β$_c$-EP	37ᵈ	6.4	42
(Pro⁵)-β$_h$-EPᶜ	23	12.9	50
(Leu⁵)-β$_h$-EP	51	27	54
(D-Leu⁵)-β$_h$-EP	58	32	56
(D-Ala²,D-Leu⁵)-β$_h$-EP	63	36	55

Note: Sar, sarcosine.

ᵃ Includes some TLC data.
ᵇ β$_c$ − EP = β-endorphin from camel glands.
ᶜ β$_h$ − EP = β-endorphin from human pituitary glands.
ᵈ Solvent system 1-butanol-pyridine-0.1 *N* NH₄OH containing 0.1% acetic acid, 2:1:3.

Packing: P1 = Sephadex® G-50.
             P2 = Bio-Gel® A-0.5 m (200—400 mesh).
Layer: L1 = Silica gel.
Solvent: S1 = 1-Butanol-pyridine-0.6 *M* ammonium acetate, 5:3:10 v/v.

Table LC^a 55 (continued)
β-ENDORPHIN ANALOGUES
MODIFIED IN POSITIONS 2 & 5

S2 = 1-Butanol-pyridine-acetic acid-water,
5:5:1:4 v/v.
Detection: D1 = Ninhydrin and Cl$_2$-toluidine.

Note: Here, $R_F = V_H/V_E$, where V_H is the hold-up volume and V_E is the elution volume of the peak.

REFERENCE

1. **Yamashiro, D., Li, C. H., Tseng, L.-E., and Loh, H. H.**, *Int. J. Pept. Prot. Res.*, 11, 251, 1978.

Reproduced by permission of Munksgaard International Publishers Ltd.

Table LC 56
ANALOGUES OF
β$_c$-ENDORPHIN^a

Packing	P1
Column	
Length (cm)	59.8
Diameter (cm)	1.18
Solvent	S1
Flow rate (mℓ/hr)	4.5
Temperature (°C)	24
Detection	D1

Peptide	$R_F \times 100$
β$_c$-EP	30
(D-Tyr^1)-β$_c$-EP	29
(D-Ala^2)-β$_c$-EP	35
(D-Phe^4)-β$_c$-EP	35
(D-Met^5)-β$_c$-EP	35

^{a} β$_c$-Endorphin (β$_c$-EP) = β-endorphin from camel glands.

Packing: P1 = Sephadex® G-50.
Solvent: S1 = 1-Butanol-pyridine-0.6 *M* NH$_4$OAc, 5:3:10, the pH of the lower phase is 8.2.
Detection: D1 = Collection of fractions and Folin-Lowry analysis.

Note: Here $R_F = V_H/V_E$, where V_H is the hold-up volume and V_E is the elution volume of the peak.

REFERENCE

1. **Yamashiro, D., Tseng, L.-F., Doneen, B. A., Loh, H. H., and Li, C. H.**, *Int. J. Pept. Prot. Res.*, 10, 159, 1977.

Reproduced by permission of Munksgaard International Publishers, Ltd.

Table LC 57
ENDORPHIN ANALOGUES

Packing	P1	P1	P1
Column			
Length (cm)	30	30	30
Diameter (mm, I.D.)	3.9	3.9	3.9
Solvent	S1	S2	S3
Flow rate (mℓ/min)	1.0	1.0	1.0
Temperature	rt	rt	rt
Detection	D1	D1	D1

Peptide		k'	
D-Ala^2-β-Endorphin (Human)	3.09	2.69	1.85
γ-Endorphin (β-Lipotropin 61-77)	2.18	1.15	1.08
α-Endorphin (β-Lipotropin 61-76)	1.94	1.15	1.00
2-MeAla^2-β-Endorphin (Human)	3.27	3.08	1.92
Arg-β-Endorphin (Human)	2.67	2.46	
β-Endorphin (Human) (β-Lipotropin 61-91)	2.26	1.82	
β-Endorphin (Camel)	2.76	2.55	

Packing: P1 = μBondapak®/Phenyl.
Solvent: S1 and S1 consist of aqueous acetonitrile containing 0.01 *M* ammonium acetate adjusted to pH 4.5 with acetic acid, the percentages of acetonitrile being 40 and 45, respectively. S3 is aqueous acetonitrile containing 0.005 *M* ammonium acetate adjusted to pH 4.5 with acetic acid, the percentage of acetonitrile being 50.
Detection: D1 = UV at 254 and 280 nm.

REFERENCE

1. **Currie, B. L., Chang, J.-K., and Cooley, R.,** *J. Liq. Chromatogr.*, 3, 513, 1980.

Reprinted by courtesy of Marcel Dekker, Inc.

Table LC 58
IODINATED AND TRITIATED ANALOGUES OF HUMAN β-ENDORPHIN (β$_h$-ENDORPHIN)

Packing	P1	P1	P1
Column			
Length (cm)	34	34	34
Diameter (cm)	1.0	1.0	1.0
Solvent	S1	S2	S3
Detection	D1	D1	D1

Analogue		R_F	
β$_h$-Endorphin	0.176	0.326	0.451
(Dit^1)-β$_h$-Endorphin	0.375	0.675	0.850
(Dit^{27})-β$_h$-Endorphin	0.375	0.677	0.854
(Dit^{1,27})-β$_h$-Endorphin	0.615	0.855	0.950
(^{3}H$_2$-Tyr^1)-β$_h$-Endorphin	0.177	0.320	0.446
(^{3}H$_2$-Tyr^{27})-β$_h$-Endorphin	0.179	0.329	0.453
(^{3}H$_2$-Tyr^{1,27})-β$_h$-Endorphin	0.174	0.328	0.451

Abbreviation: Dit, diiodotyrosine.

Table LC 58 (continued)
IODINATED AND TRITIATED ANALOGUES OF
HUMAN β-ENDORPHIN (β$_h$-ENDORPHIN)

Packing: Pl = Sephadex® G-50.
Solvent: S1, S2 and S3. Partition chromatography was carried out using the
two phases from butanol-acetic acid-water-pyrimidine, 4:1:5x, x
being 0.01, 0.05, and 0.20 for S1, S2, and S3, respectively.
Detection: Dl = counts/min or optical density at 280 nm.

Note: Here, $R_F = V_H/V_E$, wher V_h is the hold-up volume and V_E is the where
V_h is the hold-up volume and V_E is the elution volume of the peak.

REFERENCE

1. **Houghten, R. A., Chang, W.-C., and Li, C. H.,** *Int. J. Pept. Prot.
Res.*, 16, 311, 1980.

Table LC 59
β-LIPOTROPIN 61-91 FRAGMENTS

Packing	P1
Column	
Length (cm)	30
Diameter (cm)	0.39
Solvent	S1
Flow rate (mℓ/min)	2
Temperature	rt
Detection	D1

β-LPH 61-91 fragment[a]	t$_r$ (min)
80-91	2.1
62-69	4.5
70-76	4.5
62-65 (desTyr-Met-enkephalin)	5.5
61-67	8.4
61-69	8.4
61-65 (Met-enkephalin)	11.0
Leu-enkephalin	17.6
66-77	19.0
62-76 (desTyr-α-endorphin)	20.5
70-77	21.8
65-77	22.3
61-76 (α-endorphin)	25.5
62-77 (desTyr-α-endorphin)	33.5
61-77 (α-endorphin)	35.0
62-91 (desTyr-β-endorphin)	43.6
61-91 (β-endorphin)	43.6

[a] β-LPH, β-lipotropin.

Packing: Pl = Reversed phase μBondapak® C$_{18}$
(octadecyltrichlorosilane chemically
bound to 10 μm porous silica particles).
Solvent: S1 = Gradient elution with 0.01M ammonium acetate adjusted to pH 4.15 with
glacial acetic acid (A) and methanol

Table LC 59 (continued)
β-LIPOTROPIN 61-91 FRAGMENTS

(B). The ratio of the concentration of A and B was 7:3 initially and 1:3 finally.

Detection: Dl = UV at 210 rm.

REFERENCE

1. **Loeber, J. G., Verhoef, J., Burbach, J. P. H., and Witter, A.,** *Biochem. Biophys. Res. Commun.,* 86, 1288, 1979.

Reproduced by permission of Academic Press, Inc.

Table LC 60
EFFECT OF MOBILE PHASE
BUFFER COMPOSITION ON
RETENTION OF SOMATOSTATIN

Buffer	k'
0.02 *M* Ammonium acetate, pH 7.0	25.5
0.02 *M* Ammonium acetate, pH 4.4	7.0
0.1 *M* H$_3$PO$_4$, pH 2.1	9.5
0.1 *M* NaH$_2$PO$_4$, pH 2.1	4.6
0.1 *M* NaH$_2$PO$_4$ + 0.05 *M* TMA,[a] pH 2.1	2.9
0.1 *M* NaH$_2$PO$_4$ + 0.05 *M* TEA,[b] pH 2.1	2.3
TEAP,[c] pH 3.4	2.4
TEAP, pH 2.1	2.1

[a] TMA, tetramethyl ammonium chloride.
[b] TEA, triethylamine.
[c] TEAP, 0.25 NH$_3$PO$_4$, pH adjusted with TEA.

Packing:	MicroPak® CN-AQ-10, cyanopropyl phase on 10-μm diameter silica, packed in methanol.
Column:	Stainless steel, 30 cm × 4 mm.
Solvent:	Buffer — acetonitrile, 85:15.
Flow rate:	1.0 mℓ/min.
Temperature:	Ambient.
Detection:	UV at 210 nm.
Sample:	2 μg somatostatin in 10 μℓ water.

REFERENCE

1. **Wehr, C. T., Correia, L., and Abbott, S. R.,** *J. Chromatogr. Sci.,* 20, 114, 1982.

Reproduced from the Journal of Chromatographic Science by permission of Preston Publications, Inc.

Table LC 61
SOMATOSTATIN ANALOGUES

Packing	P1
Column	
Length (cm)	150
Diameter (mm, I.D.)	4.6
Material	SS
Solvent	S1
Flow rate (mℓ/min)	1.0
Detection	D1

Peptide	r
Somatostatin	1.00
Somatostatin, linear	0.54
$Ser^{3,14}$-somatostatin, linear	0.32
Des Asn^5-somatostatin	0.70
Des Ala^1,Gly^2,Asn^5-somatostatin	0.69

Packing: P1 =   LiChrosorb® RP-8, 5-μm particles.

Solvent: S1 =   25.5% of acetonitrile in phosphate buffer, pH 4.5 (ionic strength = 0.1).

Detection: S1 =   UV at 210 nm.

REFERENCE

1. **Abrahamsson, M. and Groningsson, K.,** *J. Liq. Chromatogr.*, 3, 495, 1980.

Reprinted by courtesy of Marcel Dekker, Inc.

Table LC 62
NEUROTENSIN ANALOGUES

Packing	P1	P1
Solvent	S1	S2
Flow rate (mℓ/min)	2	1.5
Detection	D1	D1

Peptide	t_r(min)	
Neurotensin	5.9	10.5
Neurotensin-(4-13)	3.9	4.6
Neurotensin-(6-13)	3.7	4.5
Neurotensin-(8-13)	ND	5.7
(Ala^{13})Neurotensin-(8-13)	3.2	6.1[a]
(Ala^{12})Neurotensin-(8-13)	3.5	11.4[a]
(Lys^8)Neurotensin-(8-13)	4.0	5.3
(Lys^9)Neurotensin-(8-13)	4.2	5.0
(Cit^8)Neurotensin-(8-13)	5.2	6.2
(Cit^9)Neurotensin-(8-13)	5.2	7.1
(Phe^{11})Neurotensin-(8-13)	5.6	7.4
Ac-Neurotensin-(8-13)	ND	7.4

[a]   In this case the mobile phase contained 15% acetonitrile.

Table LC 62 (continued)
NEUROTENSIN ANALOGUES

Abbreviations: ND, not determined; Cit, citrulline; Ac, acetyl.

Packing: P1 = Two columns of C_{18} μBondapak® in series.

P2 = Two columns of RP_{18} LiChrosorb® in series.

Solvent: S1 = 10 m*M* Triethylammonium phosphate, pH 3-methanol, 50:50 v/v.

S2 = 1% Phosphoric acid-acetonitrile, 74:26 v/v.

Detection: D1 = Absorbance at 230 nm.

REFERENCE

1. **Granier, C., Van Rietschoten, J., Kitabgi, P., Poustis, C., and Freychet, P.**, *Eur. J. Biochem.*, 124, 117, 1982.

Reproduced with permission.

Table LC 63
DIASTEREOMERS AND
ANALOGUES OF
NEUROTENSIN (NT)

Packing	P1	P1	P1
Column			
Length (cm)	30	30	30
Diameter (cm)	0.4	0.4	0.4
Solvent	S1	S1	S1
Flow rate (mℓ/min)	1	1	1
Temperature (°C)	30	40	50
Detection	D1	D1	D1

Neurotensin	t_r(min)		
NT	27.6	28.2	30.2
(D-Arg⁹)-NT	26.5	27.4	29.2
(D-Glu⁴)-NT	29.9	31.2	32.6
(Leu¹¹)-NT	25.8	26.4	27.6
(Lys⁸)-NT	30.3	31.0	32.8
(Lys⁹)-NT	30.9	31.7	33.4
(Phe¹¹)-NT	23.4	24.1	25.9
(D-Phe¹¹)-NT	16.1	17.5	18.7
(D-Pro¹⁰)-NT	21.8	23.1	24.3
(Trp¹¹)-NT	23.4	24.1	25.9
(D-Tyr¹¹)-NT	19.2	20.7	22.0

Packing: P1 = MicroPak® AX-10, a difunctional weak anion-exchange bonded phase prepared on LiChrosorb® Si-60 silica, 10 μm.

Table LC 63 (continued)
DIASTEREOMERS AND ANALOGUES OF NEUROTENSIN (NT)

Solvent: S1 = Eluent A, acetonitrile; eluent B, 0.01 *M* triethylammonium acetate, pH 6.0; gradient program linear, starting from 23% B with a linear rate of 0.3% B/min.

Detection: D1 = UV at 220 nm and amino acid analysis of samples of the eluate after drying *in vacuo* and hydrolyzing with HCl.

REFERENCE

1. **Dizdaroglu, M., Simic, M. G., Rioux, F., and St.-Pierre, S.,** *J. Chromatogr.,* 245, 158, 1982.

Table LC 64
NEUROPEPTIDES

Packing	P1
Column	
Length (cm)	30
Diameter (mm, I.D.)	3.9
Solvent	S1
Flow rate (mℓ/min)	1
Temperature	rt
Detection	D1

Peptide	t_r(min)
LPH_{88-91}	3.6
Met-enkephalin	16.9
Leu-enkephalin	18.2
α-MSH	18.9
γ-LPH	19.5
hACTH	20.4
β-LPH	21.4
β-Endorphin	22.4

Packing: P1 = μBondapak® C_{18}, particle size 10 μm.

Solvent: S1 = Gradient of 3.5 to 49% acetonitrile with 0.08% TFA in water over 20 min.

Detection: D1 = UV at 206 nm.

Table LC 64 (continued)
NEUROPEPTIDES

REFERENCE

1. McDermott, J. R., Smith, A. I.,
Biggins, J. A., Chyad Al-
Noaemi, M., and Edwardson, J.
A., *J. Chromatogr.*, 222, 371,
1981.

Reproduced by permission of Elsevier Science Publishers B.V.

Table LC 65
NEUROPEPTIDES

Packing	P1	P2
Column		
Length (cm)	10	
Diameter (mm, I.D.)	8	
Material		SS
Solvent	S1	S1
Flow rate (mℓ/min)	1.0	1.0
Detection	D1	D1

Peptide	t_r(min)	
TRH	9.17	5.56
LH-RH	19.81	18.51
Neurotensin	20.29	20.26
Substance P	22.06	21.60

Packing: P1 = Z-module radial-compression system using 10-μm irregular C_{18} capped silica cartridges.

P2 = μBondapak® 10-μm C_{18} column.

Solvent: S1: Solvent A = 11 m*M* TFA-2.6 m*M* acetic acid; solvent B = 70% acetonitrile containing 11 m*M* TFA. A 20-min linear gradient of 3 to 70% solvent B at 1 mℓ/min was established.

Detection: D1 = UV at 206 nm.

REFERENCE

1. **Smith, A. I. and McDermott, J. R.**, *J. Chromatogr.*, 306, 99, 1984.

Table LC 66
HPLC OF OPIOID PEPTIDES

Packing	P1	P1	P1
Column			
Length (mm)	250	250	250
Diameter (mm)	4.6	4.6	4.6
Solvent	S1	S2	S3
Flow rate (mℓ/hr)	40	40	40
Detection	D1	D1	D1

Peptide	t_r(min)		
Met-enkephalin	33	57	61
Leu-enkephalin	41	73	83
Met^5-β$_H$-endorphin	101	141	148
Leu^5-β$_H$-endorphin	106	150	155
Met^{65}-β-lipotropin (61-69)	31	51	51
Leu^{65}-β-lipotropin (61-69)	39	67	67

Packing: P1 = Lichrosorb® C_{18}, 10-μm particle size.
Solvent: S1 = 0.5 *M* acetic acid - 1.0 *M* pyridine, pH 5.5.
S2 = 0.5 *M* formic acid adjusted to pH 4.0 with pyridine.
S3 = 0.5 *M* formic acid adjusted to pH 3.0 with pyridine.
Detection: D1 = Fluorescamine.

REFERENCE

1. **Lewis, R. V., Stein, S., and Udenfriend, S.,** *Int. J. Pept. Prot. Res.,* 13, 493, 1979.

Reproduced by permission of Munksgaard International Publishers, Ltd.

Table LC 67
REVERSED-PHASE HPLC OF ANGIOTENSINS AND IODINATED DERIVATIVES

Packing	P1
Column	
Length (mm)	150
Diameter (mm, I.D.)	4.6
Material	SS
Solvent	S1
Flow rate (mℓ/min)	1
Temperature	rt
Detection	D1

Peptide	t_r(min)
Angiotensin I	9.6
Angiotensin II	3.3
Angiotensin III	6.0
^{125}I-Angiotensin I	19.9
^{125}I-Angiotensin II	5.4
^{125}I-Angiotensin III	16.8
Sarcosine^1, Ala^8-angiotensin II	2.5
des-Asp^1,Ile^8-angiotensin II	3.9

Table LC 67 (continued)
REVERSED-PHASE HPLC OF ANGIOTENSINS AND IODINATED DERIVATIVES

Packing: P1 = Bio-Sil® ODS-10(C_{18}-octadecylsilyl, 10-μm particle size).
Solvent: S1 = 50 mM NaH$_2$PO$_4$-25% acetonitrile, pH 6.0.
Detection: D1 = UV at 210 or 224 nm.

REFERENCE

1. Guy, M. N., Roberson, G. M., and Barnes, L. D., *Anal. Biochem.*, 112, 272, 1981.

Table LC 68
EFFECT OF TEMPERATURE ON THE HPLC OF ANGIOTENSINS

Packing	P1	P1	P1	P1
Column				
Length (cm)	30	30	30	30
Diameter (cm)	0.4	0.4	0.4	0.4
Solvent	S1	S1	S1	S1
Flow rate (mℓ/min)	1	1	1	1
Temperature (°C)	26	30	40	50
Detection	D1	D1	D1	D1
Angiotensin		t_r (min)		
A III	10.3	10.5	11.0	11.6
(Val^4)-A III	11.6	11.8	12.2	12.8
A III inhibitor	13.2	13.3	13.4	13.7
(Asn^1-Val^5)-A II	15.2	15.6	16.2	17.1
(Sar^1-Ile^8)-A II	18.1	17.8	17.1	17.1
(Sar^1-Ala^8)-A II	27.2	26.3	25.0	24.3
(Sar^1-Gly^8)-A II	27.2	26.3	26.2	26.8
(Sar^1-Thr^8)-A II	29.6	28.8	27.8	26.8
(Sar^1-Val^5-Ala^8)-A II	30.5	29.5	27.8	26.8
A II	34.4	35.5	38.3	40.8
A I	40.8	42.8	45.8	48.0
(Val^5)-A II	46.0	47.7	49.8	51.1

Packing: P1 = MicroPak® AX-10.
Solvent: S1 = A = acetonitrile; B = 0.01 M triethylammonium acetate, pH 6.0. Gradient elution was performed starting from 24% B with a rate of 0.1% B for 25 min and then 0.5% B/min.
Detection: D1 = UV at 220 nm.

REFERENCE

1. Dizdaroglu, M., Krutzsch, H. C., and Simic, M. G., *Anal. Biochem.*, 123, 190, 1982.

Reproduced by permission of Academic Press, Inc.

Table LC 69
DANSYL-MELANOTROPIN INHIBITING FACTOR (Dns-MIF) AND ITS METABOLITES

Packing	P1	P2	P1	P1	P2	P1	P2	P1	P1	P2	P1	P2	P1
Column													
Length (mm)	30	30	30	30	30	30	30	30	30	30	30	30	30
Diameter (mm, I.D.)	3.9	3.9	3.9	3.9	3.9	3.9	3.9	3.9	3.9	3.9	3.9	3.9	3.9
Solvent	S1	S1	S2	S3	S3	S4	S4	S5	S5	S6	S7	S8	S9
Flow rate (mℓ/min)	3	3	3	3	3	3	3	3	3	3	3	3	3
Detection	D1	D1	D1	D1	D1	D1	D1	D1	D1	D1	D1	D1	D1
Compound							k'						
MIF	24	3.2	30	5.0	5.3	7.0	8.0	4.5	3.2	30	0.4	30	0.6
Pro-Leu	3.4	1.0	4.7	1.2	0.6	1.4	0.6	0.6	0.4	30	0.4	5.8	0.6
Leu-Gly	1.7	1.0	3.7	0.5	0.6	0.4	0.6	0.6	0.4		0.4	3.5	0.6
Pro	1.0	1.0	3.7	0.5	0.6	0.4	0.6	0.6	0.4	30	0.4	3.	0.6
Leu	2.0	3.6	3.7	0.5	0.6	1.0	5.3	0.6	2.5		0.4	3.5	0.6
Gly	1.0	1.0	10	0.5	0.6	1.0	0.6	0.6	0.4	30	0.4	1.0	0.6
Gly-NH$_2$	5.3		30	2.6	4.0	4.2	5.3	1.8	2.8		0.4	30	0.6
OH	1.0	1.0	3.7	0.5	0.6	0.4	0.6	0.6	0.4	1.4	0.4	1.0	0.6

Packing: P1 = μBondapak® C$_{18}$, 10 μm.
P2 = μBondapak® phenyl, 10 μm.

Solvent: S1, S2, S3, S4, S5, S6, and S7 are acetonitrile-0.01 M sodium sulfate buffer, pH 7 mixtures in the respective ratios: S1, 45:55;
S2, 40:60; S3, 43.5:56.5; S4, 41.5:58.5; S5, 50:50; S6, 20:80; S7, 80:20. S8 = Methanol-sodium sulfate buffer, 50:50. S9 =
Isopropanol-1% acetic acid, 5:95.

Detection: D1 = UV at 254 nm.

REFERENCE

1. **Hui, K.-S., Salschutz, M., Davis, B. A., and Lajtha, A.,** *J. Chromatogr.,* 192, 341, 1980.

Reproduced by permission of Elsevier Science Publishers B.V.

Table LC 70
DERIVATIVES OF MIF AND ITS METABOLITES

Ethansyl derivative

Packing	P1	P2	P1	P1	P2	P1	P2	P1	P2	P2
Column Length (mm)	30	30	30	30	30	30	30	30	30	30
Diameter (mm, I.D.)	3.9	3.9	3.9	3.9	3.9	3.9	3.9	3.9	3.9	3.9
Solvent	S1	S1	S2	S3	S3	S4	S4	S5	S5	S6
Compound (k')										
MIF	>30	7.6	21	12	11	16	13	6.6	6.8	30
Pro-Leu	1.8	0.6	3.0	1.2	0.4	2.2	1.5	2.2	0.4	6.8
Leu-Gly	0.6	0.6	2.0	11	0.4	0.4	0.6	0.6	0.4	15
Pro	1.4	0.6	1.0	1	0.4	0.8	0.6	0.6	0.4	5.0
Leu	1.4	3.6	2.2	1.4	7.5	1.6	9.0	0.6	4.0	7.0
Gly	0.6	3.8	0.8	0.6	7.5	0.4	9.0	0.6	4.0	3.0
Gly-NH$_2$	>30	3.8	8	5.6	2.5	6.6	6.0	3.4	2.6	8.6
OH	0.6	0.6	0.8	0.6	0.4	0.4	0.6	0.6	0.4	1.8

Propansyl derivative

Packing	P1	P2	P1	P1	P2	P1	P2	P1	P2	P2
Column Length (mm)	30	30	30	30	30	30	30	30	30	30
Diameter (mm, I.D.)	3.9	3.9	3.9	3.9	3.9	3.9	3.9	3.9	3.9	3.9
Solvent	S1	S1	S2	S3	S3	S4	S4	S5	S5	S6
Compound (k')										
MIF	>30	>30	>30	>30	>30	30	30	14	7.8	30
Pro-Leu	4.0	1.4	7.8	4.6	1.0	5.6	3.0	2.2	0.4	30
Leu-Gly	2.6	1.0	5.8	2.6	1.0	3.8	1.0	0.6	0.4	30
Pro	2.2	2.0	3.6	2.2	2.0	2.8	1.8	0.6	1.8	13
Leu	3.0	0.8	5.8	3.0	1.2	4.2	1.8	1.4	6.0	18
Gly	1.0	0.8	1.8	1.4	1.0	1.8	1.0	0.6	6.0	7

Table LC 70 (continued)
DERIVATIVES OF MIF AND ITS METABOLITES

Gly-NH$_2$	>30	>30	>30	>30	>30	>30	>30	7.4	4.4	30
OH	1.0	0.8	2.0	1.4	1.0	1.4	1.0	0.6	0.4	6.2
Packing	P1	P2	P1	P1	P2	P1	P2	P1	P2	P2
Column										
Length (mm)	30	30	30	30	30	30	30	30	30	30
Diameter (mm, I.D.)	3.9	3.9	3.9	3.9	3.9	3.9	3.9	3.9	3.9	3.9
Solvent	S1	S1	S2	S3	S3	S4	S4	S5	S5	S6

Compound

k'

Bns derivative

Compound										
MIF	30	30	30	30	30	>30	>30	29	12	30
Pro-Leu	13	3.0	11	9.4	7.0	13	9.8	6.2	1.0	11
Leu-Gly	7.0	2.2	2.2	6.2	5.0	9.0	7.5	1.6	1.0	13
Pro	6.0	1.8	10	5.0	3.8	6.2	5.0	3.6	1.0	11
Leu	8.2	2.2	16	6.6	5.0	9.4	7.3	4.0	1.0	9.4
Gly	3.8	1.4	7.0	3.4	3.3	4.6	4.3	2.0	1.0	15
Gly-NH$_2$	>30	12	>30	>30	>30	>30	>30	15	7.2	30
OH	3.0	1.4	5.8	2.6	3.0	3.2	3.5	1.6	1.0	11
Packing	P1	P2	P1	P1	P2	P1	P2	P1	P2	P2
Column										
Length (mm)	30	30	30	30	30	30	30	30	30	30
Diameter (mm, I.D.)	3.9	3.9	3.9	3.9	3.9	3.9	3.9	3.9	3.9	3.9
Solvent	S1	S1	S2	S3	S3	S4	S4	S5	S5	S6

Compound

k'

Monoisopropansyl derivative

Compound										
MIF	6.2	3.2	11	7.0	5.3	8.2	8.0	4.0	3.2	>30

Pro-Leu	1.4	1.0	0.6	0.4	0.6	1.0	0.6	0.6	0.4	5.8
Leu-Gly	0.6	1.0	0.6	0.8	0.6	0.4	0.6	0.6	0.4	3.5
Pro	2.8	1.0	0.9	0.8	0.6	0.4	0.6	0.6	0.4	3.0
Leu	2.2	3.6	1.0	0.6	0.6	0.4	5.3	5.3	2.5	3.5
Gly	0.6	1.0	0.8	0.4	0.6	0.4	0.6	0.6	0.4	1.0
Gly-NH$_2$	2.8		4.2	3.0	4.0	3.2	5.3	2.2	2.8	>30
OH	0.6	1.0	0.8	0.4	0.6	0.4	0.6	0.4	0.4	1.0

Note: MIF, melanotropin inhibiting factor; ethansyl, diethylnaphthylene-sulfonyl; propansyl, dipropylnaphthylenesulfonyl; BnS, dibutyl-naphthylenesulfonyl; monoisopropansyl, monoisopropylaminonaphthylenesulfonyl.

Packing: P1 = μBondapak® C$_{18}$, 10-μm, prepacked column.
P2 = μBondapak® phenyl, 10-μm, prepacked column.

Solvent: S1, S2, S3, S4, and S5 are acetonitrile-0.01 M sodium sulfate buffer, pH 7, mixtures in the respective ratios: S1, 45:55; S2, 40:60; S3, 43.5:56.5; S4, 41.5:58.5; S5, 50:50. S6 = Methanol-sodium sulfate buffer, 50:50.

Detection: D1 = UV at 254 nm.

REFERENCE

1. **Hui, K.-S., Salschutz, M., Davis, B. A., and Lajtha, A.**, *J. Chromatogr.*, 192, 341, 1980.

Reproduced by permission of Elsevier Science Publishers B.V.

Table LC 71[a]
α-MELANOTROPIN PEPTIDES

Packing or layer	L1	P1	P2	P3
Column				
Length (cm)		18	50	50
Diameter (cm)		1.0	1.0	1.0
Solvent	S1	S2	S3	S3
Flow rate (mℓ/hr)		1.2	60	60
Detection	D1	D2	D2	D2

MSH[b] peptide		$R_F \times 100$		
α-MSH	61	39	62	63
3,5-diiodotyrosine²-α-MSH			33	26
3,5-ditritiotyrosine²-α-MSH	61	38		63

[a] TLC data also included
[b] α-MSH, α-melanotropin.

Layer: L1 = Silica gel G.
Packing: P1 = Sephadex® G-50.
              P2 = LiChrosorb® RP-8.
              P3 = LiChrosorb® RP-18.
Solvent: S1 = *n*-Butanol-pyridine-acetic acid-water, 5:5:1:4 v/v.
              S2 = *n*-Butanol-pyridine-0.1% acetic acid, 10:2:11 v/v.
              S3 = 17% *n*-Propanol in 1 *M* pyridine acetate, pH 5.5.
Detection: D1 = Ninhydrin.
              D2 = Fluorescamine or radioactivity.

Note: Here $R_F = V_H/V_E$ where V_H is the hold-up volume and V_E is the elution volume of the peak.

REFERENCE

1. **Buckley, D. I. and Ramachandran, J.,** *Int. J. Pept. Prot. Res.*, 17, 514, 1981.

Table LC 72
INFLUENCE OF THE MOBILE PHASE ON PEPTIDE RETENTION TIMES IN HPLC

Limiting mobile phase	t_r(min)		
	αCB-1[a]	αCB-2	αCB-3
1-Propanol	18.3	28.0	34.0
2-Propanol	21.6	34.3	39.2
70% 1-Propanol, 30% tetrahydrofuran	19.2	30.9	37.0
70% 1-Propanol, 30% acetonitrile	26.6	36.8	44.0
50% 1-Propanol, 50% tetrahydrofuran	22.3	33.0	40.0
Tetrahydrofuran	33.2	41.8	47.4
Dioxane	34.1	42.0	49.6
Ethanol	48.1	NE[b]	NE
Acetonitrile	49.0	NE	NE
Methanol	NE	NE	NE

[a] αCB-1, αCB-2, and αCB-3 are peptides produced by cyanogen bromide cleavage of human hemoglobin α chain.
[b] NE, not eluted.

Table LC 72 (continued)
INFLUENCE OF THE MOBILE PHASE ON PEPTIDE RETENTION TIMES IN HPLC

Column: LiChrosorb RP-8, 5 μm, 4.6 × 250 mm. Samples of 100 μg were dissolved in 0.1% TFA. A binary solvent system was employed. The starting solvent was 0.013 M TFA in water, and a linear gradient was established using a limiting organic solvent containing 0.013 M TFA. All solvents contained 0.1% TFA to prevent baseline shift.

Flow rate: 0.7 mℓ/min.
Rate of solvent change: 1.6%/min.
Temperature: 30°C.
Detection: UV.

REFERENCE

1. **Mahoney, W. C. and Hermodson, M. A.**, *J. Biol. Chem.*, 255, 11199, 1980.

Reproduced by permission of the American Society of Biological Chemists and the authors.

Table LC 73
HPLC OF γCB-3[a]: RETENTION TIME AS A FUNCTION OF SOLVENT

Solvent	t_r(min)	% solvent to elute peptide
Methanol	34.6	57.6
Ethanol	27.6	46.0
1-Propanol	20.3	33.8
2-Propanol	22.5	37.5
Dioxane	26.2	43.6
Tetrahydrofuran	26.1	43.4
Acetonitrile	28.1	46.8

[a] γCB-3 is the peptide containing residues 134 to 146 of the γ chain of human hemoglobin.

Packing: LiChrosorb® RP-8, 5 μm.
Column: 4.6 × 250 mm.
Solvent: Linear gradients were formed of each solvent with 0.1% of TFA in both the water and organic phases. The rate of solvent change was 1.6%/min.
Temperature: 30°C.
Flow rate: 0.7 mℓ/min.
Detection: UV at 210 nm.

REFERENCE

1. **Mahoney, W. C. and Hermodson, M. A.**, *J. Biol. Chem.*, 255, 11199, 1980.

Table LC 74
COMPARISON OF γCB-3 PREPARATIONS FROM HPLC OR GEL FILTRATION

	HPLC		Gel filtration		
Hb F sample	Gly[a]	Ala	Gly	Ala	%Gγ chains
1	0.79	2.20	0.80	2.21	79/80
2	0.80	2.37	0.81	2.25	80/81
3	0.74	2.35	0.71	2.30	74/71
4	0.79	2.34	0.66	2.40	79/66

[a]   Residues/peptide.

Note: The 13 residue COOH-terminal cyanogen bromide peptide of the γ chain of fetal globin isolated from human embryos (γ CB-3) was isolated by reversed-phase HPLC on a LiChrosorb® C_8 column and also by gel filtration on Sephadex®. The fragments were hydrolyzed and subjected to amino acid analysis to determine the ratio of $^Gγ$ to $^Aγ$ chains in the sample. The data from the samples obtained by HPLC were consistent with those obtained by gel filtration. The HPLC analyses took 26 min each while the Sephadex® columns took 24 hr or more.

REFERENCE

1.   **Mahoney, W. C. and Hermodson, M. A.,** *J. Biol. Chem.*, 255, 11199, 1980.

Reproduced by permission of the American Society of Biological Chemists and the authors.

Table LC 75
TRYPTIC PEPTIDES OF THE β-CHAIN OF HEMOGLOBIN

Packing	P1	P1	P1
Column			
Length (mm)	250	250	250
Diameter (mm)	4.6	4.6	4.6
Solvent	S1	S2	S3
Detection	D1	D1	D1
Peptide		t_r(min)	
β-T_1	18	31	18
β-T_4	56	59	50

Note: Peptides β-T_1 and β-T_4 have the amino acid compositions Val-His-Leu-Thr-Pro-Glu-Glu-Lys and Leu-Leu-Val-Val-Tyr-Pro-Trp-Thr-Gln-Arg, respectively. They did not give a very good separation using solvent system S3.

Packing: P1   = Partisil®-10-ODS-2.

Table LC 75 (continued)
TRYPTIC PEPTIDES OF THE β-CHAIN OF HEMOGLOBIN

Solvent: S1 = Acetonitrile in 0.1% H_3PO_4.
          S2 = Acetonitrile in 0.1% TFA.
          S3 = Acetonitrile in 5.4 mM H_3PO_4-49 mM KH_2PO_4. All the chromatograms were developed with a linear gradient of the organic phase from 5 to 70% acetonitrile.
Detection: D1 = UV at 210 nm.

REFERENCE

1. Acharya, A. S., Di Donato, A., Manjula, B. N., Fischetti, A., and Manning, J. M., *Int. J. Pept. Prot. Res.*, 22, 78, 1983.

Table LC 76
CYANOGEN BROMIDE PEPTIDE FRAGMENTS OF MYOGLOBIN

Packing		P1		P1
Solvent		S1		S2
Flow rate (mℓ/min)		2.0		2.0
Detection		D1		D1

Myoglobin residues	t_r(min)			
1—55	26.4[a]	26.0	45.8[a]	45.6
56—131	27.5[a]	28.3	47.9[a]	50.0
132—153	20.1[a]	20.6	33.2[a]	34.2

[a] Predicted on the basis of amino acid retention constants.

Packing: P1 = μBondapak® C_{18} column.
Solvent: S1 = The mobile phase was 0.1% TFA, pH 2, and the mobile phase modifier was acetonitrile containing 0.07% TFA. The concentration of the modifier was increased linearly at 2%/min over 60 min.
          S2 = As S1 but the rate of increase of the modifier was 1%/min.
Detection: D1 = UV at 216 nm.

REFERENCE

1. Sasagawa, T., Okuyama, T., and Teller, D. C., *J. Chromatogr.*, 240, 329, 1982.

Reproduced by permission of Elsevier Science Publishers B.V.

Table LC 77
TYROSINE-RELATED
PEPTIDES OF
THYMOPOIETIN$_{32\text{-}36}$
PENTAPEPTIDE

Packing	P1
Column	
Length (ft)	1.0
Diameter (in., O.D.)	0.25
Material	SS
Solvent	S1
Flow rate (mℓ/min)	1.0
Temperature	rt
Detection	D1

Peptide	t_r(min)
Arg-Lys-Asp-Val-Tyr	7.5
Lys-Asp-Val-Tyr	9.8
Asp-Val-Tyr	17.5
Val-Tyr	5.9
Tyr	4.3

Packing: P1 = μBondapak® C$_{18}$ reversed-phase support.

Solvent: S1 = 0.08 M TEAP, pH 4.0-acetonitrile, 96:4 v/v. The TEAP buffer is prepared by diluting 85% phosphoric acid with distilled water to give the appropriate molarity of phosphate. The pH of the solution is then adjusted by addition of triethylamine.

Detection: D1 = UV at 280 nm.

REFERENCE

1. **Tischio, J. P. and Heytei, N.,** *J. Chromatogr.,* 236, 237, 1982.

Table LC 78
LUTEINIZING HORMONE RELEASING
FACTOR (LRF) ANALOGUES

Packing	P1
Solvent	S1
Flow rate (mℓ/min)	1.0
Temperature (°C)	50
Detection	D1

Peptide	t_r(sec)
[Δ³-LPro-pClDPhe-DTrp-Ser-Tyr / βAla-Pro-Arg-NᵅCH₃Leu-DTrp]	1293
H₂N-Δ³-LPro-pClDPhe-DTrp-Ser-Tyr / HO-βAla-Pro-Arg-NᵅCH₃ Leu-DTrp]	1236
[Δ³-DPro-pClDPhe-DTrp-Ser-Tyr / βAla-Pro-Arg-NᵅMeLeu-DTrp]	1280

Table LC 77 (continued)
TYROSINE-RELATED
PEPTIDES OF
THYMOPOIETIN$_{32\text{-}36}$
PENTAPEPTIDE

Peptide	t_r(sec)
H$_2$N-Δ^3-DPro-pClDPhe-DTrp-Ser-Tyr ⌐ HO-βAla-Pro-Arg-N$^\alpha$MeLeu-DTrp ⌐	1241

Packing: P1   = PAC ^{125}I protein analysis column.
Solvent: S1   = TEAP buffer, pH 2.25-30% acetonitrile. TEAP
                      buffer is prepared by bringing the pH of 0.25
                      N H$_3$PO$_4$ to 2.25 with triethylamine redistilled
                      over *p*-toluenesulfonyl chloride.
Detection: D1   = UV at 210 nm.

REFERENCE

1. **Rivier, J. E.,** *J. Chromatogr.*, 202, 211, 1980.

Reproduced by permission of Elsevier Science Publishers B.V.

Table LC 79
INFLUENCE OF FLOW RATE ON THE
RETENTION TIME OF LRF[a]

Flow rate (mℓ/min)	Retention time (min)	Elution volume (mℓ)
1.0	26.6	26.6
1.5	17.3	25.95
2.0	12.9	25.80
3.0	8.5	25.50

[a]   LRF, luteinizing hormone releasing factor.
[b]   TEAP buffer is obtained by bringing the pH of 0.25 N
      phosphoric acid to 3—3.5 with triethylamine. Both phos-
      phoric acid and triethylamine should be free from UV
      absorbing materials.

Column: μBondapak® alkylphenyl column, 30 × 0.4 cm.
Solvent: Isocratic conditions were 15% CH$_3$CN (75% A,
         25% B: B = 60% CH$_3$CN), 85% TEAP[b] buffer.
         The A buffer was TEAP buffer, pH 3.0; the B
         buffer had a composition of 40% TEAP buffer and
         60% CH$_3$CN.

REFERENCE

1. **Rivier, J. E.,** *J. Liq. Chromatogr.*, 1, 343, 1978.

Reprinted by courtesy of Marcel Dekker, Inc.

Table LC 80
SYNTHETIC RELEASING FACTORS

Packing	P1	P1	P2
Column			
Length (cm)	25	25	30
Diameter (cm)	0.46	0.46	0.46
Solvent	S1	S2	S1
Flow rate (mℓ/min)	2	2	2
Temperature	rt	rt	rt
Detection	D1	D1	D1
Peptide[a]		t_r(sec)	
TRF	169	389	104
LRF	719	787	586
SS-14	937	1009	810
$(Met(O)^{21})CRF$	1147	1290	950
CRF	1189	1320	982

[a] TRF, thyrotropin releasing factor; LRF, luteinizing hormone releasing factor; SS-14, somatostatin-14; CRF, corticotropin releasing factor.

Packing: P1   = Vydac® C_{18}, 5 μm.
           P2   = μBondapak® CN, 10 μm.
Solvent: S1:   Buffer A = TEAP, pH 2.25, buffer B = acetonitrile, 60% in buffer A. A gradient from 0 to 80% B is established in 20 min. TEAP buffer is obtained by adjusting the pH of 0.25 N phosphoric acid with triethylamine.
           S2:   Buffer A = 0.1% TFA, buffer B = 60% acetonitrile, 40% of 0.24% TFA. A gradient from 0 to 80% B is established in 20 min.
Detection: D1   = UV at 210 nm.

REFERENCE

1. **Rivier, J., Rivier, C., Spiess, J., and Vale, W.,** *Anal. Biochem.*, 127, 258, 1982.

Reproduced by permission of Academic Press, Inc.

Table LC 81
RETENTION OF HUMAN GROWTH HORMONE (HGH) TRYPTIC PEPTIDES UNDER DIFFERENT CONDITIONS

Packing		P1		P1		P1		P1		P2		P3	
Column													
Length (cm)		30		30		30		30		30		10	
Diameter (cm, I.D.)		0.4		0.4		0.4		0.4		0.4		0.8	
Solvent		S1		S2		S3		S4		S1		S5	
Flow rate (mℓ/min)		1.0		1.0		1.0		1.0		1.0		1.5	
Temperature		rt		rt		rt		rt		rt		rt	
Detection		D1		D1		D1		D1		D1		D1	
Peptide	Sequence	k′	Order[a]	k′	Order	k′	Order	k′	Order	k′	Order	k′	Order
T1	FPTIPLSR	14.3	15	19.2	13	17.6	15	22.5	11	17.6	16	54.0	14
T2	LFDNAMLR	12.7	12	17.5	12	16.0	14	22.5	11	16.3	14	48.0	13
T3	AHR	0.2	3	3.3	4	1.9	4	15.8	6	0.1	1	8.0	1
T4	LHQLAFDTYQEFEEAYIPK	19.8	17	10.7	14	19.2	17	27.2	17	23.9	20	58.2	15
T5	EQK	0.1	1	0.05	1	0.1	1	8.9	1	0.1	1	0.3	1
T6[b]	YSFLQNPQTSLCFSESIPTPSNR	22.5	19	24.0	16	21.1	19	28.4	18	20.6	18	68.5	17
T7	EETQQK	0.3	4	0.1	2	0.1	1	10.4	3	0.4	4	0.3	1
T8	SNLQLLR	10.7	11	41.7	21	14.8	11	37.0	21	12.5	11	83.5	19
T9	ISLLIQSWLEPVQFLR	28.7	21	31.6	20	27.2	21	33.4	20	22.0	19	91.8	21
T10	SVFANSLVYGASNSDVYDLLK	20.7	18	21.2	15	20.3	18	26.4	16		21	61.2	16
T11	DLEEGIQTLMGR	18.2	16	29.1	18	18.3	16	26.0	15	16.3	14	88.0	20
T12	LEDGSPR	3.7	6	4.6	5	7.2	6	14.2	5	3.4	6	10.2	5
T13	TGQIFK	6.1	7	9.7	7	11.4	7	18.6	7	6.4	7	28.3	8
T14	QTYSK	3.4	5	5.2	7	5.3	5	13.1	4	0.9	5	11.3	6
T15	FDTNSHNDDALLK	9.4	19	31.0	19	13.2	10	20.8	8	11.1	8	25.2	7
T16[b]	NYGLLYCFR	22.5	19	24.0	16	21.1	19	28.4	18	18.9	17	68.5	17
T17	K	0.1	1	0.1	2	0.1	1	8.9	1	0.1	1	0.7	3
T18[c]	DMDK	13.1	13	15.2	10	15.7	12	23.6	13	15.6	12	41.0	11
T19[c]	VETFLR	13.1	13	15.2	10	15.7	12	23.6	13	15.6	12	41.0	11
T20[d]	IVQCR	9.4	8	12.2	8	12.5	8	20.8	8	11.4	9	32.7	9
T21[d]	SVEGSCGF	9.4	8	12.2	8	12.5	8	20.8	8	11.4	9	32.7	9

Table LC 81 (continued)
RETENTION OF HUMAN GROWTH HORMONE (HGH) TRYPTIC PEPTIDES UNDER DIFFERENT CONDITIONS

[a]   Elution order.

[b]   Not present in the Cm-(Cys)-HGH digest but replaced by peptide T-6a (YSFLQNPQTSL(Cm)CFSESIPTPSNR) and peptide T-16a(NYGLLY(Cm)CRF).

[c]   Corresponds to peptide T-18 + 19 (DMDKVETFLR). Under different digest conditions, the peptide KDMDKVETFLR can also be isolated from these RP-HPLC separations.

[d]   Not present in the Cm-(Cys)-HGH digest but replaced by peptide T-20a (IVQ(Cm)CR) and peptide T-21a(SVEGS(Cm)CGF).

Packing: P1 = μBondapak® C_{18}.
          P2 = μBondapak® alkylphenyl.
          P3 = RCM-Radial-Pak C_{18}.

Solvent: S1 = Solvent A, water-15 mM H_3PO_4, pH 2.3; solvent B, 15 mM H_3PO_4, water-acetonitrile, 50:50 v/v. Linear 60 min gradient from 0 to 100% B.

          S2 = Solvent A, water-100 mM NH_4HCO_3; solvent B, 100 mM NH_4HCO_3, water-acetonitrile, 50:50 v/v. Linear 60 min gradient from 0 to 100% B.

          S3 = Solvent A, water-0.1% TFA; solvent B, 0.1% TFA, water-acetonitrile, 50:50 v/v. Linear 60 min gradient from 0 to 100% B.

          S4 = Solvent A, water-15 mM sodium heptanesulfonate-15 mM H_3PO_4; solvent B, 15 mM sodium heptanesulfonate-15 mM H_3PO_4, water-acetonitrile, 50:50 v/v. Linear 60 min gradient from 0 to 100% B.

          S5 = Solvent A, water, 100 mM NH_4HCO_3; solvent B, 100 mM NH_4HCO_3, water-acetonitrile, 50:50 v/v. Linear 180 min gradient from 0 to 100% B.

Detection: D1 = Fractions were collected manually and adjusted to pH 7.0 with 15 mM NaOH if necessary. The organic solvent was removed under vacuum, the residue lyophilized, and amino acid analyses and N-terminal sequence determination performed.

REFERENCE

1. **Grego, B., Lambrou, F., and Hearn, M. T. W.**, *J. Chromatogr.*, 266, 89, 1983.

Reproduced by permission of Elsevier Science Publishers B.V.

Table LC 82
DEPENDENCE OF THE CAPACITY RATIO ON
ELUENT COMPOSITION

Packing	P1	P1	P1	P1
Column				
Length (mm)	250	250	250	250
Diameter (mm, I.D.)	4.6	4.6	4.6	4.6
Solvent	S1	S2	S3	S4
Flow rate (mℓ/min)	1.0	1.0	1.0	1.0
Detection	UV	UV	UV	UV

Peptide		k′		
Bovine insulin	0.17	1.01	2.69	4.23
Porcine insulin	0.18	1.95	4.10	9.10

Packing: P1 = Nucleosil® 10 C_{18}.

Solvent: S1, S2, S3, and S4 have the composition methanol-X-0.1 M $NaClO_4$, pH 2.2, 5:1:4 where X is isopropanol, tetrahydrofuran, acetonitrile, and methanol, respectively.

REFERENCE

1. **Gazdag, M. and Szepesi, G.,** *J. Chromatogr.,* 218, 603, 1981.

Reproduced by permission of Elsevier Science Publishers B.V.

Table LC 83
ACTH DERIVATIVES

Packing	P1	P1
Column		
Length (mm)	250	250
Diameter (mm, I.D.)	4.6	4.6
Solvent	S1	S2
Flow rate (mℓ/min)	1.0	1.0
Detection	UV	UV

Derivative	k′		k′, pH 2.1 (predicted)[a]
$ACTH_{15-32}$	1.5	0	2.15
$ACTH_{1-32}$	7.5	4.7	13.8
$ACTH_{1-14}$	10.1	9.1	20.15
$ACTH_{1-24}$	13.9	17.7	26.25
$ACTH_{1-28}$	10.6	10.0	20.8

[a] Predicted on the basis of retention times obtained by summing retention coefficients for each polypeptide.

Packing: P1 = Nucleosil® 10 C_{18}.

Solvent: S1 = Solutions 1 and 2, 4:6; solution 1 contains 16.7% acetonitrile in methanol, solution 2 contains 0.1 M Na_2SO_4 in 0.01 M aqueous phosphate buffer, pH 2.2.

Table LC 83 (continued)
ACTH DERIVATIVES

S2 = Solutions 3 and 4, 35:65; solution 3 contains 16.7% isopropanol in methanol, solution 4 contains 0.1 M NaClO$_4$ in 0.01 M aqueous phosphate buffer, pH 2.2.

REFERENCE

1. **Gazdag, M. and Szepesi, G.,** *J. Chromatogr.*, 218, 603, 1981.

Reproduced by permission of Elsevier Science Publishers B.V.

Table LC 84
HPLC OF KININS

Packing	P1
Column	
Length (cm)	20
Diameter (mm, I.D.)	4
Material	SS
Solvent	S1
Flow rate (mℓ/min)	1.5
Temperature (°C)	30
Detection	D1

Kinin	t_r(min)
Bradykinin	8.50— 9.10
Met-Lys-bradykinin	10.10—10.80
Kallidin	14.50—15.10

Packing: P1 = Nucleosil® 5 C$_8$ reversed phase (pre-packed column).

Solvent: S1 = 30% acetonitrile in 0.1 M sodium phosphate buffer, pH 7.0.

Detection: D1 = UV at 215 nm.

REFERENCE

1. **Geiger, R., Hell, R., and Fritz, H.,** *Hoppe-Seyler's Z. Physiol. Chem.*, 363, 527, 1982.

Table LC 85
S-PEPTIDE ANALOGUES ON C$_8$ BONDED STATIONARY PHASES

Packing	P1	P2	P3
Column			
Length (cm)	10	25	10
Diameter (cm)	0.46	0.9	0.46
Solvent	S1	S2	S3
Flow rate (mℓ/min)	1.0	4.2	1.8
Detection	D1	D1	D1

Table LC 85 (continued)
S-PEPTIDE ANALOGUES ON C$_8$ BONDED STATIONARY PHASES

Peptide		k′	
S-Peptide	3.3	2.7	
(Hse^{13})-S-peptide (1-13)lactone		2.1	
(Ile^{13},Met^{20})-S-peptide-(Gly-Gly)	6.8	6.2	
(Ile^{13},Met(0)^{20})-S-peptide-(Gly-Gly)	2.7	2.3	
(Ile^{13},Hse^{20})-S-peptide lactone		3.5	8.0
(Ile^{13},Hse^{20})-S-peptide			5.2

Abbreviation: Hse, homoserine.

Packing: P1 = Polygosil® 60-5 C$_8$.
P2 = LiChrosorb® RP-8.
P3 = Nucleosil® 5-C$_8$.
Solvents: S1, S2, and S3 are acetonitrile-0.05 *M* TFA mixtures with compositions 16:84, 17:83, and 16.5:83.5, respectively.
Detection: S1 = UV at 225 nm.

REFERENCE

1. **Hoogerhout, P. and Kerling, K. E. T.,** *Rec. Trav. Chim. Pays-Bas,* 101, 246, 1982.

Reproduced by permission of the Koninklijke Nederlandse Chemische Vereniging.

Table LC 86
PEPTIDE ANALOGUES CONTAINING 4,4-DIFLUORO-L-PROLINE AND 4-KETO-L-PROLINE

Dipeptide	Solvent	R$_F$ × 100
Z-Kpro-Pro-OMe	S1	59
Z-Kpro-Gly-OEt	S2	47
Z-Kpro-Phe-OMe	S1	54
Z-Dfpro-Gly-OEt	S3	66
Z-Dfpro-Phe-OMe	S4	41

Abbreviations: Z, *N*-benzyloxycarbonyl; Kpro, 4-keto-L-proline; Dfpro, 4,4-difluoro-L-proline.

Packing: Silica Woelm TSC.
Solvent: S1 = CH$_2$Cl$_2$-acetone, 7:1.
S2 = CH$_2$Cl$_2$-acetone, 2:1.
S3 = CH$_2$Cl$_2$-acetone, 15:1.
S4 = CH$_2$Cl$_2$-acetone, 20:1.

REFERENCE

1. **Sufrin, J. R., Balasubramanian, T. M., Vora, C. M., and Marshall, G. R.,** *Int. J. Pept. Prot. Res.,* 20, 438, 1982.

Table LC 87
CHIRAL SEPARATION OF DANSYL DIPEPTIDES

Packing	P1	P1
Column		
Length (cm)	15	15
Diameter (mm, I.D.)	4.6	4.6
Material	SS	SS
Solvent	S1	S2
Temperature (°C)	30	30
Detection	UV	UV

Dansyl derivative of	k′	
Gly-L-Ala	1.2	1.35
Gly-D-Ala	1.3	1.15
Gly-L-α-NBu	1.6	2.35
Gly-D-α-NBu	1.7	1.8
Gly-L-Nval	2.5	6.35
Gly-D-Nval	2.65	3.15
Gly-L-Nleu	4.1	8.1
Gly-D-Nleu	4.35	5.7
Gly-L-Val	2.5	3.7
Gly-D-Val	2.6	2.35
Gly-L-Leu	3.95	8.45
Gly-D-Leu	4.2	5.7
Gly-L-Met	2.2	4.05
Gly-D-Met	2.3	3.0
Gly-L-Thr	1.1	1.3
Gly-D-Thr	1.2	1.0
L-Ala-Gly	0.8	
D-Ala-Gly	0.85	
L-Leu-Gly	2.0	5.2
D-Leu-Gly	2.1	6.15

Abbreviation: Nbu, aminobutyric acid.

Packing: P1 = A chemically bonded *n*-octyl (C_8) phase was synthesized. Hypersil® 5-μm particles were slurried overnight in 6 *N* HNO_3 and subsequently washed with distilled, deionized water to neutral pH. After bonding with *n*-octyldimethyl-chlorosilane, the unreacted accessible silanol groups were capped with trimethylsilane. Chemical analysis revealed bonded phase coverage of 3.4 μmol/m^2.

Solvent: S1 = 0.8 m*M* C_3-C_8 dien*-Zn(II), 0.17 *M* ammonium acetate, pH 9, acetonitrile-water, 33:65 v/v.

   S2 = Similar to S1, with Ni(II) replacing Zn(II).

   * C_3-C_8-dien = L-2-ethyl-4-octyl-diethyltriamine. The chelating agent and metal salt are carefully weighed on an analytical balance and added to the appropriate amount of an aqueous ammonium acetate solution, pH about 7.0. The pH is adjusted with aqueous NH_4OH and the appropriate amount of organic modifier added. Where appropriate glacial acetic acid is added instead of aqueous NH_4OH. The mobile phase is then filtered and degassed with helium.

REFERENCE

1. **Lindner, W., LePage, J. N., Davies, G., Seitz, D. E., and Karger, B. L.,** *J. Chromatogr.,* 185, 323, 1979.

Reproduced by permission of Elsevier Science Publishers B.V.

Table LC 88
RESOLUTION OF DIPEPTIDE DERIVATIVES ON
CHIRAL AMIDE-BONDED SILICA GEL

Packing	P1	P2	P3	P4
Column				
Length (cm)	20	20	20	20
Diameter (cm, I.D.)	0.4	0.4	0.4	0.4
Material	SS	SS	SS	SS
Solvent	S1	S1	S1	S1
Flow rate (cm/sec)	0.090	0.104	0.095	0.096
Temperature (°C)	40	40	40	40
Detection	D1	D1	D1	D1

Racemate	k_1'	k_2'	k_1'	k_2'	k_1'	k_2'	k_1'	k_2'
Z-Leu-Leu-OMe	5.48	5.89	5.51	5.98	5.47	5.88	6.42	1.00
Z-Phe-Phe-OMe	6.23	6.54	6.51	6.85	6.29	6.46		

Abbreviation: Z, benzyloxycarbonyl.

Packing: P1 = *N*-propionyl-L-valyl-aminopropylsilanized silica.
           P2 = *N-n*-butyryl-L-valyl-aminopropylsilanized silica.
           P3 = *N-n*-valeryl-L-valyl-aminopropylsilanized silica.
           P4 = *N*-pivaloyl-L-valyl-aminopropylsilanized silica.
Solvent: S1 = 2-propanol in *n*-hexane, 2% concentration for Z-Leu-Leu-OMe
           and 3% concentration for Z-Phe-Phe-OMe.
Detection: D1 = UV at 230 and 254 nm.

REFERENCE

1. **Hara, S. and Dobashi, A.,** *J. Chromatogr.*, 186, 543, 1979.

Table LC 89
HPLC OF BLOCKED PEPTIDE INTERMEDIATES

Compound	X	Y	Solvent	t_r(min)
H-(10-14)	But	But	S1	80—94
Z-(8-14)	Acm	OH	S2	42—95
H-(8-14)	Acm	OH	S3	82.5—150
Bpoc-(8-14)	Acm	OH	S4	45—110
Z-(7-14)	But	But	S5	47.5—60
H-(7-14)	But	But	S6	75—250
Z-(6-14)	But	But	S7	135—153
Z-(3-14)	But	But	S8	33—52
Bpoc-(3-14)	Acm	OH	S9	37.5—57.5
Boc-(1-14)	Acm	OH	S10	40—75
Boc-(1-14)	Acm	OH	S11	88—100

Ala = Tyr(Bu^t)

Peptides are based on the structure

```
       1   2   3   4   5   6   7  8   9  10   11  12  13  14
Boc-Ala-Gly-Cys-Lys-Asn-Phe-Phe-Trp-Lys-Thr-Phe-Thr-Ser-Cys-OY
             |   |                    |   |       |   |   |
            Acm Boc                  Boc Bu^t   Bu^t Bu^t  X
```

Table LC 89 (continued)
HPLC OF BLOCKED PEPTIDE INTERMEDIATES

Abbreviations: Z, Benzyloxycarbonyl; Acm, acetamidomethyl; Bpoc, 2-(*p*-biphenylyl)-2-propyloxycarbonyl; Boc, tert-butyloxycarbonyl; Bu^t, tert-butyl.

Column: Heavy-wall glass column, 3.8 × 43 cm, prepacked with silica gel 60 (60Å) porosity.

Solvent:
- S1 = Column equilibration with CHCl$_3$, then an EtOH gradient.
- S2 = Column equilibration with system 1 (CHCl$_3$-iPrOH-AcOH, 88:10:2), then elution with system 1 for 50 min, followed by elution with system 1 containing 5% MeOH.
- S3 = Column equilibration with system 1, then elution with system 2 (CHCl$_3$-MeOH-AcOH, 70:20:5 v/v).
- S4 = Column equilibration with system 6 (CHCl$_3$-iPrOH-AcOH, 88:5:2), followed by elution with system 6 for 50 min, then with system 3 (CHCl$_3$-MeOH-AcOH, 85:5:2 v/v).
- S5 = Column equilibration with system 1 and elution with system 1.
- S6 = Column equilibration with system 1, then elution with system 2 for 180 min, then with system 3.
- S7 = Column equilibration with CHCl$_3$, then elution with CHCl$_3$ for 55 min, then with an MeOH gradient.
- S8 = Column equilibration with system 1, then elution with system 2.
- S9 = Column equilibration with system 4 (CHCl$_3$-MeOH-AcOH, 88:10:5 v/v) followed by elution with system 4.
- S10 = Column equilibration with system 1, then elution with system 2.
- S11 = Column equilibration with system 1, then elution with system 1 for 50 min, then with system 2.

Detection: UV at 254 nm or infrared at 5.8 μm.

REFERENCE

1. Gabriel, T. F., Jiminez, M. G., Felix, A. M., Michaelwsky, J., and Meienhofer, J., *Int. J. Pept. Prot. Res.*, 9, 129, 1977.

Reproduced by permission of Munksgaard International Publishers, Ltd.

Table LC 90
CHROMATOGRAPHY OF PROTECTED PEPTIDES IN VARIOUS MOBILE PHASES

Packing	P1	P1	P1	P1	P1	P1	P1	P1
Column								
Length (mm)	250	250	250	250	250	250	250	250
Diameter (mm, I.D.)	3	3	3	3	3	3	3	3
Solvent	S1	S2	S3	S4	S5	S6	S7	S8
Detection	UV	UV	UV	UV	UV	UV	UV	UV
TFA-dipeptide or -tripeptide derivative				k'				
TFA-Phe-Pro-OMe	11.7	15.3	15.6	12.0	9.45	10.3	8.36	8.04
TFA-L-Pro-D-Phe-OMe	9.82	12.9	13.2	11.8	7.91	9.00	8.18	7.93
TFA-L-Pro-L-Phe-OMe	11.7	13.8	14.5	13.0	8.64	9.84	8.60	8.59
TFA-Gly-D,L-Phe-OMe	6.56	7.33	7.27	6.22	4.69	5.36	4.45	4.22

Table LC 90 (continued)
CHROMATOGRAPHY OF PROTECTED PEPTIDES IN VARIOUS MOBILE PHASES

TFA-dipeptide or -tripeptide derivative				k'				
TFA-D,L-Phe-Gly-OMe	6.27	7.00	7.44	6.00	4.51	5.36	4.45	4.14
TFA-L-Ala-L-Phe-OMe	8.00	8.82	8.27	6.98	5.82	5.95	5.15	4.43
TFA-L-Phe-L-Ala-OMe	7.18	9.18	8.42	7.04	5.27	5.82	4.91	4.61
TFA-L-Tyr-Gly-OMe	4.54	5.18	4.82	4.30	3.73	3.82	3.18	2.97
TFA-Gly-L-Tyr-OMe	5.18	5.69	5.45	4.50	4.27	4.27	3.62	3.27
TFA-Gly-L-Tyr-(OMe)$_2$	10.7	12.4	11.9	9.97	9.33	8.93	7.71	7.91
TFA-L-Phe-Gly-L-Val-OMe	15.2	17.4	18.2	13.5	11.5	11.1	9.35	8.30
TFA-L-Val-Gly-L-Phe-OMe	12.9	15.3	15.9	12.2	9.18	9.69	8.29	7.43

Packing: P1 = The tripeptide L-Val-L-Ala-L-Pro was bonded to Partisil®-10 using the reagent Y-5918, 1-trimethoxysilyl-2-(4-chloromethylphenyl)-ethane. Elemental analysis showed 12.3% carbon.

Solvent: S1 to S8 are citrate buffers with the following values of pH and ionic strength.

S1 = pH 7.15, ionic strength, 1.0 M.
S2 = pH 5.0, ionic strength, 1.0 M.
S3 = pH 3.8, ionic strength, 1.0 M.
S4 = pH 2.1, ionic strength, 1.0 M.
S5 = pH 7.4, ionic strength, 0.5 M.
S6 = pH 5.2, ionic strength, 0.5 M.
S7 = pH 3.8, ionic strength, 0.5 M.
S8 = pH 2.2, ionic strength, 0.5 M.

REFERENCE

1. **Fong, G. W.-K. and Grushka, E.**, *Anal. Chem.*, 50, 1154, 1978.

Reprinted with permission. Copyright 1978 American Chemical Society.

Table LC 91
COPPER COMPLEXES OF PEPTIDES AND AMINO ACIDS

Packing	P1	P1	P1	P1	P1	P1	P1
Column							
Length (cm)	15	15	15	15	15	15	15
Diameter (cm)	1.5	1.5	1.5	1.5	1.5	1.5	1.5
Solvent	S1	S2	S3	S4	S5	S6	S7
Detection	D1	D1	D1	D1	D1	D1	D1
Ligand				$R_g{}^a$			
Gly	1	1	1	1	1	1	1
Arg	0.49	0.57	0.58	0.60	0.65	0.75	0.82
Lys	0.52	0.59	0.55	0.56	0.65	0.66	0.75
Pro	0.97	0.99	0.97	1.00	1.03		
Ser	1.34	1.17	1.07	1.05	1.05		
Thr	1.42	1.13	1.08	1.03	1.02		
Glu			10.8	5.65	3.16	1.86	1.29
Asp			10.9	5.65	2.99	1.85	1.22
Gly-Gly	5.38	2.94	2.18	1.89	1.57	1.35	1.13
Leu-Leu	4.32	2.51	1.82	1.64	1.43	1.29	
Gly-Gly-Gly	19.00	5.72	3.45	2.77	2.14	1.17	1.39
Ala-Gly-Gly	16.50		3.92	2.68	2.15	1.58	

Table LC 91 (continued)
COPPER COMPLEXES OF PEPTIDES AND AMINO ACIDS

Ligand				R_g[a]			
Leu-Leu-Leu		5.67	3.38	2.75	2.10	1.73	
tetra-Ala			6.25	3.25	2.20	1.45	1.22
tetra-Gly			5.70	3.70	2.38	1.73	1.30
penta-Gly			15.00	8.10	4.75	2.52	1.45
hexa-Gly				7.00	3.63	2.34	1.34
Tyr	7.30	3.05		2.50			

[a] $R_g = V_e/V_g$, V_e being the elution volume of the compound and V_g the elution volume of glycine in the same buffer system.

Packing: P1 = DEAE Sephadex® A-25 equilibrated in 5 mmol/ℓ sodium borate buffer, pH 8.5.

Solvents: S1, S2, S3, S4, S5, S6, and S7 are sodium borate buffers, pH 8.5, with NaCl added to give ionic strengths of 0.0075, 0.03, 0.06, 0.1, 0.16, 0.26, and 0.5, respectively.

Detection: D1 = UV at 570 nm.

Copper complexes were formed by adding an excess of basic copper carbonate to the aqueous solution of peptide. Sodium hydroxide was added to maintain the pH at 9.0 to 9.5 and the tube shaken for 10 to 15 min. The tube was then centrifuged and the solution of the copper complex aspirated into a clean tube.

REFERENCE

1. **Sampson, B. and Barlow, G. B.**, *J. Chromatogr.*, 183, 9, 1980.

Reproduced by permission of Elsevier Science Publishers B.V.

Table LC 92
COMPARISON OF PREDICTED AND OBSERVED CAPACITY RATIOS

Packing	P1	P1	P1	P1
Column				
Length (mm)	250	250	250	250
Diameter (mm, I.D.)	4.6	4.6	4.6	4.6
Solvent	S1	S2	S3	S4
Flow rate (mℓ/min)	1.0	1.0	1.0	1.0
Detection	UV	UV	UV	UV

Peptide		k'		k' (predicted)
Aprotinine		4.55		42.5[c]
Insulin[a]			3.61	71.9[d]
Insulin[b]			2.53	62.0[d]
Proinsulin[a]			12.90	95.9[d]
Proinsulin[b]			4.45	87.0[d]
Oxytocin	1.31		6.50	13.3[d]
Oxytocein	2.14			5.12[d]

[a]   Porcine.
[b]   Bovine.
[c]   pH 7.25.
[d]   pH 2.1.

Table LC 92 (continued)
COMPARISON OF PREDICTED AND OBSERVED CAPACITY RATIOS

Packing: P1 = Nucleosil® 10 C₁₈.
Solvent: S1 = Solutions 1 and 2, 4:6 v/v. Solution 1 contains 16.7% acetonitrile in methanol; solution 2 contains 0.1 M Na₂ SO₄ in 0.01 M aqueous phosphate buffer, pH 2.2.

S2 = Solutions 1 and 3, 3:7 v/v. Solution 3 contains 0.05 M Na₂SO₄ in 0.01 M aqueous phosphate buffer, pH 6.5.

S3 = Solutions 4 and 5, 35:65 v/v. Solution 4 contains 16.7% isopropanol in methanol; solution 5 contains 0.1 M NaClO₄ in 0.01 M aqueous phosphate buffer, pH 2.2.

S4 = Solution 6 and 5, 4:6 v/v. Solution 6 contains 33% isopropanol in acetonitrile.

REFERENCE

1. **Gazdag, M. and Szepesi, G.,** *J. Chromatogr.*, 218, 603, 1981.

Table LC 93
MEDIUM-SIZED PEPTIDES

Packing	P1
Column	
Length (cm)	25
Diameter (cm)	0.46
Solvent	S1
Flow rate (mℓ/min)	1.5
Temperature	rt
Detection	D1

Peptide	t_r(min)
Angiotensin I	112
Angiotensin II	100
Dynorphin	118
Leu-enkephalin	81
Met-enkephalin	73
Glucagon	119
Spermidine	54
Spermine	50
Vasopressin	73

Packing: P1 = LiChrosorb® RP-18, 5 μm (Merck).
Solvent: S1 = Elution was carried out with acetonitrile-heptafluorobutyric acid-water, 10:0.1:89.9 v/v isocratically for 30 min followed by a linear acetonitrile gradient of 0.4% min at a flow rate of 1.5 mℓ/min.
Detection: D1 = UV at 210 nm, or fluorescamine.

REFERENCES

1. **Mabuchi, H. and Nakahashi, H.,** *J. Chromatogr.*, 233, 107, 1982.

Table LC 94
PEPTIDE RETENTION ON REVERSED-PHASE SUPPORTS

Packing	P1	P2	P3	P4	P5	P6	P7
Peptide			% acetonitrile giving k' = 2				
Leu-enkephalin	23.2	23.0	10.4	22.2	19.5	16.0	5.5
α-Endorphin	22.0	22.5	12.3	20.0	20.2	16.0	7.0
Oxytocin	29.5	28.7	13.5	27.7	29.5	26.0	18.0
Ranatensin	41.5	39.8	31.7	36.4	42.5	33.0	32.5

Packing:
P1 = MicroPak® MCH-10, a reversed-phase column with a monomeric octadecyl (C_{18}) phase on 10-μm diameter silica at approximately 12% carbon by weight.

P2 = MicroPak® MCH-N-CAP-10, which is a modification of MCH-10 in which residual silanols are end-capped by a secondary silanization reaction.

P3 = C_{18}-diol, a mixed phase containing a low loading of octadecyl groups on 10-μm diameter silica, with the remainder of the silica surface carrying a hydrophilic glycerylpropyl phase.

Pr = Hexyl phase bonded to 10-μm diameter silica.

P5 = Phenylethyl phase bonded to 10-μm diameter silica.

P6 = MicroPak® CN-AQ-10-EC, an experimental end-capped version of packing P7.

P7 = MicroPak® CN-AQ-10, a cyanopropyl phase on 10-μm diameter silica, which is supplied packed in methanol. All the reversed-phase columns contained stationary phases bonded to LiChrosorb® base material, packed in 4 mm I.D. stainless steel tubes 15 or 30 cm long.

Solvent: 0.02 M ammonium acetate, pH 7.0/acetonitrile.
Flow rate: 1.0 mℓ/min.
Temperature: 30°C.
Detection: Fluorescence.

REFERENCE

1. **Wehr, C. T., Correia, L., and Abbott, S. R.**, *J. Chromatogr. Sci.*, 20, 114, 1982.

Reproduced from the *Journal of Chromatographic Science* by permission of Preston Publications, Inc.

Table LC 95
EFFECT OF MOBILE PHASE COMPOSITION ON RETENTION OF PEPTIDES

Packing	P1	P1	P1	P1	P1	P1	P1	P1
Column								
Length (cm)	25	25	25	25	25	25	25	25
Diameter (mm)	4.0	4.0	4.0	4.0	4.0	4.0	4.0	4.0
Solvent	S1	S2	S3	S4	S5	S6	S7	S8
Flow rate (mℓ/min)	1.0	1.0	1.0	1.0	1.0	1.0	1.0	1.0
Detection	UV	UV	UV	UV	UV	UV	UV	UV

Table LC 95 (continued)
EFFECT OF MOBILE PHASE COMPOSITION ON RETENTION OF PEPTIDES

Peptide	t_r							
Trityrosine	100	104	92	98	98	94	87	97
Met-enkephalin	100	102	94	99	99	95	88	97
Neurotensin	100	98	102	118	99	96	80	95
Somatostatin	100	100	103	119	102	100	84	95
Bradykinin	100	97	95	122	101	97	77	93
Substance P	100	102	107	129	100	99	83	95
Insulin	100	94	105	—[a]	99	98	93	—[a]

[a] Not eluted.

Retention t_r: Values are expressed as percent relative to the conditions using solvent S1.

Packing: P1 = Bio-Rad® ODS, 10 μm.
Solvent: S1 = 0.1 M NaClO₄, 0.1% H₃PO₄ with an acetonitrile gradient of 2.5%/min.
           S2 = 0.1 M NaClO₄, 0.1% H₃PO₄ with a methanol gradient of 3.5%/min.
           S3 = 0.1 M NaClO₄, 0.1% H₃PO₄ with a propanol gradient of 1.8%/min.
           S4 = 0.1% H₃PO₄ with an acetonitrile gradient of 2.5%/min.
           S5 = 0.6 M NaClO₄, 0.1% H₃PO₄ with an acetonitrile gradient of 2.5%/min.
           S6 = 0.06 M HClO₄ with an acetonitrile gradient of 2.5%/min.
           S7 = 0.1 M NaH₂PO₄, 0.1% H₃PO₄, with an acetonitrile gradient of 2.5%/min.
           S8 = 0.1% TFA with an acetonitrile gradient of 2.5%/min.

$$REFERENCE$$

1. **Meek, J. L. and Rossetti, Z. L.**, *J. Chromatogr.*, 211, 15, 1981.

Reproduced by permission of Elsevier Science Publishers B.V.

Table LC 96
CHROMATOGRAPHY OF PEPTIDES ON DIFFERENT STATIONARY PHASES

Packing	P1	P1	P2	P2	P3	P3	P4	P4	P5	P5
Solvent	S1	S2	S1	S2	S1	S2	S1	S2	S1	S2
Flow rate (mℓ/mm)	0.8	0.8	0.8	0.8	0.8	0.8	0.8	0.8	0.8	0.8
Temperature (°C)	40	40	40	40	40	40	40	40	40	40
Detection	D1	D1	D1	D1	D1	D1	D1	D1	D1	D1
Peptide	t_r									
Leu-Enk-Arg-Lys[a]	25.2	27.6	19.3	27.4	32.7	36.9	20.2	24.7	14.2	20.0
Leu-Enk-Arg-Arg	26.8	29.1	22.3	29.2	35.2	39.0	22.7	26.9	15.9	21.5
Lys-bradykinin	30.0	31.1	27.1	31.4	41.2	42.6	27.6	30.0	22.0	24.2
β-Melanotropin	31.3	31.8	29.8	31.8	43.3	45.9	30.8	32.7	25.1	25.5
Met-enkephalin	31.8	32.4	26.7	31.3	33.9	37.9	25.1	28.1	17.1	22.9
LHRH	34.3	32.6	31.2	34.8	36.5	40.1	25.9	28.6	22.2	24.7
Dynorphin	35.4	33.7	32.5	34.7	50.4	48.6	31.5	34.8	29.0	28.0
Leu-enkephalin	39.0	36.7	33.2	35.7	41.9	43.3	29.1	32.1	25.1	27.7
Angiotensin	39.8	38.7	37.6	39.8	57.3	55.5	38.8	39.0	34.0	32.4
Neurotensin	49.6	49.0	49.8	51.3	60.4	62.0	48.2	47.5	41.9	41.2

[a] Enk, enkephalin.

Table LC 96 (continued)
CHROMATOGRAPHY OF PEPTIDES ON DIFFERENT STATIONARY PHASES

Packing: P1 = μBondapak® C_{18}.

            P2 = Partisil® ODS-3.

            P3 = LiChrosorb® RP-8.

            P4 = MicroPak® MCH-5.

            P5 = Vydac 218TP.

Solvent: S1 = Linear gradient from 0 to 60% solvent B in 60 min. Solvent A = methanol-water, 25:75 v/v, containing 0.1% TFA. Solvent B = acetonitrile-water, 80:20 v/v containing 0.1% TFA.

            S2 = Linear gradient from 0 to 60% solvent B in 60 min. Solvent A = acetonitrile-water, 10:90 v/v, containing 0.1% TFA. Solvent B = acetonitrile-water, 80:20 v/v, containing 0.1% TFA.

Detection: D1 = UV at 215 nm.

REFERENCE

1. **Tan, L.,** *J. Chromatogr.*, 266, 67, 1983.

Reproduced by permission of Elsevier Science Publishers B.V.

Table LC 97
HIGH PERFORMANCE GEL PERMEATION CHROMATOGRAPHY OF PEPTIDES

Packing	P1	P2
Column		
Length (mm)	300	300
Diameter (mm, I.D.)	7.5	7.5
Solvent	S1	S1
Flow rate (mℓ/min)	1.0	1.0
Temperature (°C)	23	23
Detection	D1	D1

Peptide	t_r(sec)	
Actinomycin C	403.3 ± 0.88	1032 ± 3.0
Bacitracin	396.6 ± 0.73	1010.4 ± 3.2
Insulin A chain		942 ± 3.4
ACTH 1-24	365.6 ± 1.08	
Trypsin inhibitor		878 ± 3.4
Porcine β-lipotropin	314.7 ± 0.33	
Cytochrome c	309.9 ± 0.53	
Lysozyme		813 ± 3.2

Packing: P1   = Bio-Sil® TSK 125 (silica modified with silanoyl groups), two columns being used.

              P2   = Bio-Gel® TSK 20 (silica with hydroxylated polyethers).

Solvent: S1   = 6 *M* guanidine hydrochloride.

Detection: D1   = At 279 nm.

REFERENCE

1. **Richter, W. O., Jacob, B., and Schwandt, P.,** *Anal. Biochem.*, 133, 288, 1983.

Table LC 98
GEL PERMEATION CHROMATOGRAPHY OF PEPTIDES IN SEVERAL SOLVENTS

Packing	P1	P1	P1	P1	P1	P1	P1	P1	P1	P1
Column										
Length (cm)	60	60	60	60	60	60	60	60	60	60
Diameter (mm, I.D.)	7.5	7.5	7.5	7.5	7.5	7.5	7.5	7.5	7.5	7.5
Solvent	S1	S2	S3	S4	S5	S6	S7	S8	S9	S10
Temperature (°C)	22	22	22	22	22	22	22	22	22	22
Detection	D1	D1	D1	D1	D1	D1	D1	D1	D1	D1

Peptide					k'					
Cytochrome	0.28	0.80	0.59	0.90	0.78	0.64	0.80	0.90	0.76	0.40
Substance P	0.48	0.97	3.28	2.26	1.25	3.01	1.18	1.22	1.27	0.68
Angiotensin I	3.0	0.96	2.16	1.04	1.19	2.20	1.15	1.21	1.27	1.13
(Leu^5)-enkephalin	0.50	1.04	1.82	1.05	1.25	1.76	1.16	1.22	1.28	1.33
Gly-Gly-Leu	0.90	0.99	1.20	1.31	1.26	1.16	1.16	1.23	1.27	1.33
Norleucine	0.48	0.99	1.25	1.04	1.35	1.32	1.27	1.33	1.35	1.38

Packing: P1 = TSK GEL G2000SW, 10 μm.
Solvent: S1 = Distilled water.
      S2 = 0.15 M phosphate buffer, pH 7.4.
      S3 = 0.15 M phosphate buffer, pH 7.4, 1 M NaCl.
      S4 = 0.15 M phosphate buffer, pH 7.4, 1 M NaCl, 20% methanol.
      S5 = 0.15 M phosphate buffer, pH 7.4, 1 M NaCl, 20% methyl Cellosolve®.
      S6 = 0.15 M phosphate buffer, pH 7.4, 1 M NaCl, 1% Brij-35.
      S7 = 0.15 M phosphate buffer, pH 7.4, 1 M NaCl, 10% methanol, 1% Brij-35.
      S8 = 0.15 M phosphate buffer, pH 7.4, 1 M NaCl, 20% methanol 1% Triton® X-100.
      S9 = 0.15 M phosphate buffer, pH 7.4, 1 M NaCl, 2% methyl Cellosolve,® 1% Triton® X-100.
      S10 = 0.15 M phosphate buffer, pH 7.4,1 M NaCl, 20% methyl Cellosolve,® 1% sodium dodecyl sulfate.
Detection: D1 = *o*-Phthalaldehyde.

REFERENCE

1. **Shioya, Y., Yoshida, H., and Nakajima, T.,** *J. Chromatogr.,* 240, 341, 1982.

Reproduced by permission of Elsevier Science Publishers B.V.

Table LC 99
POLYPEPTIDES AND PROTEINS

Packing	P1
Column	
Length (mm)	100
Diameter (mm, I.D.)	5
Material	SS
Solvent	S1
Flow rate (mℓ/min)	1—1.5
Temperature	rt
Detection	D1

Peptide or protein	t_r (min)
β-LPH$_{34-39}$ (ovine)	8.5
L-Tryptophan	10.0

Table LC 99 (continued)
POLYPEPTIDES AND PROTEINS

Peptide or protein	t_r (min)
Arginine vasotocin	12.0
Lysine vasopressin	13.0
Arginine vasopressin	14.0
$ACTH_{5-10}$	17.0
Phenylalanylphenylalanine(Phe 2)	18.0
$ACTH_{1-18}$	18.5
Met-enkephalin	19.0
Oxytocin	19.5
$ACTH_{4-10}$	20.5
$ACTH_{1-24}$	21.5
α-Endorphin	22.0
Leu-enkephalin	22.0
Insulin A (bovine)	22.0
Angiotensin II	23.0
Cobra neurotoxin 3	24.0
Neurotensin	24.5
α-Melanotrophin	26.0
Bombesin	26.0
RNAse	27.5
Phe 3	28.0
Gastrin 1	28.5
Substance P	29.0
Substance P_{4-11}	30.0
$ACTH_{1-39}$ (human)	30.5
$ACTH_{18-39}$	30.5
$ACTH_{34-39}$	31.0
Somatostatin	32.0
Insulin (bovine)	32.0
$ACTH_{1-39}$(porcine)	33.0
Insulin B (bovine)	33.5
β-Endorphin (ovine)	34.0
β-Lipotrophin (human)	34.5
Calcitonin (human)	34.5
Cytochrome c	35.0
Glucagon	36.0
Phe 4	36.5
Calcitonin (Salmon)	37.0
Lysozyme	37.5
Bovine serum albumin	43.0
Myoglobin	45.0
Melittin	46.0

Packing: P1 = Hypersil®-ODS, 5 μm.

Solvent: S1 = The primary solvent is 0.1 *M* NaH$_2$PO$_4$-H$_3$PO$_4$, pH 2.1, total phosphate concentration 0.2 *M*, the secondary solvent being acetonitrile. The gradient profile is composed of three linear segments, the acetonitrile concentration being increased to 10% in 5 min, to 40% in 45 min, and to 60% after 50 min when the gradient is terminated.

Detection: D1 = UV and endogenous fluorescence, the latter being usable only with compounds containing tryptophan.

Table LC 99 (continued)
POLYPEPTIDES AND PROTEINS

REFERENCE

1.   **O'Hare, M. J. and Nice, E. C.**, *J. Chromatogr.*, 171,
209, 1979.

Reproduced by permission of Elsevier Science Publishers B.V.

Table LC 100
POLYPEPTIDES AND PROTEINS ON SHORT AND ULTRA-
SHORT ALKYLSILANE-BONDED REVERSED-PHASE PACKINGS

Packing	P1	P2	P3	P4
Column				
Length (mm)	150	150	150	150
Diameter (mm, I.D)	5	5	5	5
Material	SS	SS	SS	SS
Solvent	S1	S1	S1	S1
Flow rate (mℓ/min)	1.0	1.0	1.0	1.0
Temperature (°C)	45	45	45	45
Detection	D1	D1	D1	D1
Polypeptide or protein		$t_r(min)^a$		
Ribonuclease A	−10.2	−10.7	−11.3	−12.0
Epidermal growth factor (mouse)		−7.6	−8.0	
Calcitonin (synthetic human)	−4.1	−4.4	−4.1	−5.0
Cytochrome *c*	−2.5	−3.1	−2.3	−3.8
Parathyroid hormone (bovine)	−2.4	−2.3	−2.1	−2.6
Lysozyme	0	0	0	0
Bovine serum albumin	3.2	3.3	5.8	7.2
Human serum albumin	3.8	3.5		
α-Lactalbumin (bovine)	4.6	7.4	6.6	7.2
α-Lactalbumin (rat)	4.9		7.9	
Myoglobin	7.1	9.0	9.5	8.9
α-Hemoglobin (bovine)	8.2	9.2	9.6	11.3
Carbonic anhydrase (bovine erythrocytes)	9.8	9.5	9.6	10.0
β-Lactoglobulin A (bovine milk)	10.1	10.0		8.8
β-Lactoglobulin B (bovine milk)	10.1	10.0		8.8
Elastase	11.1	10.0	11.7	NR^b
Prolactin (ovine P-S-13)	13.1	16.7	16.6	16.6
Human placental lactogen	14.3	20.2	17.6	18.5
Prolactin (rat PRL-B-2)	17.3	25.0	20.0	29.1
Growth hormone (bovine GH-B-18)	17.3	29.0		39.4
Ovalbumin	NR	NR	NR	NR

a   Retention time relative to lysozyme.
b   NR, not recovered using CH_3CN as organic modifier.

Packing: P1   = LiChrosorb® RP-2 (C_1).
            P2   = Ultrasphere® SAC (experimental) (C_3).
            P3   = Spherisorb® S5C6 (C_6).
            P4   = LiChrosorb® RP-8 (C_8).
Solvent: S1   = Linear gradient elution between an aqueous primary solvent of 0.155 *M*
                  NaCl solution adjusted to pH 2.1 with HCl, and a secondary solvent of
                  acetonitrile.
Detection: D1   = UV at 215 nm.

Table LC 100 (continued)
POLYPEPTIDES AND PROTEINS ON SHORT AND ULTRA-SHORT ALKYLSILANE-BONDED REVERSED-PHASE PACKINGS

REFERENCE

1. Nice, E. C., Capp, M. W., Cooke, N., and O'Hare, M. J., *J. Chromatogr.*, 218, 569, 1981.

Reproduced by permission of Elsevier Science Publishers B.V.

Table LC 101
THE EFFECT OF TEMPERATURE ON THE RETENTION OF POLYPEPTIDES AND PROTEINS

Packing	P1	P1	P1	P1	P1	P1
Column						
Length (cm)	7.5	7.5	7.5	7.5	7.5	7.5
Diameter (mm, I.D.)	4.6	4.6	4.6	4.6	4.6	4.6
Material	SS	SS	SS	SS	SS	SS
Solvent	S1	S1	S1	S1	S1	S1
Flow rate (mℓ/min)	1.0	1.0	1.0	1.0	1.0	1.0
Temperature (°C)	4	18	25	35	45	60
Detection	D1	D1	D1	D1	D1	D1

Peptide or protein	t_r(min)					
Tryptophan	10.8	8.0	6.0	4.5	3.4	2.6
ACTH 1-24	23.8	22.4	22.0	20.9	20.0	18.8
RNase	31.6	31.4	31.4	31.4	30.2	29.2
Lysozyme	39.8	40.4	40.8	40.6	40.2	39.2
Bovine serum albumin	41.6	42.8	44.0	44.2	44.4	44.4
Prolactin (rat)	51.0	55.6	57.6	59.6	60.4	59.8

Packing: P1 = C3, 10 μm, 30 nm pore-diameter support, a short alkyl chain bonded silica. The packing was prepared by reacting spherical silica particles with dimethylalkylchlorosilane and capping with trimethylchlorosilane.

Solvent: S1 = A linear gradient between 0.15 M NaCl, pH 2.1, and acetonitrile.

Detection: D1 = UV at 215 or 280 nm.

REFERENCE

1. O'Hare, M. J., Capp, M. W., Nice, E. C., Cooke, N. H. C., and Archer, B. G., *Anal. Biochem.*, 126, 17, 1982.

Reproduced by permission of Academic Press, Inc.

Table LC 102
GEL PERMEATION CHROMATOGRAPHY OF PEPTIDES AND PROTEINS

Packing	P1
Solvent	S1
Flow rate (ml/min)	1.0
Temperature (°C)	50
Detection	D1

Peptide or protein	t_r(sec)
Acetic acid	1370 ± 9
Indole	1449 ± 1
Thyrotropin releasing factor (TRF)	1365 ± 2
(Met^5)-enkephalin	1321 ± 5
LRF(5-10) Ac-Tyr-Gly-Leu-Arg-Pro-Gly-NH$_2$	1296 ± 3
Oxytocin	1302 ± 5
Bradykinin	1269 ± 3
Luteinizing hormone releasing factor (LRF)	1278 ± 3
LRF antagonist	1318 ± 2
Substance P	1243 ± 2
Dynorphin	1203 ± 5
α-Melanocyte stimulating hormone	1238 ± 4
Neurotensin	1221 ± 3
Bombesin	1250 ± 2
Somatostatin	1266 ± 7
CLIP (human)	1172 ± 2
β-Melanocyte stimulating hormone	1198 ± 4
ACTH(1-24)	1166 ± 8
Gastrin releasing peptide	1147 ± 2
Somatostatin (1-28)	1164 ± 5
Vasoactive intestinal peptide (VIP)	1148 ± 10
Glucagon	1155 ± 7
β-Endorphin (human)	1118 ± 4
Insulin	1116 ± 3
Cytochrome *c*	898 ± 8
Trypsin inhibitor (soybean)	857 ± 3
Growth hormone (human)	896 ± 2
Trypsin (bovine)	897 ± 1
Chymotrypsinogen (bovine)	869 ± 9
Carbonic anhydrase	775 ± 2
Albumin (egg)	753 ± 4
γ-Globulins	732 ± 5

Packing: P1 = PACI-125 protein analysis column.
Solvent: S1 = Dilute TEAP buffer, pH 2.25-acetonitrile, 70:30. TEAP buffer is made by bringing the pH of 0.25 N H$_3$PO$_4$ to 2.25 with triethylamine that has been redistilled over *p*-toluenesulfonyl chloride. The buffer is then diluted 1:1 with water.
Detection: D1 = UV at 210 nm.

REFERENCE

1. **Rivier, J. E.**, *J. Chromatogr.*, 202, 211, 1980.

Reproduced by permission of Elsevier Science Publishers B.V.

Table LC 103
THE EFFECT OF TRIETHYLAMMONIUM PHOSPHATE (TEAP) CONCENTRATION ON RETENTION OF PEPTIDES AND PROTEINS

Packing	P1	P1	P1
Solvent	S1	S2	S3
Flow rate (mℓ/min)	1.0	1.0	1.0
Temperature (°C)	50	50	50
Detection	D1	D1	D1

Peptide or protein	t_r(sec)		
Bovine serum albumin	729	710	700
Cytochrome *c*	861	818	810
β_h-endorphin	1019	1046	1048
Clip	1075	1108	1112
LRF	1167	1206	1214

Note: β_h-Endorphin, human β-endorphin; Clip, adrenocorticotropic hormone, (18-39); LRF, luteinizing hormone releasing factor.

Packing: P1 = PAC I-125 column.
Solvent: S1, S2, and S3 consist of aqueous buffer-acetonitrile, 7:3. For S1 the aqueous buffer is TEAP, pH 2.25, for S2, TEAP, pH 2.25-water, 1:1 v/v and for S3 TEAP, pH 2.25-water, 1:3 v/v. TEAP buffer is obtained by bringing the pH of 0.25 *N* phosphoric acid to 2.25 with triethylamine which has been redistilled over *p*-toluenesulfonyl chloride. Several liters are made at one time and are filtered over a C_{18} cartridge. As a result any hydrophobic UV-absorbing material is eliminated.
Detection: D1 = UV at 210 nm.

REFERENCE

1. **Rivier, J. E.,** *J. Chromatogr.,* 202, 211, 1980.

Reproduced by permission of Elsevier Science Publishers B.V.

Table LC 104
SPECIFIC RESOLUTION OF SOME PAIRS OF PROTEINS ON TSK-GEL SW

Packing		P1	P2	P3
Column				
Length (cm)		65	65	65
Diameter (mm, I.D.)		7.5	7.5	7.5
Solvent		S1	S1	S1
Flow rate (mℓ/min)		1.0	1.0	1.0
Temperature (°C)		25	25	25
Detection		D1	D1	D1

Peptide	Compounds separated	Specific resolution (R$_s$)[a]		
1. Thyroglobulin	1 and 2		3.37	1.84
2. γ-Globulin	2 and 3	4.31	7.76	1.95
3. Bovine serum albumin	3 and 4	6.28	7.52	3.30
4. β-Lactoglobulin	4 and 5	7.28	7.03	3.20
5. Myoglobin	5 and 6	6.67	5.56	1.78
6. Cytochrome *c*	6 and 7	3.40	2.24	1.13
7. Glycylglycylglycylglycine				

[a] $R_s = 2(V_2 - V_1)/(W_2 + W_1)(\log M_1 - \log M_2)$ where V, W, and M are the elution volumes, peak widths at the base, and the molecular weights of two components.

Packing: P1 = G2000SW.
            P2 = G3000SW.
            P3 = G4000SW.
Solvent: S1 = 0.1 *M* phosphate buffer containing 0.3 *M* NaC1, pH 7.
Detection: D1 = UV at 220 nm.

REFERENCE

1. **Kato, Y., Komiya, K., Sasaki, H., and Hashimoto, T.,** *J. Chromatogr.,* 190, 297, 1980.

Table LC 105
GEL PERMEATION CHROMATOGRAPHY OF PROTEINS

Packing	P1	P1
Column		
Length (cm)	30	30
Diameter (mm)	7.8	7.8
Solvent	S1	S2
Flow rate (mℓ/min)	2	2
Temperature	rt	rt
Detection	D1	D1

Protein	k′	
Bovine serum albumin	0.11	0.13
Ovalbumin	0.17	0.20
Chymotrypsinogen A	0.70	0.44

Table LC 105 (continued)
GEL PERMEATION
CHROMATOGRAPHY OF
PROTEINS

Protein	k'	
Soybean trypsin inhibitor	0.34	0.37
Ribonuclease A	0.63	0.46
Cytochrome c	1.15	0.59
Insulin	1.70	1.67
Glucagon		1.20

Packing: P1   = Waters I-125 protein analysis columns (two columns in series each 30 cm long).

Solvent: S1   = 0.1 M ammonium acetate, pH 7.0.

　　　　 S2   = 0.1 M potassium phosphate, pH 7.0.

Detection: D1   = UV at 230 nm (acetate) or 220 nm (phosphate).

REFERENCE

1. **Jenik, R. A. and Porter, J. W.**, *Anal. Biochem.*, 111, 184, 1981.

Reproduced by permission of Academic Press, Inc.

Table LC 106
RESOLVING POWER R_S OF GEL
PERMEATION COLUMNS AS A
FUNCTION OF THE THROUGH-
FLOW RATE

Through-flow rate (mℓ/hr·cm²)	R_S	
	Type F	Type S
3.6	2.49	3.07
7.3	2.04	2.49
13	1.73	2.22
19	1.48	1.85
25.8	1.31	1.74
32.4		1.57

$$\text{Resolving power } R_S = \frac{2(V_{E2} - V_{E1})}{W_1 + W_2}$$

where V_E = eluted volume; W = peak width at base.

Note: Bovine serum albumin and myoglobin were separated on 600 × 22 mm columns of Fractogel® TSK HW-55, Types F and S. Type F has a particle size (moist with water) of 0.032—

Table LC 106 (continued)
RESOLVING POWER R$_S$ OF GEL PERMEATION COLUMNS AS A FUNCTION OF THE THROUGH-FLOW RATE

0.063 mm and type S a particle size (moist with water) of 25—40 μm. Using type S it is possible to achieve greater resolving power than with type F at similar elution rates (necessitating a slightly higher pressure) on account of the smaller particle size. Conversely, type S allows separations to be performed at a much higher elution rate with roughly identical resolving power (provided a correspondingly higher pressure is used).

REFERENCE

1. **Merck and Co.,** Fractogel® TSK pamphlet.

Table LC 107
PREDICTED AND OBSERVED RETENTION TIMES OF PEPTIDES

Packing	P1
Solvent	S1
Flow rate (mℓ/min)	2.0
Detection	D1

	t$_r$(min)		
Peptide	Observed		Predicted
GGG	1.8	1.3[a]	1.9[b]
PG	2.5	1.3	2.3
ARKM*	2.8	1.0	2.0
TEEQ	5.0	4.7	5.2
GL-NH$_2$	5.3	5.5	5.2
Ac-AAA	6.1	6.5	6.8
MTAK	6.5	8.0	6.4
NLC*	6.6	7.0	6.9
MARKM*	7.5	10.4	9.3
MAR	7.8	5.3	7.2
YK	8.0	6.0	7.5
TPGSR	8.1	7.6	7.3
KYE	8.2	7.7	9.4
GY	8.5	9.9	8.8
TEAEMK	9.2	10.7	10.3
EY	9.6	9.3	9.1
HLK	9.8	12.9	10.6
FK	9.9	11.4	10.8
IRE	10.3	12.2	11.7
PL	10.3	13.8	11.2
IAE	10.9	13.1	10.9
GF	11.5	14.1	11.9
KMKDTDSEEE	11.5	9.9	13.7
AFR	12.0	13.5	12.9
DIAAK	12.0	11.7	13.1

Table LC 107 (continued)
PREDICTED AND OBSERVED RETENTION TIMES OF PEPTIDES

Peptide	t_r(min) Observed		Predicted
QIAE	12.0	14.2	11.0
NIPC*	12.4	11.5	10.6
ASEDLK	13.0	11.9	12.7
EAFR	13.5	14.5	14.3
FDR	13.8	12.6	12.8
VFDKDGNGY	14.8	19.0	19.7
FKE	15.0	12.5	12.5
KVFGR	15.6	15.5	15.6
SLGQNPTEAE	15.8	17.4	16.9
GW	16.3	15.5	15.3
MIRE	16.5	16.9	15.3
SHPETLEK	16.7	19.2	17.5
HGLDNYR	17.0	18.0	17.6
WY-NH$_2$	17.1	16.6	17.4
Ac-ADQL	18.2	16.6	17.8
LFK	18.2	19.4	16.2
IAEFK	19.5	20.7	19.6
ADIDGDGQVNYEE	19.8	21.1	22.4
VFDKDGNGYI	20.2	19.0	19.7
ISAAELR	20.3	21.0	20.1
FESNFNTQATNR	20.3	19.0	19.4
ELGTVMR	21.2	21.0	19.2
GHHEAELK	21.3	18.9	17.2
WWC*NDGR	21.4	22.1	24.9
LQDMINE	22.0	21.0	20.7
FVQMMTAK	22.5	23.6	22.8
WWC*	23.5	23.4	24.5
QIAEFK	23.8	21.3	20.9
Ac-ADQLTEEQIAE	24.0	25.7	26.0
RSLGQNPTEAELQDM*	24.0	25.2	25.1
MIREADIDGDGQVNYEE	24.8	27.6	29.7
FLTMMAR	25.1	25.3	24.9
VDADGNGTIDEPE	25.3	24.5	23.8
HVMTNLGEK*LTDEEVDEM*	25.7	28.1	27.7
LGTVMRSLGQNPTEAE	25.8	27.5	27.4
SALLSSDITASVNG*	26.0	26.4	25.9
NTDGSTDYGILQINSR	26.9	27.6	27.9
VEADVAGHGQDILIR	26.9	29.0	28.7
FLTMMARKMKDTDSEEE	27.0	29.1	30.7
LRHVMTNLGEK*LTDE	27.0	24.7	24.8
VTVPLVSDAEC**R	27.3	27.0	26.1
VFDKDGNGYISAAELR	27.5	28.5	30.1
AFRVFDKDGNGYISAAE	28.6	29.3	31.2
LRHVMTNLGEK*LTDEEVDE	28.6	29.6	30.8
VFDKDGNGYISAAEL	29.0	28.5	29.5
GYSLGNWVC**	29.1	29.1	29.6
IREADIDGDGQVNYEEFVQM*	29.2	28.8	30.5
Ac-ADQLTEEQIAEFK	29.2	27.7	28.4
EAFSLFDKDGDGTITTK	30.0	31.4	31.5
ALELFR	30.2	25.4	24.6
AFSLFDKDGDGTITTKE	30.4	31.4	31.5
HVMTNLGEK*LTDEEVDEMIR	33.5	31.0	32.5
NKALELFRKDIAAKYKELGYQG	34.2	34.0	37.3
PGYPGVYTEVSYHVDWIK	34.8	34.5	36.7

Table LC 107 (continued)
PREDICTED AND OBSERVED RETENTION TIMES OF PEPTIDES

Peptide	t_r(min) Observed		Predicted
DDYGADEIFDSMIC** AGVPEGGK	35.9	34.8	37.1
EADIDGDGQVNYEEFVQMMTAK	37.2	31.2	33.5
INEVDADGNGTIDFPEFLTM*	37.5	33.9	34.1
KDTDSEEEIREAFRVFDKDGNGYISAAELRHVMTNLGEK*			
LTDEEVDEM*	37.8	40.6	45.8
IILHENFDYDLLDNDISLLK	38.5	38.1	40.8
ASSTNLKDILADLIPKEQARIKTFRQQHGNTVVGQITVDM*	39.0	40.3	43.7
HGVTVLTALGAILK	40.5	32.1	30.6
Ac-ADQLTEEQIAEFKEAFSLFDKDGDGTITTKELGYVMR	42.1	41.3	44.6
SQLSAAITALNSESNFARAYAEGIHRTKYWELIYEDC** M*	42.3	42.9	47.5
Ac-ADQLTEEQIAEFKEAFSLFDKDGDGTITTKELGTVM*	42.8	41.4	45.4
SLGQNPTEAELQDMINEVDADGNGTIDFPEFLTM	44.0	39.0	42.3
YLEFISEAIIHVLHSR	45.0	36.5	38.2
MARKMKDTDSEEEIREAFRVFDKDGNGYISAAELRHVMTNLGEK*			
LTDEEVDEMIREADIDGDGQVNYEEFVQMMTAK	45.5	46.0	53.8
NGLAGPLHGLANQEVLVWLTQLQKEVGKDVSDEKLRDYIWNTLN			
SGRVVPGYGHAVLRKTDPRYIC** QREFALKHLPHDPM*	45.5	50.3	53.8
VLSEGEWQLVLHVWAKVEADVAGHGQDILIRLFKSHPETLEK			
FDRFKHLKTEAEM*	45.6	47.2	54.1
SLGQNPTEAELQDMINEVDADGNGTIDFPEFLTMMAR	45.8	39.8	43.6
KMKDTDSEEEIREAFRVFDKDGNGYISAAELRHVMTNLGEK*			
LTDEEVDEMIREADIDGDGQVNYEEFVQMMTAK	46.2	45.5	53.1
VDADGNGTIDFPEFLTMMARKMKDTDSEEEIREAFRVFDKDGNGY			
ISAAELRHVMTNLGEK* LTDEEVDEMIREADIDGDGQVNYEEFVQMMTAK	48.3	50.1	58.8
FKEAFSLFDKDGDGTITTKELGTVMRSLGQNPTEAELQDMINEVD			
ADGNGTIDFPEFLTMMARKMKDTDSEEEIREAFRVFDKDGNGYISAAE	48.9	50.7	59.0
KASEDLKKHGVTVLTALGAILKKKGHHEAELKPLAQSHATKHKIPI			
KYLEFISEAIIHVLHSRHPGNFGADAQVAM*	50.0	49.6	56.2
Ax-ADQLTEEQIAEFKEAFSLFDKDGDGTITTKELGTVMRSLGQNPT			
EAELQDMINEVDADGNGTIDFPEFLTMMARKMKDTDSEEEIREAFRVFDK			
GNGYISAAELRHVMTNLGEK* LTDEEVDEMIREADIDGDGQVNYEEF			
VQMMTAK	51.0	56.0	66.7

Note: M*, Homoserine or its lactone; C*, aminoethylcysteine; C**, carboxymethylcysteine; K*, trimethyllysine.

[a] Predicted on the basis of modified Rekker's constants.
[b] Calculated from weighted fit parameters.

Packing: P1 = μBondapak® C_{18} column.
Solvent: S1 = The mobile phase was 0.1% TFA, pH 2. The mobile phase modifier was acetonitrile containing 0.07% of TFA. The concentration of the mobile phase modifier was increased linearly from 0 to 60% over 60 min.
Detection: D1 = UV at 216 nm.

REFERENCE

1. **Sasagawa, T., Okuyama, T., and Teller, D. C.,** *J. Chromatogr.*, 240, 329, 1982.

Reproduced by permission of Elsevier Science Publishers B.V.

Table LC 108
PREDICTED AND OBSERVED RETENTION TIMES OF PEPTIDES

Packing	P1
Column	
Length (mm)	250
Diameter (mm)	4.6
Solvent	S1
Flow rate (mℓ/min)	1.0
Temperature	rt
Detection	D1

Peptide	Elution point[a,b] Actual	Calculated
AF	27.70	31.30
FL	38.00	34.06
GF	28.50	30.29
GW	31.00	30.43
GY	8.50	29.27
LL	33.50	33.58
LM	27.70	31.86
VF	31.00	33.05
ETY	28.50	28.65
EWE	28.50	30.16
KWK	17.00	32.39
LWL	42.50	37.95
LWM	43.00	36.23
KDVY	28.00	32.98
YPFP	41.00	37.33
FDASV	31.50	32.96
HPPGF	35.00	33.68
IISMR	32.50	33.02
KPPGF	34.50	35.10
RKDVY	29.00	31.43
RPPGF	34.50	32.56
TKPPR	11.50	28.82
YGGFL	44.00	37.27
YGGFM	39.50	35.55
DAWALR	39.50	34.61
EVIRVL	44.50	37.42
FFKFSR	41.00	37.14
HSEFTF	27.80	28.18
LAHAIH	27.00	34.72
MKRPPG	28.00	31.35
NDYNPR	26.00	25.62
YGGFLR	39.50	35.72
AAEEVGL	33.00	34.32
AIHQVIK	26.00	31.58
RVYIHPF	49.00	39.96
YPEPGPI	47.00	43.00
DRVYIHPF	41.70	39.92
RPPGFSPF	44.00	37.65
SENNFQPK	27.80	25.96
VGANFLQR	38.10	32.22
MKRPPGFSP	36.50	36.44
RPPGFSPFR	42.00	36.10
RPPGFSPYR	35.00	35.08

Table LC 108 (continued)
PREDICTED AND OBSERVED RETENTION TIMES OF PEPTIDES

Peptide	Elution point[a,b]	
	Actual	Calculated
WAGGNASGE	29.50	29.25
DAGLELQAYR	40.00	35.02
DRVYIHPFHL	47.00	43.25
GSQDLANQYK	30.00	27.82
MKRPPGFSPF	43.00	40.68
RGSQDLANQY	29.50	25.28
SASLPNASVK	29.70	32.32
HNAGYPAAMPF	42.50	39.96
KSEGKNVVSAL	34.50	34.08
TITGSEAPNEW	24.50	32.81
EEQWKELASVVK	45.00	39.16
IASLILNTPELR	48.50	42.57
YVMGHFRWDRFG	48.00	43.34
KEWLVEVKGMADR	45.00	42.87
SLMKRPPGFSPFR	40.70	41.83
YGGFLRRIRPKLK	41.00	44.02
AGCKNFFWKTFTSC	9.50	46.27
NVVSALQLDMTNYK	45.70	38.73
DTNSKKMNLGVGAYR	34.80	33.70
GGFMTSEKSQTPLVTL	46.60	39.73
YGGFMTSEKSQTPLVT	40.50	39.18
DVYLPKPSWGNHTPIFR	45.50	49.32
YADSGEGDFLAEGGGVR	43.50	39.09
YGGFMTSEKSQTPLVTL	48.00	42.95
ASAELALGENSEAFKSGR	39.00	35.70
HTTQDTLANSDPTGSHAK	28.00	25.64
SDHASWHNAGYPAAMPFESK	37.70	42.53
SYSMEHFRWGKPVGKKRRPVKVYP	39.00	53.74
HFIEQGIDVVLSQSYAKNMGLYGER	51.00	50.81
YGGFMTSEKSQTPLVTLFKNAIIKNAYKKGE	50.50	59.76
CNBr peptide 1-32[a]	43.00	57.41
CNBr peptide 33-76[a]	51.50	72.28
CNBr peptide 77-141[a]	62.50	110.95
G-D,L-Nle	12.50	29.82
FLamide	33.00	32.17
LMamide	18.40	29.97
Y-D-AGF-D-L	50.20	38.27
YPFPamide	38.00	35.44
Y-D-AGF-D-Lamide	51.00	36.38
Y-D-AGFMamide	39.00	34.66
GEGFLG-D-FL	59.50	41.92
HPFHL-D-LVY	55.50	44.84
AGAFTVICR	43.00	36.78
DYMGWMDFamide	49.50	39.99
RPPG-p-chloro-FSP-p-chloro-FR	51.30	36.10
CYFQNCPRGamide	35.50	32.13
CYIQNCPRGamide	32.50	31.66
WAGG-iso-D-NASGE	29.40	27.27
WAGG-B-D-DASGE	29.50	31.16
pyro-EGLPPRPKIPP	39.50	44.33
RPKPQQFFGLMamide	49.00	37.57
pyro-EADPNKFYGLMamide	46.10	41.05
pyro-EPSKDAFIGLMamide	51.00	42.53

Table LC 108 (continued)
PREDICTED AND OBSERVED RETENTION TIMES OF PEPTIDES

Note: Actual elution points are measured values expressed as a percentage of buffer B in A + B buffer mixture. Calculated elution points are derived from peptide hydrophobicity values, calculated using Rekker's constants.

a   Aminoethyl-(α-chain of human hemoglobin).

Nonstandard abbreviations: Nle, norleucine; iso-D, isoaspartic acid; pyro-E, pyroglutamic acid.

Packing: P1   = LiChrosorb® RP-8 or RP-18, 5- or 10-μm particle sizes.
Solvent: S1   = Linear gradient of 0 to 50% of buffer B in A + B buffer mixture. Buffer A, 0.125 M pyridine-formate, pH 3.0; buffer B, 1.0 M pyridine-acetate, pH 5.5—60% propan-1-ol (final concentrations based on pyridine).
Detection: D1   = Fluorescamine.

REFERENCE

1. **Wilson, K. J., Honegger, A., Stötzel, R. P., and Hughes, G. J.,** *Biochem. J.,* 199, 31, 1981.

Reproduced by permission of the Biochemical Society.

Table LC 109
ALANINE-CONTAINING ANALOGUES OF DYNORPHIN-(1-13)

Packing	P1
Column	
Length (cm)	30
Diameter (cm)	2.2
Solvent	S1
Flow rate (mℓ/min)	4
Detection	D1

Peptide	t_r (min)
Dyn-(1-13)[a]	54
(Ala[1])-Dyn-(1-13)	69
(Ala[2])-Dyn-(1-13)	90
(Ala[3])-Dyn-(1-13)	91
(Ala[4])-Dyn-(1-13)	78
(Ala[5])-Dyn-(1-13)	70
(Ala[6])-Dyn-(1-13)	77
(Ala[7])-Dyn-(1-13)	76
(Ala[8])-Dyn-(1-13)	53
(Ala[9])-Dyn-(1-13)	72
(Ala[10])-Dyn-(1-13)	62
(Ala[11])-Dyn-(1-13)	71

a   Dyn-(1-13), dynorphin-(1-13).

Packing: P1   = Nucleosil® C_{18}, 30 μm.

Table LC 109 (continued)
ALANINE-CONTAINING
ANALOGUES OF DYNORPHIN-(1-13)

Solvent: S1 =   Chromatography was started with
                  a solution of 15% acetonitrile in
                  0.1% TFA, a linear gradient of
                  acetonitrile from 15 to 45% in
                  0.1% TFA being carried out be-
                  tween 10 and 90 min.

Detection: D1 =   UV at 280 nm.

REFERENCE

1. **Turcotte, A., Lalonde, J.-M., St. Pierre, S., and Lemaire, S.,** *Int. J. Pept. Prot. Res.,* 23, 361, 1984.

Reproduced by permission of Munksgaard International Publishers, Ltd.

Table LC 110
DIPEPTIDES

Dipeptide	K	pH	Solvent
Tyr-Tyr	29.3	2.0	S1
	6.35	5.8	S1
	35	8.6	S1
	0.04	11.6	S1
	3.38	2.0	S1
Tyr-Glu	0.20	5.8	S1
	0.48	8.6	S1
		11.6	S1
Tyr-Val	15.4	2.0	S1
	2.03	5.8	S1
	19.3	8.6	S1
	0.48	11.6	S1
Tyr-Arg	1.67	2.0	S1
	0.91	5.8	S1
	6.18	8.6	S1
	0.31	11.6	S1
Glu-Tyr	7.88	2.0	S1
	0.31	5.8	S1
	0.24	8.6	S1
	0	11.6	S1
Val-Tyr	15.4	2.0	S1
	1.65	5.8	S1
	16.5	8.6	S1
	0.43	11.6	S1
Arg-Tyr	4.31	2.0	S1
	0.82	5.8	S1
	2.54	8.6	S1
	0.19	11.6	S1
Gly-Phe	6.44	2.1	S2
	1.05	5.9	S2
	2.34	10.0	S2
Phe-Gly	3.00	2.1	S2
	1.40	5.9	S2
	3.29	10.0	S2
Phe-Phe	3.39	2.2	S3
	1.46	6.4	S3
	3.95	10.05	S3

Table LC 110 (continued)
DIPEPTIDES

Dipeptide	k'	pH	Solvent
	0.765	2.4	S4
	0.333	6.4	S4
	0.602	10.4	S4
Tyr-Gly	5.60	2.0	S5
	1.94	5.9	S5
	0.10	11.1	S5
Gly-Tyr	11.0	2.0	S5
	1.10	5.9	S5
	0	11.1	S5
Trp-Tyr	28.1	2.0	S6
	7.43	5.9	S6
	15.9	9.5	S6
	1.31	11.1	S6

Solvent: S1 = Acetonitrile-water, 1:99.
 S2 = Acetonitrile-water, 5:95.
 S3 = Acetonitrile-water, 1:4.
 S4 = Acetonitrile-water, 3:7.
 S5 = Water.
 S6 = Acetonitrile-water, 1:9.

Note: The mobile phase pH was maintained with HCl, NaOH, and phosphate buffer, while ionic strength, 0.10 M, was controlled with added NaCl. The column, 150 × 4.1 mm, contained PRP-1, a spherical 10 μm, PSDV copolymer of large surface area and porosity (Hamilton Co.). The column temperature was controlled at 25°C, the flow rate was 1.0 mℓ/min, and detection was at 254 or 208 nm.

REFERENCE

1. **Pietrzyk, D. J., Smith, R. L., and Cahill, W. R., Jr.,** *J. Liq. Chromatrogr.*, 6, 1645, 1983.

Table LC 111
TRIPEPTIDES

Tripeptide	k'	pH	Solvent
Phe-Gly-Gly	2.59	2.1	S2
	1.33	5.9	S2
	4.13	10.0	S2
Gly-Gly-Phe	8.07	2.1	S2
	1.10	5.9	S2
	1.78	10.0	S2
Phe-Phe-Gly	1.80	2.2	S3
	1.56	6.4	S3
	3.99	10.05	S3

Table LC 111 (continued)
TRIPEPTIDES

Tripeptide	k'	pH	Solvent
Gly-Phe-Phe	5.49	2.2	S3
	1.68	6.4	S3
	2.02	10.05	S3
Phe-Gly-Phe	4.86	2.2	S3
	1.32	6.4	S3
	2.11	10.05	S3
Phe-Phe-Phe	2.96	2.4	S4
	1.63	6.4	S4
	2.82	10.2	S4
Gly-Gly-Tyr	15.6	2.0	S5
	1.46	5.9	S5
	0	11.1	S5
Gly-Tyr-Gly	10.5	2.0	S5
	2.68	5.9	S5
	0.18	11.1	S5
Tyr-Gly-Gly	5.34	2.0	S5
	2.28	5.9	S5
	0.16	11.1	S5
Trp-Gly-Tyr	42.8	2.0	S6
	9.48	5.9	S6
	13.0	9.5	S6

For conditions, see Table LC 110.

REFERENCES

1. **Pietrzyk, D. J., Smith, R. L., and Cahill, W. R., Jr.,** *J. Liq. Chromatogr.*, 6, 1645, 1983.

Table LC 112
TETRAPEPTIDES

Tetrapeptide	k'	pH	Solvent
Phe-Gly-Phe-Gly	3.09	2.2	S3
	1.69	6.4	S3
	2.77	10.05	S3
Phe-Gly-Gly-Phe	3.24	2.2	S3
	1.08	6.4	S3
	2.79	10.05	S3
Gly-Phe-Phe-Gly	3.69	2.2	S3
	2.84	6.4	S3
	2.52	10.05	S3
Phe-Phe-Gly-Gly	1.45	2.2	S3
	1.57	6.4	S3
	3.94	10.05	S3
Phe-Phe-Phe-Gly	2.91	2.4	S4
	1.50	6.4	S4
	2.72	10.2	S4
Gly-Gly-Gly-Tyr	18.8	2.0	S5
	1.99	5.9	S5
	0.11	11.1	S5
Gly-Gly-Tyr-Gly	12.8	2.0	S5

Table LC 112 (continued)
TETRAPEPTIDES

Tetrapeptide	k′	pH	Solvent
	3.83	5.9	S5
	0.13	11.1	S5
Gly-Tyr-Gly-Gly	10.9	2.0	S5
	4.74	5.9	S5
	0.31	11.1	S5
Tyr-Gly-Gly-Gly	5.96	2.0	S5
	2.97	5.9	S5
	0.24	11.1	S5
Trp-Gly-Gly-Tyr	34.8	2.0	S6
	9.92	5.9	S6
	11.6	9.5	S6
	1.41	11.1	S6
Tyr-Gly-Gly-Trp	4.62	2.0	S6
	4.43	5.9	S6
	4.27	9.5	S6

For conditions, see Table LC 110.

REFERENCES

1. **Pietrzyk, D. J., Smith, R. L., and Cahill, W. R., Jr.,** *J. Liq. Chromatogr.*, 6, 1645, 1983.

Table LC 113
PENTAPEPTIDES

Pentapeptide	k′	pH	Solvent
Phe-Gly-Gly-Gly-Gly	2.40	2.1	S2
	1.63	5.9	S2
	5.27	10.0	S2
Gly-Phe-Gly-Gly-Gly	5.03	2.1	S2
	2.81	5.9	S2
	5.95	10.0	S2
Gly-Gly-Phe-Gly-Gly	6.50	2.1	S2
	3.13	5.9	S2
	4.18	10.0	S2
Gly-Gly-Gly-Phe-Gly	6.75	2.1	S2
	2.60	5.9	S2
	3.60	10.0	S2
Gly-Gly-Gly-Gly-Phe	9.27	2.1	S2
	1.42	5.9	S2
	1.87	10.0	S2
Gly-Phe-Gly-Phe-Gly	3.64	2.2	S3
	1.87	6.4	S3
	2.31	10.05	S3
Gly-Phe-Phe-Gly-Gly	3.20	2.2	S3
	2.34	6.4	S3
	2.40	10.05	S3
Gly-Phe-Gly-Gly-Phe	4.36	2.2	S3
	1.31	6.4	S3
	1.39	10.05	S3

Table LC 113 (continued)
PENTAPEPTIDES

Peptide	k′		
Phe-Phe-Phe-Gly-Gly	1.24	2.4	S4
	1.28	6.4	S4
	2.43	10.2	S4
Phe-Phe-Gly-Gly-Phe	1.59	2.4	S4
	1.00	6.4	S4
	1.61	10.2	S4
Phe-Phe-Gly-Phe-Gly	2.02	2.4	S4
	1.76	6.4	S4
	2.37	10.2	S4
Trp-Gly-Gly-Gly-Tyr	29.6	2.0	S6
	7.29	5.9	S6
	11.0	9.5	S6
	1.56	11.1	S6
Trp-Gly-Gly-Gly-Tyr	27.1	2.0	S6
	7.10	5.9	S6
	10.8	9.5	S6
	1.77	11.1	S6

For conditions, see Table LC 110.

REFERENCE

1. **Pietrzyk, D. J., Smith, R. L., and Cahill, W. R., Jr.**, *J. Liq. Chromatogr.*, 6, 1645, 1983.

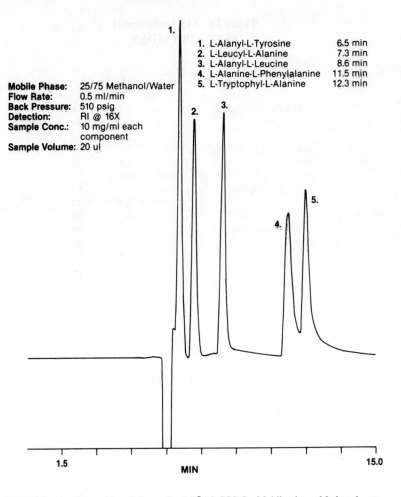

Mobile Phase: 25/75 Methanol/Water
Flow Rate: 0.5 ml/min
Back Pressure: 510 psig
Detection: RI @ 16X
Sample Conc.: 10 mg/ml each component
Sample Volume: 20 ul

1. L-Alanyl-L-Tyrosine	6.5 min
2. L-Leucyl-L-Alanine	7.3 min
3. L-Alanyl-L-Leucine	8.6 min
4. L-Alanine-L-Phenylalanine	11.5 min
5. L-Tryptophyl-L-Alanine	12.3 min

FIGURE LC 1. Dipeptides. Column: Partisil® 10 CCS/C₈. Mobile phase: Methanol-water, 25:75 v/v. Flow rate: 0.5 mℓ/min. Detection: RI. Reference: Whatman Liquid Chromatography Product Guide, 1982, 27. (Reproduced by permission of Whatman Chemical Separation, Inc.)

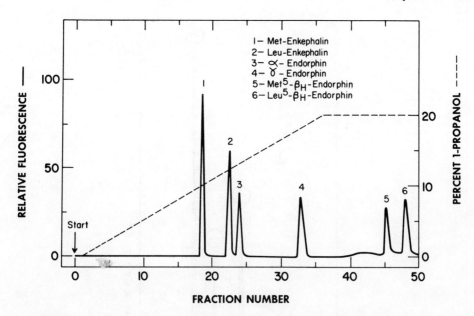

FIGURE LC 2. Opioid peptides. Column: LiChrosorb® C_{18}, 10-μm particle size, 4.6 × 250 mm. The enkephalins (10 nmol) and the endorphins (2 nmol) were loaded on the column in 1 mℓ of buffer. Solvent: 0.5 *M* formic acid adjusted to pH 4.0 with pyridine. The gradient was 0—20% *n*-propanol (-—-). Detection: Fractions were collected at 3-min intervals and peptides detected with fluorescamine. Reference: Lewis, R. V., Stein, S., and Udenfriend, S., *Int. J. Pept. Prot. Res.*, 13, 493, 1979. (Reproduced by permission of Munksgaard International Publishers, Ltd.)

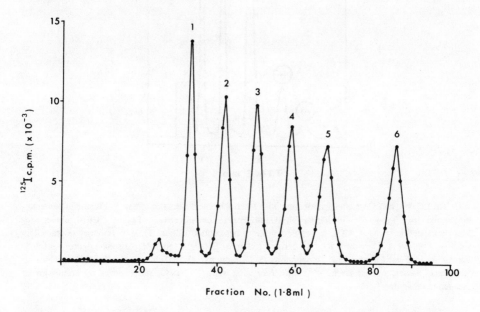

FIGURE LC 3. β-Endorphin-related peptides. Radioiodinated peptides; 3 μg of peptide labeled with 1 mCi of ^{125}I, 4 × 10^4 cpm of each peptide (*N*-acetyl endorphin $_{1-26}$, *N*-acetyl β-endorphin $_{1-27}$, β-endorphin$_{1-26}$, β-endorphin$_{1-27}$, *N*-acetyl β-endorphin$_{1-31}$, and β-endorphin$_{1-31}$) were separated. Column: SP-Sephadex® C25 (pyridinium form), 70 × 0.7 cm. Solvent: Linear gradient (100 mℓ mixer) with 50% acetic acid to 1 *M* pyridine in 50% acetic acid. Detection: Gamma counter assay of eluted fractions. Reference: Smyth, D. G., *Anal. Biochem.*, 136, 127, 1984. (Reproduced by permission of Academic Press, Inc.)

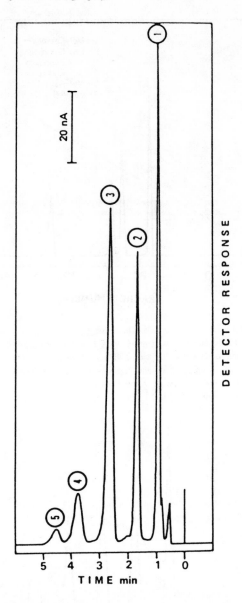

FIGURE LC 4.   HPLC of cholecystokinins with electrochemical detection. Peak 1: FK33-824, an enke-phalin analogue, Tyr-D-Ala-Gly-MePhe-Met(O)-ol. Peak 2: Leu-enkephalin. Peak 3: CCK-4, cholecysto-kinin tetrapeptide. Peak 4: CCK-8s, cholecystokinin octapeptide sulfate. Peak 5: Impurity contained in CCK-4. Column: LiChrosorb® RP-18, 2.1 × 150 mm. Solvent: 150 mM phosphate buffer, pH 5.5, containing 11.5 vol% 1-propanol. Temperature: 40°C. Flow rate: 0.6 mℓ/min. Detection: Electrochemical. Reference: Sauter, A. and Frick, W., *Anal. Biochem.*, 133, 307, 1983. (Reproduced by permission of Academic Press, Inc.)

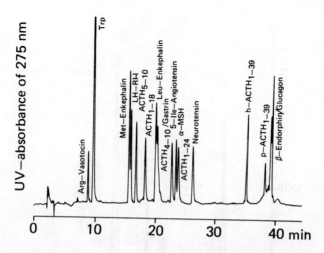

FIGURE LC 5.   Biogenic peptides. Column: RP-18, 5-μm particles, 250 × 4.6 mm, I.D. Solvent: Solvent A was 0.05 *M* TFA; solvent B was acetonitrile. The gradient was increased stepwise from 5 to 80% B within 60 min. Flow rate: 1.3 mℓ/min. Temperature: 40°C. Detection: UV at 275 and 254 nm. Reference: Schöneshöffer, M. and Fenner, A., *J. Chromatogr.*, 224, 472, 1981. (Reproduced by permission of Elsevier Science Publishers B.V.)

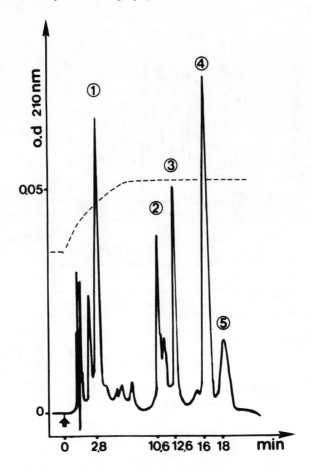

FIGURE LC 6.   The secretin-glucagon family of peptides. (1) Vasoactive intestinal polypeptide (VIP). (2) Gastric inhibitory polypeptide
(GIP). (3) Secretin. (4) Glucagon. (5) Pancreatic polypeptide. The
chromatography revealed the presence of impurities in the natural peptides VIP and GIP and in synthetic secretin. Column: μBondapak® C-
18, 0.39 × 30 cm, at room temperature. Solvent: Triethylammonium
phosphate (TEAP) buffer, 0.25 N, pH 3.5-acetonitrile. A gradient of
26 to 31% acetonitrile was established in 7 min. Flow rate: 2 mℓ/min.
Detection: UV at 210 nm. Reference: Fourmy, D., Pradayrol, L., and
Ribet, A., *J. Liq. Chromatogr.*, 5, 2123, 1982. (Reproduced by courtesy of Marcel Dekker, Inc.)

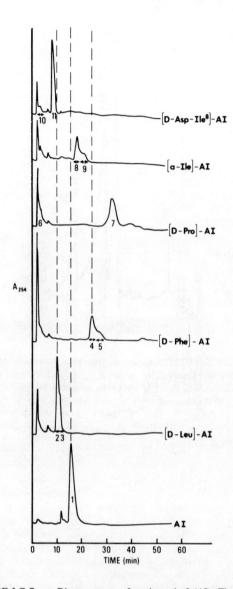

FIGURE LC 7. D-Diastereomers of angiotensin I (AI). The D-Pro, D-Leu, and D-Phe diastereomers were completely resolved from each other and from angiotensin I. The alloisoleucine diastereomer was retained on the column longer than angiotensin I and was not completely resolved from angiotensin I. Each diastereomer preparation contained at least one peptide impurity, the amino acid composition of the impurity differing from that of the expected peptide. Column: ODS (octadecylsilane). Solvent: 0.1 *M* TEAP, pH 3.5-acetonitrile, 81:19. Flow rate: 1 mℓ/min. Detection: Fractions absorbing at 254 nm were collected and analyzed for amino acid content. Reference: Margolis, S. A. and Konash, P. L., *Anal. Biochem.*, 134, 163, 1983. (Reproduced by permission of Academic Press, Inc.)

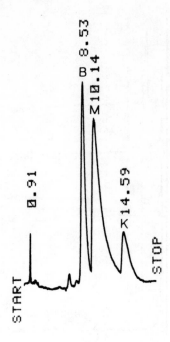

FIGURE LC 8.   HPLC of kinins. B, Bradykinin; K, kallidin; M, Met-Lys-bradykinin. Column: Nucleosil® 5 C$_8$. Solvent: Isocratic elution with 3% acetonitrile in 0.1 *M* sodium phosphate buffer, pH 7.0. Temperature: 30°C. Flow rate: 1.5 mℓ/min. Detection: UV at 215 nm. Reference: Geiger, R., Hell, R., and Fritz, H., *Hoppe-Seyler's Z. Physiol. Chem.*, 363, 527, 1982. (Reproduced by permission of Walter de Gruyter & Co.)

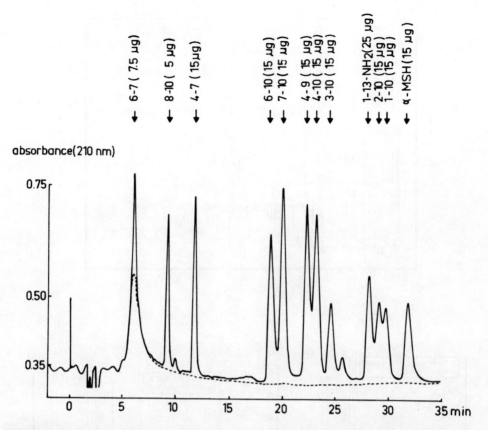

FIGURE LC 9.   N-Terminal fragments of ACTH. The abbreviations refer to ACTH 1-39 as the basic sequence. Column: μBondapak® C$_{18}$, 10 μm 300 × 3.9 mm, I. D. Solvent: Elution carried out with a 30-min convex gradient of 0.01 *M* ammonium acetate, pH 4.2 (X) and methanol containing 1.5 mℓ/ℓ acetic acid (Y). Initial conditions, X/Y = 95:5, final conditions X/Y = 50:50. Flow rate: 2 mℓ/min. Temperature: Ambient. Detection: UV at 210 nm. The baseline in the UV profile of the gradient used is shown by the interrupted line. Reference: Verhoef, J., Codd, E. E., Burbach, J. P. H., and Witter, A., *J. Chromatogr.*, 233, 317, 1982. (Reproduced by permission of Elsevier Science Publishers B.V.)

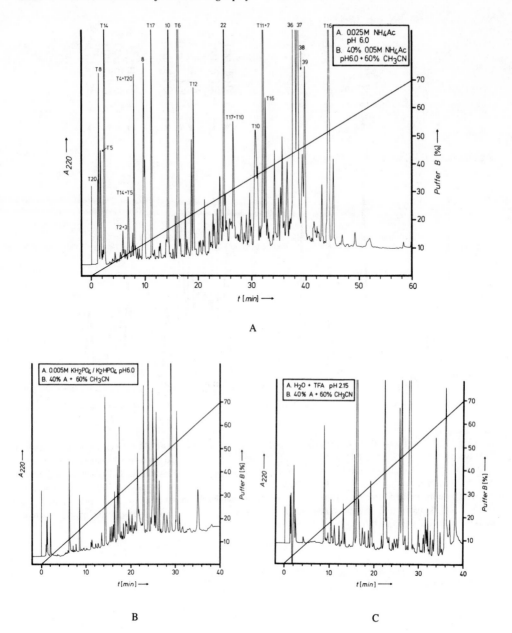

FIGURE LC 10.   Separation of the tryptic digest of the Bence-Jones protein den using different buffer systems. Column: ODS-Hypersil® 5μm, 250 × 4.6 mm. Solvent: (A) Buffer A, 25 mM ammonium acetate, pH 6.0, buffer B, 40% 50 mM ammonium acetate, pH 6.0, 60% acetonitrile. A linear gradient of buffer B from 0 to 70% in 60 min was established. (B) Buffer A, 5 mM KH$_2$PO$_4$/K$_2$HPO$_4$, pH 6.0, buffer B, 40% buffer A, 60% acetonitrile. A linear gradient of buffer B from 0 to 70% was established in 40 min. (C) Buffer A, water brought to pH 2.15 with TFA, buffer B, 40% buffer A, 60% acetonitrile. A linear gradient of buffer B from 0 to 70% was established in 40 min. Detection: UV at 220 nm. Note: These experiments showed the suitability of different buffer systems for the primary separation of peptide mixtures. The systems could also be effectively used for the rechromatography of fractions obtained during a primary separation. Reference: Yang, C. Y., Pauly, E., Kratzin, H., and Hilschmann, N., *Hoppe-Seyler's Z. Physiol. Chem.*, 362, 1131, 1981. (Reproduced by permission of Walter de Gruyter & Co.)

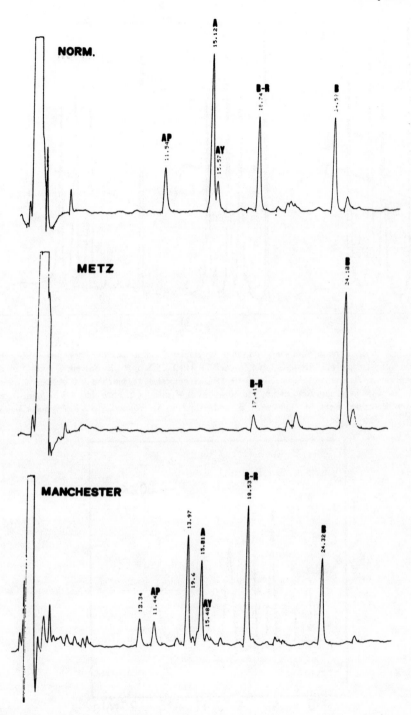

FIGURE LC 11. Fibrinopeptides. Fibrinopeptides released by thrombin from (A) normal human fibrinogen, (B) fibrinogen Metz, and (C) fibrinogen Manchester. Column: Hibar RT 250-4, LiChrosorb® RP-18 with 5-μm particles. Solvent: A gradient between 6 and 17% acetonitrile in 0.025 M ammonium acetate, pH 6.0 with H_2PO_4, in 40 min. Detection: UV at 210 nm. Note: In fibrinogen Metz the arginine residue of the thrombin cleavage site, i.e., position 16 of the A chain, is homozygously replaced by cysteine. In fibrinogen Manchester the arginine residue in position 16 of the A α chain is substituted by histidine. References: (1) Henschen, A., Lottspeich, F., Kehl, M., and Southan, C., *Ann. N.Y. Acad. Sci.*, 408, 28, 1983. (2) Kehl, M., Lottspeich, F., and Henschen, A., *Hoppe-Seyler's Z. Physiol. Chem.*, 362, 1661, 1981.

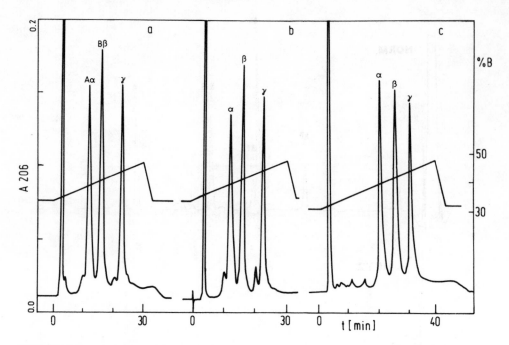

FIGURE LC 12. Derivatives of fibrinogen and fibrin. (a) Human S-(carboxymethyl)fibrinogen. (b) S-(carboxy-methyl)fibrin. (c) Mercaptolysed fibrin. Column: Vydac® TP RP10, 25 × 0.46 cm. Solvent: Solvent A was 0.1% TFA in water; solvent B, acetonitrile containing 0.1% TFA to compensate for baseline drift. A linear gradient was established, 36—48% in 30 min for a and b, and 32—48% in 40 min for c. Flow rate: 1.5 mℓ/min. Detection: UV at 206 nm. Reference: Kehl, M., Lottspeich, F., and Henschen, A., *Hoppe-Seyler's Z. Physiol. Chem.*, 363, 1501, 1982.

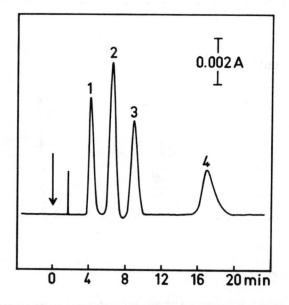

FIGURE LC 13. Phosphopeptides. Peak 1: P-M66, Arg-Arg-Ala-Ser(P)-Val-Ala. Peak 2: P-M57, Leu-Arg-Arg-Ala-Ser(P)-Val-Ala. Peak 3: P-M87, Val-Leu-Arg-Arg-Ala-Ser(P)-Val-Ala. Peak 4: P-M07, Gly-Val-Leu-Arg-Arg-Ala-Ser(P)-Val-Ala. Column: 250 × 4 mm column of Spherisorb® C_{18}, 10 μm preceded by a short precolumn of Bondapak C_{18}/Corasil®. Solvent: Phosphate buffer, pH 3.2 ethanol, 78:22. Counter-ion = 1-hexane sulfonic acid, 0.015 M. Flow rate: 1.5 ℓ/min. Reference: Fransson, B., Ragnarsson, U., and Zetterqvist, O., *Anal. Biochem.*, 126, 174, 1982. (Reproduced by Academic Press, Inc.)

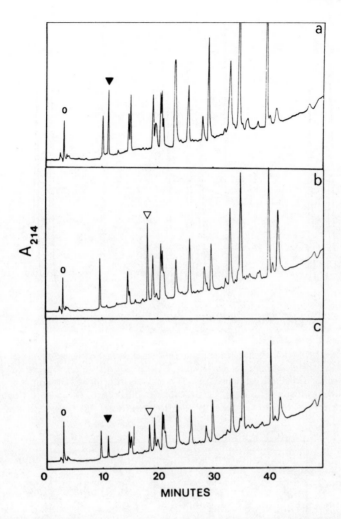

FIGURE LC 14.   Peptide mapping of closely homologous proteins. Liquid chromatographic maps of rabbit myosin light chains. Myosin light chains were incubated in 50 mM sodium phosphate buffer, pH 7.8, with *S. aureus* V$_8$ protease (1:500 mol/mol) for 16 hr. A 50-min linear gradient from 0 to 50% of acetonitrile in 20 mM potassium phosphate buffer, pH 5.9, was begun immediately on injection of the digest. The flow rate was 1.0 mℓ/min. A column of μBondapak® C$_{18}$ was used. The peak marked 0 is present in both the maps and is due to the digestion buffer. Peptides indicated by ▼ are unique to the LC1F map and that indicated by ▽ is unique to the LC3F map. (a) LC1F, (b) LC3F, (c) LC1F + LC3F. *Note:* The LC1F rabbit myosin light chain differs from the LC3F in two respects; it contains an additional 41 residues at its N-terminal end, and the 8 N-terminal residues of LC3F contain 5 amino acid replacements compared with the corresponding sequence of LC1F. Apart from these differences, the amino acid sequences of the two light chains appear to be identical. It is probable that the additional fragments in the map of LC1F have their origin in the "difference peptide" of this particular light chain, while the distinctive peptide seen in the digest of LC3F could arise from the N-terminal part of the molecule. Reference: Dalla Libera, L., Betto, R., and Biral, D., *J. Chromatogr.*, 264, 164, 1983. Reproduced by permission of Elsevier Science Publishers B.V.)

Section I.III.

PAPER CHROMATOGRAPHY TABLES

Wherever possible, tables are arranged according to classes of chemical compounds. This was not always possible when different chemical compounds were chromatographed under the same experimental conditions. The reader is referred to the compound index for specific compounds which may appear in different tables.

Table PC 1
POLYPEPTIDES CONTAINING ARGININE AND GLYCINE

Paper	P1	P1
Solvent	S1	S2
Peptide	**$R_F \times 100$**	
Arg-Gly-Arg	28	53
Arg-Gly$_2$-Arg	22	76
Arg-Gly$_3$-Arg	33	73
Arg-Gly$_4$-Arg	23	73
Arg-Gly$_5$-Arg	9	74
Arg-Gly$_6$-Arg	16	74
Arg-Gly$_7$-Arg	10	76
Arg-Gly$_6$-Phe-Arg	37	90
Arg-Gly$_3$-Phe-Gly$_3$-Arg	37	90
Arg-Gly$_3$-Phe-Gly$_2$-Phe-Arg	25	91

Paper: P1 = Whatman No. 4.
Solvent: S1 = 1-Butanol-pyridine-acetic acid-water, 30:20:6:24 v/v.
          S2 = 1-Propanol-ammonia (d = 0.88)-water, 3:6:1 v/v.
Detection: Chlorine-starch-potassium iodine, ninhydrin or Sakaguchi's reagent.
Technique: Descending development.

REFERENCE

1. **Abramson, F. B., Elliott, D. F., Lindsay, D. G., and Wade, R.,** *J. Chem. Soc. C.,* p. 1048, 1970.

Reproduced by permission of the Royal Society of Chemistry.

Table PC 2
GASTRIN-LIKE PEPTIDES AND DERIVATIVES

Paper	P1
Solvent	S1
Detection	D1

Peptide	$R_F \times 100$
Trp-Leu-Asp-Phe-NH$_2$	91
Deoxyfructosyl-Trp-Leu-Asp-Phe-NH$_2$	82
Glucosyl-deoxyfructosyl-Trp-Leu-Asp-Phe-NH$_2$	70
Thr-Gly-Trp-Leu-Asp-Phe-NH$_2$	71
Deoxyfructosyl-Thr-Gly-Trp-Leu-Asp-Phe-NH$_2$	65
Glucosyl-deoxyfructosyl-Thr-Gly-Trp-Leu-Asp-Phe-NH$_2$	39
Ala-Tyr-Gly-Trp-Leu-Asp-Phe-NH$_2$	89
Deoxyfructosyl-Ala-Tyr-Gly-Trp-Leu-Asp-Phe-NH$_2$	80
Glucosyl-deoxyfructosyl-Ala-Tyr-Gly-Trp-Leu-Asp-Phe-NH$_2$	69
Glu-Ala-Tyr-Gly-Trp-Nle-Asp-Phe-NH$_2$	90
Deoxyfructosyl-Glu-Ala-Tyr-Gly-Trp-Nle-Asp-Phe-NH$_2$	73
Glucosyl-deoxyfructosyl-Glu-Ala-Tyr-Gly-Trp-Nle-Asp-Phe-NH$_2$	45

Paper: P1 = Whatman No. 1.
Solvent: S1 = Butanol-acetic acid-water, 4:1:5 v/v.
Detection: D1 = *p*-Dimethylaminobenzaldehyde reagent.

REFERENCE

1. **Previero, A., Mourier, G., Bali, J.-P., Lignon, M. R., and Moroder, L.,** *Hoppe-Seyler's Z. Physiol. Chem.,* 363, 813, 1982.

Reproduced by permission of Walter de Gruyter & Co.

Section I.IV.

THIN-LAYER CHROMATOGRAPHY (TLC) TABLES

Wherever possible, tables are arranged according to classes of chemical compounds. This was not always possible when different chemical compounds were chromatographed under the same experimental conditions. The reader is referred to the compound index for specific compounds that may appear in different tables.

Table TLC 1
SOAP TLC OF PEPTIDES

Layer	L1	L2	L3	L4	L5	L6
Solvent	S1	S1	S1	S1	S1	S1
Detection	D1	D1	D1	D1	D1	D1
Technique	T1	T1	T1	T1	T1	T1

Peptide			$R_F \times 100$			
Gly-Gly	96	76	60	88	68	44
Gly-Ala	96	68	55	85	62	39
Gly-Ser	96	81	71	93	75	53
Gly-Ile	83	22	14	38	22	9
Gly-Leu	81	20	12	32	20	8
Gly-Pro	96	ES	51	75	56	28
Gly-Tyr	90	44	33	59	44	22
Gly-Phe	72	18	12	29	18	7
Ala-Ala	95	60	48	74	58	36
Ala-Gly	96	74	60	87	67	43
Ala-Ser	96	77	70	92	71	50
Ala-Val	90	37	25	52	37	20
Ala-Ile	77	21	14	33	21	9
Ala-Pro	92	48[a]	43	67	49	24
Ala-Tyr	90	40	30	54	41	20
Ala-His	96	16	6	38	12	2
Asp-Gly	96	ND	92	95	78	58
Asp-Ala	96	ND	78	88	68	49
Phe-Gly	76	19	13	32	19	8
Phe-Ala	75	17	12	26	15	7
Arg-Gly	96	6	3	21	5	2
Arg-Asp	96	11	5	29	9	3
Ile-Gly	90	30	24	47	35	14
Leu-Leu	55	6	5	9	5	3
Leu-Val	71	13	11	20	12	7
Leu-Tyr	72	16	14	24	15	8
Gly-Gly-Gly	96	75	56	83	66	42
Gly-Gly-Ala	96	64	49	78	59	38
Gly-Ala-Gly	96	67	49	78	59	38
Gly-Ala-Ala	96	60	46	74	56	36
Ala-Ala-Ala	96	61	51	74	57	38
Gly-Gly-Phe	76	24	16	29	17	7
Gly-Leu-Tyr	63	13	12	17	12	6
Leu-Gly-Phe	47	5	4	9	5	2
Gly-Gly-Gly-Gly	96	75	56	83	64	39
Ala-Ala-Ala-Ala	96	63	55	75	59	41
Leu-Trp-Met-Arg	49	0	0	0	0	0
Leu-Trp-Met-Arg-Phe	11	0	0	0	0	0

Note: ES, elongated spot. ND, not determined.

[a]   Tailing

Layer: L1 = SiO_2.
L2 = SiO_2 + 4% DBS (triethanolamine dodecylbenzenesulfonate).
L3 = SiO_2 + 4% Na-DSS (sodium dioctylsulfosuccinate).
L4 = SiO_2 + 1% H-DBS (dodecylbenzenesulfonic acid).
L5 = SiO_2 + 2% H-DBS.
L6 = SiO_2 + 4% H-DBS.
Layers are prepared by mixing 20 g of silanized silica gel 60 HF (C_2) (Merck) in 50 mℓ of 95% ethanol with a known concentration of detergent. Detergent concentration refers to the ethanolic solution in which the silica gel is suspended.

Table TLC 1 (continued)
SOAP TLC OF PEPTIDES

Solvent: S1 = Water-methanol-acetic acid, 64.3:30:5.7 v/v.

Detection: D1 = The wet layers are sprayed with 1% ninhydrin in pyridine-glacial acetic acid, 5:1 and the layers are then heated at 100°C for 5 min.

Technique: T1 = Temperature, 25°C; migration distance, 11 cm.

REFERENCE

1. **Lepri, L., Desideri, P. G., and Heimler, D.,** *J. Chromatogr.*, 195, 187, 1980.

Reproduced by permission of Elsevier Science Publishers B. V.

Table TLC 2
PEPTIDES ON SILANIZED SILICA GEL IMPREGNATED WITH H-DBS (DODECYLBENZENESULFONIC ACID)

Layer	L1	L1	L1	L1	L1	L1
Solvent	S1	S2	S3	S4	S5	S6
Detection	D1	D1	D1	D1	D1	D1
Technique	T1	T1	T1	T1	T1	T1
Peptide			$R_F \times 100$			
Gly-Gly	68	60	70	70	69	83
Gly-Ala	60	55	64	64	71	85
Gly-Ser	73	70	75	76	75	85
Gly-Ile	17	15	19	13	32	67
Gly-Leu	14	12	16	11	28	62
Gly-Pro	42	36	49	49	59	79
Gly-Tyr	31	30	40	30	50	78
Gly-Phe	12	8	12	11	24	52
Ala-Ala	56	52	62	62	70	83
Ala-Gly	66	59	69	69	73	85
Ala-Ser	72	68	74	74	73	85
Ala-Val	ES[a]	27	36	34	52	78
Ala-Ile	15	13	19	16	30	71
Ala-Pro	37	32	43	43	55	78
Ala-Tyr	31	30	39	32	46	78
Ala-His	10	4	15	14	15	76
Ala-Gly	74	71	75	77	77	87
Asp-Ala	62	61	68	65	67	84
Phe-Gly	13	11	13	9	15	37
Phe-Ala	12	10	15	8	18	49
Arg-Gly	6	3	6	4	5	56
Arg-Asp	8	4	9	7	11	69
Ile-Gly	23	21	28	20	31	56
Leu-Leu	8	4	6	3	7	31
Leu-Val	11	9	13	8	21	56
Leu-Tyr	12	11	16	9	21	55
Gly-Gly-Gly	66	58	67	66	62	82
Gly-Gly-Ala	59	52	63	61	64	82
Gly-Ala-Gly	62	54	64	62	61	82
Gly-Ala-Ala	59	51	62	59	63	82
Ala-Ala-Ala	60	54	63	61	69	82
Gly-Gly-Phe	13	10	13	9	22	55
Gly-Leu-Tyr	9	8	10	6	14	40
Leu-Gly-Phe	4	3	3	2	4	20
Gly-Gly-Gly-Gly	62	56	67	65	59	78

Table TLC 2 (continued)
PEPTIDES ON SILANIZED SILICA GEL IMPREGNATED WITH H-DBS (DODECYLBENZENESULFONIC ACID)

Peptide	$R_F \times 100$					
Ala-Ala-Ala-Ala	62	57	69	66	63	80
Leu-Trp-Met-Arg	0	0	0	0	0	0
Leu-Trp-Met-Arg-Phe	0	0	0	0	0	0

[a]   ES, elongated spot.

Layer: L1 = Silica gel + 4% H-DBS. Layers are prepared by mixing 20 g of silanized silica gel 60 HF(C$_2$) Merck, in 50 mℓ of 95% ethanol containing 4% H-DBS.
Solvent: S1 = 0.1 *M* HCl + 1 *M* CH$_3$COOH in 30% CH$_3$OH, pH 1.25.
          S2 = 0.05 *M* HCl + 1 *M* CH$_3$COOH in 30% CH$_3$OH, pH 1.55.
          S3 = 0.1 *M* NaCl + 1 *M* CH$_3$COOH in 30% CH$_3$OH, pH 2.75.
          S4 = 0.1 *M* NaCl + 0.1 *M* CH$_3$COOH in 30% CH$_3$OH, pH 3.30.
          S5 = 0.1 *M* CH$_3$COONa + 0.1 *M* CH$_3$COOH in 30% CH$_3$OH, pH 5.10.
          S6 = 1 *M* CH$_3$COONa in 30% CH$_3$OH, pH 8.15.
Detection: D1 = The wet layers are sprayed with 1% ninhydrin in pyridine-glacial acetic acid, 5:1 and the layers are then heated at 100°C for 5 min.
Technique: T1 = Temperature, 25°C; migration distance, 11 cm.

REFERENCE

1. **Lepri, L., Desideri, P. G., and Heimler, D.,** *J. Chromatogr.,* 195, 187, 1980.

Reproduced by permission of Elsevier Science Publishers B. V.

Table TLC 3
PEPTIDES ON SILANIZED SILICA GEL IMPREGNATED WITH ANIONIC OR CATIONIC DETERGENTS

	L1	L1	L2	L3	L4
Layer	L1	L1	L2	L3	L4
Solvent	S1	S2	S3	S4	S5
Detection	D1	D1	D1	D1	D1
Technique	T1	T1	T1	T1	T1
Peptide	$R_F \times 100$				
Gly-Gly	46	68	96	97	96
Gly-Ala	38	66	96	97	96
Gly-Ser	54	75	96	97	96
Gly-Ile	12	44	93	93	87
Gly-Leu	10	40	90	90	86
Gly-Pro	27	58	96	97	95
Gly-Tyr	27	69	94	84	77
Gly-Phe	10	41	78	81	69
Ala-Ala	35	64	96	97	96
Ala-Gly	43	70	96	97	96
Ala-Ser	51	72	96	97	96
Ala-Val	19	52	96	97	96
Ala-Ile	10	41	93	94	87
Ala-Pro	24	54	96	95	95
Ala-Tyr	26	62	96	88	81
Ala-His	3	18	96	97	96
Asp-Gly	58	79	96	94	88
Asp-Ala	46	71	96	94	89

Table TLC 3 (continued)
PEPTIDES ON SILANIZED SILICA GEL IMPREGNATED WITH ANIONIC OR CATIONIC DETERGENTS

Peptide	$R_F \times 100$				
Phe-Gly	10	42	79	86	74
Phe-Ala	8	41	81	89	81
Arg-Gly	2	13	96	97	96
Arg-Asp	2	18	96	97	96
Ile-Gly	15	48	94	96	95
Leu-Leu	3	26	57	80	71
Leu-Val	6	35	82	92	85
Leu-Tyr	10	49	74	79	69
Gly-Gly-Gly	39	67	96	97	96
Gly-Gly-Ala	34	67	96	97	96
Gly-Ala-Gly	35	67	96	97	96
Gly-Ala-Ala	34	67	96	97	96
Ala-Ala-Ala	36	68	96	97	96
Gly-Gly-Phe	7	43	80	78	66
Gly-Leu-Tyr	6	48	62	71	45
Leu-Gly-Phe	2	25	47	70	46
Gly-Gly-Gly-Gly	39	68	96	97	96
Ala-Ala-Ala-Ala	36	73	96	97	96
Leu-Trp-Met-Arg	2	5	32	79	58
Leu-Trp-Met-Arg-Phe	0	3	7	51	19

Layer: L1 = SiO$_2$ + 4% H-DBS.
L2 = SiO$_2$.
L3 = SiO$_2$ + 4% N-DPC (*N*-dodecylpyridinium chloride). Layers are prepared by mixing 20 g of silanized silica gel 60 HF (C$_2$) (Merck) in 50 mℓ of 95% ethanol with a known concentration of detergent. Detergent concentration refers to the ethanolic solution in which the silica gel is suspended.

Solvent: S1 = Water-acetic acid, 7:3 v/v.
S2 = Water-acetic acid, 1:1 v/v.
S3 = 0.1 *M* CH$_3$COOH + 0.1 *M* CH$_3$COONa in 30% CH$_3$OH.
S4 = 1 *M* CH$_3$OH.

Detection: D1 = The wet layers are sprayed with 1% ninhydrin in pyridine-glacial acetic acid, 5:1 and the layers are then heated at 100°C for 5 min.

Technique: T1 = temperature, 25°C; migration distance, 11 cm.

REFERENCE

1. **Lepri, L., Desideri, P. G., and Heimler, D.**, *J. Chromatogr.*, 195, 187, 1980.

Reproduced by permission of Elsevier Science Publishers B. V.

Table TLC 4
DIPEPTIDES ON SILANIZED SILICA GEL IMPREGNATED WITH H-DBS

Layer	L1	L2	L3	L4	L5
Solvent	S1	S1	S1	S1	S1
Detection	D1	D1	D1	D1	D1
Technique	T1	T1	T1	T1	T1

Peptide	$R_F \times 100$				
Gly-Thr	96	86	69	59	46
Gly-Val	91	51	37	27	17
Gly-Met	93	51	36	25	15

Table TLC 4 (continued)
DIPEPTIDES ON SILANIZED SILICA GEL IMPREGNATED WITH H-DBS

Peptide	$R_F \times 100$				
Gly-Trp	70	22	12	6	5
Gly-Glu	96	93	74	65	48
Gly-His	96	47	18	8	3
Ala-Thr	96	90	70	58	44
Ala-Trp	68	20	12	7	5
β-Ala-Ala	96	75	56	45	30
β-Ala-His	96	44	16	6	2
Ser-Gly	96	90	72	62	45
Ser-Ala	96	78	63	50	34
Ser-Leu	84	31	19	14	7
Ser-Phe	78	28	17	13	6
Ser-His	96	44	16	7	2
Thr-Gly	96	89	72	60	43
Val-Gly	94	63	49	40	25
Val-Ala	94	57	43	37	24
Val-Val	84	34	22	17	11
Val-Leu	69	16	9	8	5
Val-Phe	64	14	8	7	4
Val-Tyr	83	40	27	22	16
Leu-Gly	90	45	30	22	15
Leu-Ala	89	40	28	21	15
Leu-β-Ala	85	35	24	18	13
Leu-Ser	95	55	39	30	18
Leu-Ile	60	10	7	6	4
Leu-Trp	47	6	3	2	2
Leu-Phe	48	6	3	3	2
Pro-Gly	96	71	55	44	27
Pro-Ala	94	65	50	40	24
Pro-Leu	74	19	11	8	4
Pro-Phe	66	17	9	7	4
Met-Gly	95	54	40	30	17
Met-Val	80	25	16	12	7
Met-Leu	62	13	7	5	4
Met-Met	78	23	16	13	8
Met-Phe	55	9	6	5	3
Met-Tyr	77	30	22	17	11
Met-His	96	15	5	3	0
Trp-Gly	77	24	14	8	5
Trp-Ala	74	22	13	8	5
Trp-Leu	39	4	3	2	2
Trp-Phe	36	3	3	2	2
Phe-Ser	88	40	25	19	10
Phe-Val	65	13	7	6	4
Phe-Pro	57	16	7	6	4
Phe-Trp	40	3	3	2	2
Phe-Phe	42	4	3	2	2
Phe-Tyr	64	17	10	7	4
Tyr-Gly	93	61	43	34	20
Tyr-Ala	93	54	39	31	19
Tyr-Phe	59	12	9	6	4
His-Gly	96	43	14	7	2
His-Ala	96	35	10	5	2
His-Ser	96	47	18	10	3
His-Leu	90	4	2	2	0
His-Pro	95	31	9	4	1
His-Met	93	10	4	3	0
His-Phe	82	3	2	2	0
His-Tyr	93	13	5	3	0

Table TLC 4 (continued)
DIPEPTIDES ON SILANIZED SILICA GEL IMPREGNATED WITH H-DBS

Layer: L1 = SiO$_2$.
            L2 = SiO$_2$ + 1% H-DBS.
            L3 = SiO$_2$ + 2% H-DBS.
            L4 = SiO$_2$ + 3% H-DBS.
            L5 = SiO$_2$ + 4% H-DBS. Layers are prepared by mixing 20 g of silanized silica gel 60 HF (C$_2$)(Merck) in 50 mℓ of 95% ethanol with a known concentration of H-DBS. H-DBS concentration refers to the ethanolic solution in which the silica gel is suspended.
Solvent: S1 = Water-methanol-acetic acid, 64.3:30:5.7 v/v.
Detection: D1 = The wet layers are sprayed with 1% ninhydrin in pyridine-glacial acetic acid, 5:1 and the layers are then heated at 100°C for 5 min.
Technique: T1 = Temperature, 25°C, migration distance 11 cm.

REFERENCE

1. **Lepri, L., Desideri, P. G., and Heimler, D.,** *J. Chromatogr.*, 207, 412, 1981.

Reproduced by permission of Elsevier Science Publishers B.V.

Table TLC 5
DIPEPTIDES ON SILANIZED SILICA GEL IMPREGNATED WITH 4% H-DBS

Layer	L1	L1	L1	L1	L1	L1
Solvent	S1	S2	S3	S4	S5	S6
Detection	D1	D1	D1	D1	D1	D1
Technique	T1	T1	T1	T1	T1	T1
Peptide			$R_F \times 100$			
Gly-Thr	71	64	73	72	73	83
Gly-Val	30	28	34	33	55	81
Gly-Met	25	24	28	28	49	78
Gly-Trp	7	7	8	6	16	45
Gly-Glu	74	69	75	74	77	90
Gly-His	8	5	12	15	18	81
Ala-Thr	71	64	72	71	73	85
Ala-Trp	7	7	8	6	17	45
β-Ala-Ala	56	52	56	56	66	83
β-Ala-His	8	5	12	16	19	79
Ser-Gly	74	67	73	72	77	85
Ser-Ala	64	58	63	62	72	86
Ser-Leu	15	14	15	11	30	66
Ser-Phe	14	13	13	10	25	57
Ser-His	8	5	10	11	16	81
Thr-Gly	73	67	72	70	73	85
Val-Gly	45	42	45	42	50	76
Val-Ala	42	40	44	43	56	81
Val-Val	21	20	22	22	43	77
Val-Leu	12	11	12	10	20	55
Val-Phe	9	7	9	5	14	37
Val-Tyr	25	22	27	21	43	75
Leu-Gly	23	20	25	21	31	54
Leu-Ala	23	20	25	21	36	70
Leu-β-Ala	22	18	25	16	21	59
Leu-Ser	33	31	37	32	41	73
Leu-Ile	8	6	6	4	9	48
Leu-Trp	3	3	3	2	5	19

Table TLC 5 (continued)
DIPEPTIDES ON SILANIZED SILICA GEL IMPREGNATED
WITH 4% H-DBS

Peptide	$R_F \times 100$					
Leu-Phe	3	3	3	2	4	16
Pro-Gly	54	47	55	53	59	79
Pro-Ala	46	39	49	47	58	81
Pro-Leu	9	7	11	7	22	58
Pro-Phe	7	6	9	6	16	43
Met-Gly	34	31	37	32	41	68
Met-Val	14	12	17	12	30	70
Met-Leu	7	6	7	5	13	46
Met-Met	13	11	13	10	24	58
Met-Phe	5	4	5	3	9	27
Met-Tyr	17	16	17	14	31	67
Met-His	2	2	3	4	5	54
Trp-Gly	11	9	10	7	12	30
Trp-Ala	11	9	11	7	17	40
Trp-Leu	3	2	3	2	3	15
Trp-Phe	2	2	2	2	3	9
Phe-Ser	18	17	23	17	26	56
Phe-Val	7	7	9	6	15	44
Phe-Pro	6	4	7	5	13	32
Phe-Trp	2	0	2	2	3	12
Phe-Phe	2	2	3	2	3	11
Phe-Tyr	9	8	10	6	16	40
Tyr-Gly	38	35	45	33	43	67
Tyr-Ala	33	32	41	31	49	75
Tyr-Phe	6	5	8	4	12	33
His-Gly	6	5	10	12	19	81
His-Ala	3	3	8	9	18	81
His-Ser	6	5	11	16	20	81
His-Leu	1	0	2	2	4	ES[a]
His-Pro	3	2	6	7	15	ES
His-Met	2	2	3	2	8	ES
His-Phe	0	0	0	0	3	ES
His-Tyr	2	1	3	2	11	ES

[a] Elongated spot.

Layer: L1 = The layer is prepared by mixing 20 g of silanized silica gel HF (C$_2$) (Merck) in 50 mℓ
                of 95% ethanol containing 4% H-DBS.
Solvent: See Table TLC 2.
Detection: D1 = The wet layers are sprayed with 1% ninhydrin in pyridine-glacial acetic acid, 5:1 and
                the layers are then heated at 100°C for 5 min.
Technique: T1 = Temperature, 25°C; migration distance, 11 cm.

REFERENCE

1. **Lepri, L., Desideri, P. G., and Heimler, D.**, *J. Chromatogr.*, 207, 412,1981.

Reproduced by permission of Elsevier Science Publishers B. V.

Table TLC 6
DIPEPTIDES ON SILANIZED SILICA GEL IMPREGNATED WITH ANIONIC AND CATIONIC DETERGENTS

Layer	L1	L1	L2	L3	L3	L3
Solvent	S1	S2	S3	S4	S5	S6
Detection	D1	D1	D1	D1	D1	D1
Technique	T1	T1	T1	T1	T1	T1
Peptide			$R_F \times 100$			
Gly-Thr	46	67	96	97	96	96
Gly-Val	19	46	96	97	95	93
Gly-Met	18	56	96	96	90	89
Gly-Trp	7	36	75	68	39	32
Gly-Glu	56	75	96	97	96	96
Gly-His	2	13	96	97	96	81
Ala-Thr	46	66	96	97	96	94
Ala-Trp	7	35	75	68	41	35
β-Ala-Ala	34	56	96	97	96	94
β-Ala-His	2	13	96	97	96	94
Ser-Gly	49	68	96	97	96	95
Ser-Ala	41	62	96	97	96	95
Ser-Leu	8	32	91	91	85	78
Ser-Phe	9	36	83	82	70	56
Ser-His	2	13	96	97	96	81
Thr-Gly	49	65	96	97	96	95
Val-Gly	26	52	96	97	96	95
Val-Ala	25	50	96	97	96	95
Val-Val	14	40	93	96	92	86
Val-Leu	6	28	84	90	82	69
Val-Phe	6	43	71	81	70	42
Val-Tyr	18	60	90	87	77	63
Leu-Gly	11	44	96	96	94	86
Leu-Ala	10	43	94	96	96	90
Leu-β-Ala	10	40	95	96	96	86
Leu-Ser	17	50	96	96	96	92
Leu-Ile	4	23	69	85	75	46
Leu-Trp	3	20	47	58	25	7
Leu-Phe	3	18	48	74	48	18
Pro-Gly	27	56	96	97	96	94
Pro-Ala	25	53	96	97	96	94
Pro-Leu	5	28	88	90	84	76
Pro-Phe	5	29	74	82	71	58
Met-Gly	20	51	96	96	96	92
Met-Val	10	37	88	92	90[a]	76
Met-Leu	4	26	74	82	73	49
Met-Met	9	37	85	87	78	63
Met-Phe	4	25	58	77	56	24
Met-Tyr	15	53	81	84	72	48
Met-His	0	9	96	97	95	70
Trp-Gly	9	41	73	77	59	44
Trp-Ala	8	39	79	81	62	38
Trp-Leu	3	21	43	57	26	10
Trp-Phe	3	22	35	48	15	4
Phe-Ser	14	44	89	94	85	72
Phe-Val	6	27	75	84	75	34[a]
Phe-Pro	5	23	58	82	68	53
Phe-Trp	2	20	41	47	14	2
Phe-Phe	2	19	43	66	39	6

Table TLC 6 (continued)
DIPEPTIDES ON SILANIZED SILICA GEL IMPREGNATED WITH ANIONIC AND CATIONIC DETERGENTS

Peptide	$R_F \times 100$					
Phe-Tyr	9	43	69	79	54	15
Tyr-Gly	28	64	91	94	85	81
Tyr-Ala	26	62	93	96	89	81
Tyr-Phe	8	40	62	75	44	16
His-Gly	3	13	96	97	96	81
His-Ala	2	11	96	97	96	78
His-Ser	3	13	96	97	96	81
His-Leu	0	3	88	97	96	ES[b]
His-Pro	1	11	96	97	96	74
His-Met	0	4	96	97	96	50[a]
His-Phe	0	2	76	93	89	ES
His-Tyr	0	8	96	97	96	ES

[a]   Slightly elongated spot.
[b]   Elongated spot.

Layer:	L1 =	SiO$_2$ + 4% H-DBS.
	L2 =	SiO$_2$.
	L3 =	SiO$_2$ + 4% N-DPC (*N*-dodecylpyridinium chloride). Layers are prepared by mixing 20 g of silanized silica gel 60 HF(C$_2$)(Merck) in 50 mℓ of 95% ethanol with a known concentration of detergent. Detergent concentration refers to the ethanolic solution in which the silica gel is suspended.
Solvent:	S1 =	Water-acetic acid, 7:3 v/v.
	S2 =	Water-acetic acid, 1:1 v/v.
	S3 =	0.1 *M* CH$_3$COOH + 0.1 *M* CH$_3$COONa in 30% CH$_3$OH.
	S4 =	1 *M* CH$_3$COOH in 30% CH$_3$OH.
	S5 =	1 *M* CH$_3$COONa in 30% CH$_3$OH.
Detection:	D1 =	The wet layers are sprayed with 1% ninhydrin in pyridine-glacial acetic acid, 5:1 and the layers are then heated at 100°C for 5 min.

REFERENCE

1. **Lepri, L., Desideri, P. G., and Heimler, D.**, *J. Chromatogr.*, 207, 412, 1981.

Reproduced by permission of Elsevier Science Publishers B. V.

Table TLC 7
CLOSELY RELATED POLYPEPTIDES

Layer	L1	L1	L1	L2	L2	L2	L2
Solvent	S1	S2	S3	S3	S4	S5	S6
Detection	D1	D1	D1	D1	D1	D1	D1
Technique	T1	T1	T1	T1	T1	T1	T1
Compound	$R_F \times 100$						
Angiotensin III inhibitor	47	86	76	75	89	0	81
Angiotensin III	16	69	55	53	74	0	71
Angiotensin II	37	83	75	73	88	2	79
Des-Asp angiotensin I	24	79	63	59	81	1	75
Angiotensin I	22	79	63	60	83	1	76

Table TLC 7 (continued)
CLOSELY RELATED POLYPEPTIDES

Compound	$R_F \times 100$						
Melittin	0	6	0	0	0[a]	0	40[a]
					3[a]		44[a]
					6[b]		52[b]
Glucagon	0	44	33	26	68	0	72
Insulin-B	0	41	36	31	73	0	72
Actinomycin C_1	0	4	3	2	13	21	25
Actinomycin C	0	4	3	2	10[b]	14[b]	19[b]
$(C_1 + C_2 + C_3)$					13[b]	18[b]	22[b]
					22[a]	25[a]
Actinomycin V	0	5	4	3	15	18	22
Actinomycin I	0	9	8	5	21	30	32

[a] Secondary spot.
[b] Main spot.

Layer:	L1 =	Homemade layers of silanized silica gel.
	L2 =	RP-2 plates (Merck).
Solvent:	S1 =	1 *M* acetic acid in methanol-water (30:70 v/v).
	S2 =	1 *M* acetic acid in methanol-water (50:50 v/v).
	S3 =	1 *M* acetic acid + 3% KCl in methanol-water (50:50 v/v).
	S4 =	1 *M* acetic acid + 3% KCl in methanol-water (60:40 v/v).
	S5 =	Water-methanol-tetrahydrofuran (40:30:30 v/v).
	S6 =	3% KCl in water-methanol-tetrahydrofuran (40:30:30 v/v).
Detection:	D1 =	The spots of actinomycins are yellow; the other compounds are made visible with a 1% solution of ninhydrin in pyridine-acetic acid, 5:1.
Technique:	T1 =	Chromatography is performed at 25°C in a thermostatic chamber. The migration distance is 11 cm for the silanized silica gel and 6 cm for the RP-2 plates.

REFERENCE

1. **Lepri, L., Desideri, P. G., and Heimler, D.,** *J. Chromatogr.*, 211, 29, 1981.

Reproduced by permission of Elsevier Science Publishers B.V.

Table TLC 8
POLYPEPTIDES ON SILANIZED SILICA GEL IMPREGNATED
WITH *N*-DODECYLPYRIDINIUM CHLORIDE

Layer	L1	L2	L2	L2	L2
Solvent	S1	S1	S2	S3	S4
Detection	D1	D1	D1	D1	D1
Technique	T1	T1	T1	T1	T1
Polypeptide	$R_F \times 100$				
Angiotensin III inhibitor	83	86	83	85	79
Angiotensin III	69	82	65	84	63
Angiotensin II	71	84	55	33	33
Des-Asp angiotensin I	71	84	60	81	58
Angiotensin I	69	82	51	51	40
Melittin	0	76	0	0	0
Glucagon	0	72	0	0	0
Insulin-B	0	ES[a]	0	0	0

Table TLC 8 (continued)
POLYPEPTIDES ON SILANIZED SILICA GEL IMPREGNATED WITH *N*-DODECYLPYRIDINIUM CHLORIDE

[a]   ES, elongated spot.

Layer:	L1 and L2 consist of homemade layers of silanized silica gel impregnated with 1% and 4% of *N*-dodecylpyridinium chloride, respectively.
Solvent:	S1 = 1 *M* acetic acid in methanol-water (30:70 v/v).
	S2 = 0.1 *M* acetic acid + 0.1 *M* KCl in methanol (30:70 v/v) (pH 3.30).
	S3 = Water.
	S4 = Water-acetic acid, 99.5:0.5 v/v.
Detection:	D1 = 1% solution of ninhydrin in pyridine-acetic acid, 5:1.
Technique:	T1 = Chromatography is performed at 25°C in a thermostatic chamber. The migration distance is 11 cm.

REFERENCE

1. **Lepri, L., Desideri, P. G., and Heimler, D.,** *J. Chromatogr.*, 211, 29, 1981.

Reproduced by permission of Elsevier Science Publishers B.V.

Table TLC 9
SMALL PEPTIDES ON LAYERS OF AMMONIUM TUNGSTOPHOSPHATE

Layer	L1	L1	L1	L1
Solvent	S1	S2	S3	S4
Detection	D1	D1	D1	D1
Technique	T1	T1	T1	T1

Peptide	$R_F \times 100$			
Gly	66	75	88	91
Gly$_2$	44	59	70	75
Gly$_3$	25	36	51	60
Gly$_4$	13	22	35	43
Gly$_5$	5	11	22	31
Gly$_6$	2	5	10	15
Gly-Ala	46	61	72	77
Gly-Ser	48	63	74	80
Gly-Leu	46	58	70	75
Gly-Ile	41	53	65	71
Gly-Val	46	61	73	78
Gly-Thr	52	66	77	83
Gly-Met	27	39	48	53
Gly-Trp	10	19	24	30
Gly-Glu	39	53	63	70
Gly-His	8	18	27	35
Gly-Pro	28	42	50	56
Gly-Tyr	35	48	57	62
Gly-Phe	28	40	48	53
Ala	68	78	91	93
Ala$_2$	52	68	80	85
Ala$_3$	34	50	65	72
Ala$_4$	31	45	57	63

Table TLC 9 (continued)
SMALL PEPTIDES ON LAYERS OF AMMONIUM
TUNGSTOPHOSPHATE

Peptide	$R_F \times 100$			
Ala$_5$	21	31	45	50
Ala-Gly	49	64	76	82
Ala-Ser	56	71	83	89
Ala-Thr	57	72	83	89
Ala-Trp	16	26	31	36
Ala-Val	50	64	76	81
Ala-Ile	45	58	68	73
Ala-Tyr	40	53	63	68
Ala-His	11	22	33	41
β- Ala-Ala	36	48	59	65
β-Ala-His	7	13	18	23
Ser-Gly	51	65	77	83
Ser-Ala	52	66	78	84
Ser-Leu	51	64	76	80
Ser-Phe	36	48	57	62
Ser-His	11	23	34	42
Val-Gly	50	65	77	83
Val-Ala	52	66	78	84
Val-Val	47	61	72	77
Val-Leu	51	64	73	78
Val-Phe	33	45	53	59
Val-Tyr	40	53	62	69
Leu-Gly	47	61	72	77
Leu-Ala	49	64	75	79
Leu-β-Ala	36	49	59	65
Leu-Ser	54	69	80	86
Leu-Val	49	64	74	78
Leu-Leu	53	63	68	71
Leu-Ile	49	62	71	74
Leu-Tyr	31	46	55	61
Leu-Trp	11	19	23	27
Leu-Phe	28	40	47	52
Ile-Gly	44	58	70	76
Thr-Gly	55	70	81	88
Met-Gly	26	40	53	59
Met-Val	25	36	51	57
Met-Leu	28	38	52	58
Met-Met	13	23	34	41
Met-Phe	13	22	32	37
Met-Tyr	16	26	39	46
Met-His	3	9	17	23
Met-Arg	2	8	17	25
Trp-Gly	10	20	32	40
Trp-Ala	14	24	35	42
Trp-Leu	12	22	32	38
Trp-Phe	4	9	14	18
Phe-Gly	28	40	52	59
Phe-Ala	29	41	54	61
Phe-Ser	34	47	59	64
Phe-Val	31	42	54	60
Phe-Trp	3	6	11	13
Phe-Phe	14	24	33	37
Phe-Tyr	19	29	40	46
Tyr-Gly	32	45	57	64
Tyr-Ala	36	49	61	69

Table TLC 9 (continued)
SMALL PEPTIDES ON LAYERS OF AMMONIUM TUNGSTOPHOSPHATE

Peptide	$R_F \times 100$			
Tyr-Phe	14	23	33	37
Asp-Gly	47	62	76	83
Asp-Ala	48	63	77	84
Arg-Gly	7	16	30	38
Arg-Phe	3	7	14	19
Arg-Asp	7	18	32	42
His-Gly	10	20	34	40
His-Ala	12	22	36	43
His-Ser	12	22	36	42
His-Leu	10	20	32	37
His-Pro	5	11	21	29
His-Met	5	11	18	24
His-Phe	4	9	15	19
His-Tyr	5	11	18	25
Gly-Gly-Ala	25	36	51	60
Gly-Ala-Gly	29	42	57	66
Gly-Ala-Ala	31	44	59	67
Gly-Gly-Phe	16	21	30	34
Gly-Leu-Tyr	17	26	36	42
Leu-Gly-Phe	12	17	25	29

Layer: L1 = Ammonium tungstophosphate is prepared by dissolving 50 g of phosphotungstic acid in 200 mℓ of 2 *M* nitric acid. 200 mℓ of 2 *M* ammonium nitrate are added with stirring, the suspension is filtered and the solid rinsed with 1 *M* ammonium nitrate solution, washed with water, and air-dried. Layers are prepared using an automatic apparatus. Ammonium tungstophosphate, 4 g, is suspended in 50 mℓ of water. To the suspension, shaken with a magnetic stirrer, is added 2 g of calcium sulfate hemihydrate which has first been passed through a 200-mesh sieve. After 10 min the aqueous slurry is sprayed on the plates.

Solvents: S1, S2, S3, and S4 are solutions of ammonium nitrate of concentrations 1, 2, 3, and 4 *M*, respectively.

Detection: D1 = The wet layers are sprayed with a solution of 1% ninhydrin in pyridine-glacial acetic acid, 5:1, and are then heated at 100°C for 5 min.

Technique: T1 = Chromatography is carried out at 25°C. The migration distance is 10 cm.

REFERENCE

1. **Lepri, L., Desideri, P. G., and Heimler, D.,** *J. Chromatogr.,* 243, 339, 1982.

Reproduced by permission of Elsevier Science Publishers B.V.

Table TLC 10
PEPTIDE DIASTEREOMERS ON SILANIZED
SILICA GEL LAYERS

Layer	L1	L1	L1	L2	L3	L3
Solvent	S1	S2	S3	S4	S5	S6
Detection	D1	D1	D1	D1	D1	D1
Technique	T1	T1	T1	T1	T1	T1

Peptide	$R_F \times 100$					
L-Ala-L-Ala	94	95	91	96	75	80
D-Ala-L-Ala	83	83	80	96	68	80
L-Ala-D-Ala	83	83	80	96	68	80
D-Ala-D-Ala	94	95	91	96	74	80
L-Ala-L-Ala-L-Ala	87	87	82	96	75	81
L-Ala-L-Ala-D-Ala	74	77	72	96	68	80
L-Ala-D-Ala-L-Ala	63	69	57	96	64	79
D-Ala-D-Ala-D-Ala	87	87	82	96	75	81
Gly-L-Ala	94	95	92	96	76	84
Gly-D-Ala	94	95	92	96	76	84
L-Leu-Gly	54	57	41	82	25	55
D-Leu-Gly	54	57	41	82	25	55
L-Ala-L-Leu	45	52	53	80	16	35
D-Ala-L-Leu	33	35	30	71	12	29
L-Leu-L-Leu	16	18	10	56	5	17
D-Leu-L-Leu	6	6	4	34	0	6
L-Leu-D-Leu	6	6	4	34	0	6
D-Leu-D-Leu	16	18	10	56	5	17
L-Leu-L-Tyr	26	29	18	70	13	33
D-Leu-L-Tyr	23	23	12	66	8	30
L-Tyr-L-Arg	47	58	19	82	1	36
L-Tyr-D-Arg	45	54	13	79	1	33

Layer: L1 = Silanized silica gel, OPTI-UP C_{12} (Antec, Bennwil, Switzerland).

L2 = Silanized silica gel, Sil C_{18}-50 (Macherey, Nagel & Co., Düren, G.F.R.).

L3 = Sil C_{18}-50 impregnated with 4% H-DBS.

Solvent: S1 = 1 M acetic acid in water.

S2 = 1 M acetic acid + HCl-water (3:97 v/v).

S3 = 0.5 M sodium acetate in water.

S4 = 1 M acetic acid in water-methanol (30:70 v/v).

S5 = 1 M acetic acid + 0.1 M HCl in water-methanol (20:80 v/v).

S6 = 1 M acetic acid + 1 M HCl in water-methanol (20:80 v/v).

Detection: D1 = Spots were visualized by spraying with 1% ninhydrin in pyridine-acetic acid, 5:1, and heating the plates at 100°C for 5 min.

Technique: T1 = The migration distance was 6 cm, chromatography being performed at 25°C.

REFERENCE

1. **Lepri, L., Desideri, P. G., Heimler, D., and Giannessi, S.,** *J. Chromatogr.*, 265, 328, 1983.

Reproduced by permission of Elsevier Science Publishers B.V.

Table TLC 11
PEPTIDE DIASTEREOMERS ON AMMONIUM TUNGSTOPHOSPHATE LAYERS

Layer	L1	L1	L1
Solvent	S1	S2	S3
Detection	D1	D1	D1
Technique	T1	T1	T1

Peptide	$R_F \times 100$		
L-Ala-L-Ala	52	78	25
D-Ala-L-Ala	54	78	29
L-Ala-D-Ala	54	78	29
D-Ala-D-Ala	52	78	25
L-Ala-L-Ala-L-Ala	34	63	9
L-Ala-L-Ala-D-Ala	32	63	7
L-Ala-D-Ala-L-Ala	34	62	13
D-Ala-D-Ala-D-Ala	34	63	9
Gly-L-Ala	46	70	17
Gly-D-Ala	46	68	17
L-Leu-Gly	47	70	13
D-Leu-Gly	47	70	13
L-Ala-L-Leu	50	72	28
D-Ala-L-Leu	55	73	36
L-Leu-L-Leu	53	67	27
D-Leu-L-Leu	53	67	28
L-Leu-D-Leu	53	67	28
D-Leu-D-Leu	53	67	28
L-Leu-L-Tyr	31	54	10
D-Leu-L-Tyr	37	54	17
L-Tyr-L-Arg	4	18	0
L-Tyr-D-Arg	6	21	0
1st front	75	90	

Layer: L1 = Ammonium tungstophosphate (see Table TLC 9).

Solvent: S1 = 1 *M* NH_4NO_3.
             S2 = 3 *M* NH_4NO_3.
             S3 = 0.5 *M* HNO_3.

Detection: D1 = Spots were visualized by spraying with 1% ninhydrin in pyridine-acetic acid, 5:1, and heating the plates at 100°C for 5 min.

Technique: T1 = The migration distance was 10 cm, chromatography being performed at 25°C.

REFERENCE

1. **Lepri, L., Desideri, P. G., Heimler, D., and Giannessi, S.,** *J. Chromatogr.,* 265, 328, 1983.

Reproduced by permission of Elsevier Science Publishers B.V.

Table TLC 12
DIPEPTIDES

Layer	L1
Solvent	S1

Dipeptide	$R_F \times 100$
Gly-Phe	76
Gly-Tyr	94
Gly-Trp	68
Ala-Phe	62
Ala-Tyr	91
Leu-Phe	20
Leu-Tyr	70
Phe-Gly	74
Phe-Ala	69
Phe-Val	54
Phe-Leu	36
Val-Phe	58
Val-Tyr	91

Layer: L1   = A 1:4 cellulose: XAD-2 plate. XAD-2 is a polystyrene-divinyl benzene copolymer.
Solvent: S1   = EtOH-H$_2$O (20:80 v/v).
Technique: Ascending chromatography in a pre-equilibrated, closed paper lined chromatographic tank.
Detection: Ninhydrin.

REFERENCE

1. **Pietrzyk, D. J., Rotsch, T. D., and Chan Leuthauser, S. W.,** *J. Chromatogr. Sci.,* 17, 555, 1979.

Reproduced from the Journal of Chromatographic Science by permission of Preston Publications, Inc.

Table TLC 13
SMALL PEPTIDES

Layer	L1	L1	L1	L1
Solvent	S1	S2	S3	S4
Detection	D1	D1	D1	D1
Technique	T1	T1	T1	T1

Peptide	$R_F \times 100$			
Ala-Ala	65	26		
Ala-Asp	56	1		
Ala-Glu	64	5		
Ala-Gly	30	17		
Ala-Gly-Gly	47	13		
Ala-Phe	94	52		
Ala-Pro	61	24		
Ala-Ser	52	17		
Glu-Ala	68	1		
Gly-Ala	52	18		
Gly-Asp	43	0		

Table TLC 13 (continued)
SMALL PEPTIDES

Peptide	$R_F \times 100$			
Gly-Gly	34	13		
Gly-Gly-Gly	32	8		
Gly-Ile	81	49		
Gly-Leu	82	51		
Gly-Lys	14	11		
Gly-Pro	47	17		
Gly-Ser	32	13		
Gly-Tyr	68	28		
Gly-Val	72	36		
Leu-Ala	97	58		
Leu-Gly	82	52		
Leu-Val	100	77		
Pro-Gly	58	24		
Ser-Gly	43	16		
Val-Ala	80	49		
Val-Gly	67	38		
Val-Gly-Gly	65	28		
Val-His	23	19		
Val-Leu	100	80		
Val-Met	90	61		
Val-Phe	96	67		
Val-Pro	69	42		
Val-Ser	68	34		
Val-Tyr	90	57		
Val-Tyr-Val	96	73		
Val-Trp	95	67		
Val-Val	95	60		
Ile-Ala	90	68	56	
Ile-Gly	84	53	46	
Ile-Glu	94	13	13	
Ile-Leu	100	88	86	
Ile-Lys	58	50	41	
Ile-Met	94	79	80	
Ile-Phe	100	81	89	
Ile-Pro	89	60	63	
Ile-Ser	87	53	38	
Ile-Trp	100	83	83	
Ile-Val	100	80	83	
Leu-Leu	99	66	83	
Leu-Met	100	71	75	
Leu-Phe	99	73	79	
Leu-Ser	88	61	41	
Leu-Trp	100	73	77	
Leu-Tyr	97	67	66	
His-Ala	20	26		29
His-Gly	24	12		20
His-Leu	70	54		63
His-Phe	57	50		63
His-Ser	12	15		18
His-Tyr	57	52		40
Lys-Ala	35	9		40
Lys-Asp	28	2		2
Lys-Gly	19	8		27
Lys-Leu	80	42		75
Lys-Lys	14	9		48
Lys-Phe	66	42		75

Table TLC 13 (continued)
SMALL PEPTIDES

Peptide	$R_F \times 100$		
Lys-Val	54	28	70
Pro-Hyp	69	21	19
Pro-Ile	100	53	74
Pro-Leu	100	57	77
Pro-Met	100	50	72
Pro-Phe	92	52	75
Pro-Trp	91	50	67
Pro-Tyr	93	47	65
Pro-Val	100	42	66
Asp-Gly	48	2	
Glu-Glu	85	0	
Glu-Gly	64	3	
Glu-Val	86	4	
Glutathione (oxidized)	35	0	
Glutathione (reduced)	68	0	
Met-Ala	79	50	36
Met-Ala-Ser	64	40	21
Met-Glu	79	13	0
Met-Gly	64	44	28
Met-Leu	97	71	69
Met-Met	91	65	51
Met-Phe	95	70	62
Met-Pro	73	52	38
Met-Ser	67	43	25
Phe-Ala[a]	91/100	45	58
Phe-Gly[a]	81/100	52	54
Phe-Leu	100	85	84
Phe-Phe	100	83	86
Phe-Pro	91	75	62
Phe-Trp[a]	100/100	72	79
Phe-Tyr[a]	100/100	78	65
Phe-Val	100	81	77
Ser-Ala	54	26	7
Ser-Leu	87	63	39
Tyr-Ala[a]	79/100	40	38
Tyr-Glu[b]	95/80	23/75	5/75
Tyr-Gly[a]	68/100	34	27
Tyr-Leu[a]	93/100	60	80
Tyr-Phe[a]	92/100	58	70
Tyr-Tyr[a]	91/100	68	52
Tyr-Val	92	72	58

[a] Major/minor spot R_F values.
[b] Two spots of similar concentrations even in different solvents.

Layer: L1 = Cellulose powder, MN 300, without binder. Before use it is purified as follows. Cellulose powder, 50 g, is slurried with a mixture of methanol and water (4:1 v/v, 200 mℓ) and the slurry is poured into a large Buchner funnel and washed in the funnel with solvents (a) to (e) in the following order: (a) 2-propanol-water-acetic acid (60:20:20 v/v, 300 mℓ); (b) methanol-water (25:75 v/v, 200 mℓ); (c) methanol-1 N HCl (60:40 v/v, 200 mℓ); (d) water (200 mℓ); (e) methanol (200 mℓ). The powder is then dried overnight *in vacuo* before use.

Table TLC 13 (continued)
SMALL PEPTIDES

Solvent: S1 = 2-Propanol-butanone-1 *N* HCl (60:15:25 v/v).

S2 = 2-Methyl-2-butanol-butanone-propanone-methanol-water-0.88 ammonia solution (50:20:10:5:15:5 v/v).

S3 = *n*-Butanol-butanone-water-0.88 ammonia solution (80:5:17:3 v/v).

S4 = 2-Methyl-2-butanol-butanone-propanone-methanol-0.88 ammonia solution (25:20:35:5:20 v/v).

Detection: D1 = Ninhydrin-cadmium acetate.

Technique: T1 = Development proceeds until the solvent front has traveled 13 cm from the origin.

REFERENCES

1. **Heathcote, J. G., Washington, R. J., Keogh, B. J., and Glanville, R. W.,** *J. Chromatogr.,* 65, 397, 1972.
2. **Heathcote, J. G., Keogh, B. J., and Washington, R. J.,** *J. Chromatogr.,* 79, 187, 1973.
3. **Heathcote, J. G., Washington, R. J., and Keogh, B. J.,** *J. Chromatogr.,* 92, 355, 1974.
4. **Heathcote, J. G., Washington, R. J., and Keogh, B. J.,** *J. Chromatogr.,* 104, 141, 1975.

Reproduced by permission of Elsevier Science Publishers B.V.

Table TLC 14
TETRAPEPTIDES CONTAINING
ALANINE AND PROLINE

Layer	L1	L1	L1	L1
Solvent	S1	S2	S3	S4
Detection	D1	D1	D1	D1
Tetrapeptide		$R_F \times 100$		
Ala-Ala-Ala-Ala	10	75	28	51
Pro-Ala-Ala-Ala	10	52	31	40
Ala-Pro-Ala-Ala	9	70	19	52
Ala-Ala-Pro-Ala	8	67	21	49
Ala-Ala-Ala-Pro	6	67	19	49

Layer: L1 = Silica gel plates (Silufol).

Solvent: S1 = Butanol-acetic acid-water (4:1:1 v/v).

S2 = Isopropanol-water (1:1 v/v).

S3 = Propanol-water (7:3 v/v).

S4 = Butanol-acetic acid-water (2:1:2 v/v).

Detection: D1 = Ninhydrin.

REFERENCE

1. **Siemion, I. Z., Sobczyk, K., and Nawrocka, E.,** *Int. J. Pept. Prot. Res.,* 19, 439, 1982.

Table TLC 15
GLYCYL AND ALANYL
HOMO-OLIGOPEPTIDES

Layer	L1
Solvent	S1
Detection	D1

Compound	$R_F \times 100$
Glycine	22
Di-glycine	21
Tri-glycine	26
Tetra-glycine	32
Penta-glycine	50
Hexa-glycine	63
Alanine	44
Di-alanine	56
Tri-alanine	69
Tetra-alanine	80
Penta-alanine	94
Hexa-alanine	—[a]

[a] Did not migrate.

Layer:	L1	= Silica gel 60.
Solvent:	S1	= Phenol-0.1% ammonia, 10:1 v/v.
Detection:	D1	= Ninhydrin.

REFERENCE

1. **Oshima, G., Shimabukuro, H., and Nagasawa, K.,** *J. Chromatogr.*, 152, 579, 1978.

Reproduced by permission of Elsevier Science Publishers B.V.

Table TLC 16
OLIGOPEPTIDES

Layer	L1	L1
Solvent	S1	S2
Detection	D1	D1
Technique	T1	T1

Peptide	$R_F \times 100$	
Lys-Tyr	1	0
Lys-Lys-Lys-Lys	4	16
Lys-Ala	5	11
Lys-Lys-Lys	6	21
Ala-Ala-Ala-Ala-Ala-Ala	7	0
Arg-Asp	11	7
Gly-Gly-Gly	11	25
γ-Glu-Glu	12	6
Gly-Gly	16	32

Table TLC 16 (continued)
OLIGOPEPTIDES

Peptide	$R_F \times 100$	
Glu-Ala	21	21
Gly-dehydro Ala	23	43
Met-Asn	28	38
DL-Ala-DL-Ala	32	53
Ala-Ala-Ala-Ala	37	63
Glu-Tyr	38	22
Pro-Phe-Gly-Lys	38	86
Lys-Leu	40	60
Bradykinin triacetate	44	52
γ-Glu-Leu	47	40
Met-Ala	47	62
Trp-Glu	48	28
DL-Ala-DL-Norval[a]	50	69
Ala-Val	50	70
Phe-Gly-Phe-Gly	68	78
Tyr-Tyr	70	53
Trp-Tyr	73	65
Trp-Trp	77	71
Trp-Leu	80	84

[a] Norval, Norvaline.

Layer: L1 = TLC plates precoated with 250-μm thick, MN 300 cellulose.

Solvent: S1 = Isoamyl alcohol-ethyl alcohol-glacial acetic acid-pyridine-distilled water, 17.5:5:1.3:17.5:15 v/v.

S2 = Isopropyl alcohol-ethyl alcohol-ammonium hydroxide, 20:20:15 v/v.

Detection: D1 = Ninhydrin.

Technique: T1 = Development in glass tanks at 23 to 27°C, without pre-equilibration of the tanks.

REFERENCE

1. **Wainwright, I. M. and Shapsak, P.,** *J. Chromatogr. Sci.,* 17, 535, 1979.

Reproduced from the Journal of Chromatographic Science by permission of Preston Publications, Inc.

Table TLC 17
TRYPTOPHANYL PEPTIDES

Layer	L1	L1	L1	L1
Solvent	S1	S2	S3	S4
Detection	D1	D1	D1	D1
Technique	T1	T1	T1	T1
Peptide		$R_F \times 100$		
Trp-Ala	65	44	43	23
Trp-β-Ala	62	30	31	13
Trp-Gly	54	32	35,41[b]	17
Trp-Glu	51,55		26,22[b]	7,0
Trp-Ile	76	69	56,60[b]	45,52[b]
Trp-Leu	76	66	55,60[b]	40,47[b]
Trp-α-Lys	46,[a]	12,[a]	2	0
	47,51	15,7		
Trp-Phe	72	59	55	42
Trp-Trp	74	62	54	42
Trp-Tyr	69	56	52	43
Trp-Val	73,[a] 69	62	49	36,42[b]
Trp-Gly-Gly	49	22	25	8
Trp-Met-Asp-Phe-NH₂·HCl	67,[a]	47	50,	40,33,[b]
	64,[b]		30,34[b]	25,[b] 16,[b]
	54[b]			8[b]

[a] Main.
[b] Trace.

Layer: L1   = The following pre-coated glass chromatoplates, 20 × 20 cm, 0.25-mm
                    layer without fluorescent indicator were used; silica gel 60 TLC plates (E.
                    Merck) and silica gel K5 plates (Whatman).
Solvent: S1   = *n*-Butanol-acetic acid-water, 5:2:3 v/v.
            S2   = *n*-Butanol-acetic acid-water, 4:1:5 v/v, upper phase.
            S3   = Chloroform-isopropanol-water, 2:7:1 v/v.
            S4   = Ethyl acetate-*n*-propanol-water, 5:4:1 v/v.
Detection: D1   = The plates are dried at 110°C for 5 min, sprayed with 0.2 *M* sodium borate
                    buffer, pH 9.0, re-dried at 110°C for 5 min, and dipped for 30 min in
                    acetone-*n*-hexane, 1:4 v/v, containing fluorescamine, 10 mg/100 mℓ. The
                    plates are then sprayed with 40% perchloric acid for 5 sec and fluorescence
                    is observed within a few minutes under a long wave (366 nm) UV lamp.
                    Plates developed with acidic soluents are sprayed with 0.5 *M* sodium borate
                    buffer, pH 9.0, and sprayed with 40% perchloric acid for about 10 sec.
Technique: T1   = Ascending chromatography at ambient temperature (about 20°C) in glass
                    tanks.

REFERENCE

1. **Nakamura, H. and Pisano, J. J.**, *J. Chromatogr.*, 152, 153, 1978.

Reproduced by permission of Elsevier Science Publishers B.V.

Table TLC 18
TRYPTOPHANYL PEPTIDES

Layer	L1	L1	L1	L1	L1	L1
Solvent	S1	S2	S3	S4	S5	S6
Detection	D1	D1	D1	D1	D1	D1
Technique	T1	T1	T1	T1	T1	T1
Peptide				$R_F \times 100$		
Gly-Trp	18	59	9	1	0	17
Pro-Trp	10	59	5	2	1	17
Leu-Trp	49	73	33	13	2	33
Phe-Trp acetate	52	72	41	15	4	31
Val-Trp	40	71	23	11(8)[a]	0	29
Gly-Trp-Gly	11	53	4	1	0	14
Lys-Trp-Lys	0	23	0	0	0	2
Leu-Trp-Leu	63	85	59(46)	27(33)	5	55,62
Gly-Gly-Trp	10	53	3	1	0	12

[a] Values in parentheses indicate minor fluorescent spots.

Layer: L1 = Pre-coated silica gel glass plates (silica gel 60, 20 × 20 cm, 0.25-mm layer, without fluorescence indicator).

Solvent: S1 = Chloroform-isopropanol-water, 2:7:1 v/v.
S2 = *n*-Butanol-acetic acid-water, 5:2:3 v/v.
S3 = Ethyl acetate-isopropanol-water, 4:5:1 v/v.
S4 = Chloroform-ethanol-acetic acid, 13:6:1 v/v.
S5 = Benzene-dioxane-methanol, 5:3:2 v/v.
S6 = *n*-Hexane-*n*-propanol-ammonia solution (29.5% as NH₃), 2:7:1 v/v.

Detection: D1 = After brief drying with a hair dryer, the plate is sprayed with 70% perchloric acid for 5 sec and the fluorescence is observed under longwave UV light. After development with alkaline solvents the plate is sprayed for 10 sec.

Technique: T1 = Ascending chromatography in a glass chromatographic tank at room temperature.

REFERENCE

1. **Nakamura, H. and Pisano, J. J.,** *J. Chromatogr.*, 152, 167, 1978.

Reproduced by permission of Elsevier Science Publishers B.V.

Table TLC 19
FLUORESCAMINE DERIVATIVES OF TRYPTOPHANYL PEPTIDES

Layer	Ll	Ll	Ll	Ll
Solvent	S1	S2	S3	S4
Detection	D1	D1	D1	D1
Technique	T1	T1	T1	T1

Peptide	$R_F \times 100$			
Trp-Ala	35(81)	36(74)	70	37(68)
Trp-β-Ala	29(85)	38(83)	70	43(70)
Trp-Gly	31(68)	26(60)	69	31(66)
Trp-Glu	6(66)	20(38)	56	26
Trp-Ile	59(88)	73(84)	75	65(70)
Trp-Leu	(88)	75(86)	75	62(70)
Trp-α-Lys	13,30(80)	12,38	64,67	30,48(65)
Trp-Phe	(85)	39(80)	73	61(69)
Trp-Tyr	(86)	39(82)	70	57(68)
Trp-Val	(86)	40(82)	72	55(69)
Trp-Gly-Gly	32(72)	22,40	65	31
Trp-Met-Asp-Phe·NH₂·HCl	18(81)	39,43(72)	66,77	31,53(68)

Note: Spots which were not detected after TLC under longwave UV light and were detected by the later perchloric acid spray are shown in parentheses.

Solvent:	S1	= Dioxane-triethanolamine-methanol, 6:1:1 v/v.
	S2	= Chloroform-isopropanol-water, 2:8:1 v/v.
	S3	= Ethanol-chloroform-28% ammonia solution-water, 5:2:1:1 v/v.
	S4	= Ethyl acetate-n-hexane, methanol-water, 60:20:25:10 v/v.

Detection: D1 = Compounds are derivatized at the origin by dipping the lower 2 cm of the plate in fluorescamine solution for 15 min. The fluorescamine solution contains 10 mg of fluorescamine in a mixture of 20 mℓ of acetone and 80 mℓ of n-hexane. The plate is dried without heating and then developed with the appropriate solvent. Air-dried plates are sprayed with 40% perchloric acid for 5 sec.

Technique: T1 = Ascending chromatography at ambient temperature (about 20°C) in glass tanks.

REFERENCE

1. **Nakamura, H. and Pisano, J. J.,** *J. Chromatogr.*, 152, 153, 1978.

Reproduced by permission of Elsevier Science Publishers B.V.

Table TLC 20
TRYPTOPHAN DIPEPTIDE METHYL ESTERS AND HYDRAZIDES

Layer	L1
Solvent	S1

Tryptophan dipeptide methyl ester	$R_F \times 100$
Boc-Trp-OSu	67
Boc-Trp-Ala-OMe	66
Boc-Trp-Cys-OMe	62
Boc-Trp-His-OMe	55
Boc-Trp-Ile-OMe	82
Boc-Trp-Leu-OMe	73
Boc-Trp-Met-OMe	72
Boc-Trp-Phe-OMe	63
Boc-Trp-Trp-OMe	61
Boc-Trp-Tyr-OMe	59
Boc-Trp-Val-OMe	77

Tryptophan dipeptide hydrazide	
Boc-Trp-Trp-N_2H_3	51
Boc-Trp-Leu-N_2H_3	50
Boc-Trp-Met-N_2H_3	62
Boc-Trp-Phe-N_2H_3	53
Boc-Trp-Cys-N_2H_3	24
Boc-Trp-Tyr-N_2H_3	53
Boc-Trp-Ala-N_2H_3	54
Boc-Trp-Ile-N_2H_3	68
Boc-Trp-Val-N_2H_3	54

Abbreviation: OSu, *N*-hydroxysuccinimidyl ester.

Layer: L1 = Baker-flex silica gel 1B-F flexible TLC sheets.
Solvent: S1 = Benzene-methanol, 4:1 v/v.

REFERENCE

1. **Pettit, G. R., Krupa, T. S., and Reynolds, R. M.,** *Int. J. Pept. Prot. Res.,* 14, 193, 1979.

Reproduced by permission of Munksgaard International Publishers, Ltd.

Table TLC 21
HISTIDYL PEPTIDES AND RELATED COMPOUNDS

Layer	L1	L1	L1	L1	L1	L1
Solvent	S1	S2	S3	S4	S5	S6
Detection	D1	D1	D1	D1	D1	D1
Technique	T1	T1	T1	T1	T1	T1
Peptide			$R_F \times 100$			
L-Histidine	65	49	76	84	62	81
Histamine dihydrochloride	74	70	86	97	78(69)	80
DL-2-Methylhistidine dihydrochloride hydrate	63(75)	40(46)	75,62	76(86)	24	79
L-3-Methylhistidine	54(66)	23	14(11)	32	6(3)	49
DL-4-Methylhistidine	63(52)	44(18,24,31)	78	87(6,10)	57	81
L-Histidine methyl ester dihydrochloride	64,81	47,78	76,87	86,99	63,87	82,88
L-Histidine ethyl ester hydrochloride	65,85(70,78)	48,86	76,90	85,99	62,89	83,88
L-Histidine hydroxamate	65	47	76	86	63	82
L-Histidinol dihydrochloride	76(53)	76(21)	86(2)	97	79(70)	87
L-Histidinol phosphate	57	26	45	12	11	71
1,4-Methylhistamine dihydrochloride	55(65)	22(40)	2(14,18)	30	1	27
L-His-L-Leu	80	75	78	95	85	88
L-His-Gly	66	52	52	86	72	81
L-His-L-Lys hydrobromide	56(72)	26(58)	18	20	5	75(88)
L-His-L-Tyr	75	70	79	92	19	87
L-His-L-Ala	68	57	53	90	75	84
DL-His-DL-His	56,58	27,29	44	22,35	6	71,74
L-His-L-Ser	62(58)	43(34)	50	76	59(49)	82
L-His-L-Phe	76	70	80	96	83	88
Glucagon	60	18	7(0,44,46)	7(1)	2	92

Note: Values in parentheses are for trace spots, usually with transient reddish-orange fluorescence.

Layer: L1 = Pre-coated silica gel TLC plate (silica gel 60, 20 × 20 cm, 0.25-mm layer, E. Merck).
Solvent: S1 = *n*-Butanol-acetic acid-water, 5:2:3 v/v.
          S2 = *n*-Butanol-acetic acid-water, 4:1:5 v/v, upper phase.
          S3 = Ethyl acetate-methanol-water, 60:25:10 v/v.

Table TLC 21 (continued)
HISTIDYL PEPTIDES AND RELATED COMPOUNDS

S4 = Chloroform-methanol-acetic acid, 6:2:2 v/v.

S5 = Benzene–dioxan-acetic acid, 2:5:3 v/v.

S6 = Isopropanol-acetic acid-water, 6:2:2 v/v.

Detection: D1 = Histidine and related compounds are first derivatized with fluorescamine and then converted into new fluorophores. To 10 μℓ of 10 mM aqueous sample in a disposable polyethylene micro test tube, plus 40 μℓ of 0.2 M sodium borate buffer, pH 9.0, is added rapidly 50 μℓ of fluorescamine solution (20 mg in 100 mℓ of acetonitrile) with vigorous shaking in a vortex-type mixer. After standing for 5 min at room temperature, 50 μℓ of 2.0 N HCl are added and the tube is tightly capped. The mixture is incubated in a water bath at 80°C for 1 hr and 1 μℓ of the cooled reaction mixture applied to the TLC plate. After chromatography and brief drying under a stream of cold air from a hair-dryer, the plate is examined in the dark under a long wave (366 nm) UV lamp.

Technique: T1 = Ascending chromatography in a glass chromatographic tank.

REFERENCE

1. **Nakamura, H.,** *J. Chromatogr.,* 131, 215, 1977.

Reproduced by permission of Elsevier Science Publishers B.V.

Table TLC 22
DEHYDROTYROSINE PEPTIDES

Layer	L1	L1
Solvent	S1	S2
Detection	D1	D1

Peptide	$R_F \times 100$	
Boc-ΔTyr(Cl$_2$Bzl)-D-Ala-NH$_2$	32	71
Boc-ΔTyr(Cl$_2$Bzl)-Gly-NH$_2$	24	76
Boc-ΔTyr(Cl$_2$Bzl)-Met-NH$_2$	45	80
Boc-ΔTyr(Cl$_2$Bzl)-Gly-Gly-Phe-Leu-OBzl	50	87
Boc-ΔTyr(Cl$_2$Bzl)-Gly-Gly-Phe-ΔPhe-Leu-OBzl	52	98
Boc-ΔTyr(Cl$_2$Bzl)-Gly-Gly-Phe-ΔLeu-OBzl	39	83
Boc-ΔTyr(Cl$_2$Bzl)-D-Ala-Gly-Phe-Leu-OBzl	56	89
Boc-Tyr(Cl$_2$Bzl)-Gly-Gly-Phe-Leu-OBzl	61	98

Abbreviations: Bzl, benzyl; Boc, tert-butyloxycarbonyl.

Layer:	L1	=	Silica gel K66F (Whatman).
Solvent:	S1	=	CHCl$_3$-MeOH-AcOH, 95:5:1 v/v.
	S2	=	CHCl$_3$-MeOH, 5:1 v/v.
Detection:	D1	=	UV or 10% H$_2$SO$_4$ spray and heat.

REFERENCE

1. **Shimohigashi, Y., Dunning, J. W., Jr., Kolar, A. J., and Stammer, C. H.,** *Int. J. Pept. Prot. Res.,* 21, 202, 1983.

Reproduced by permission of Munksgaard International Publishers, Ltd.

Table TLC 23
CYCLIC DIPEPTIDES AND
DEHYDRODIPEPTIDES

Layer	L1
Solvent	S1

Peptide	$R_F \times 100$
Cyclo(-Gly-L-Ala-)	40
Cyclo(-ΔPhe-L-Ala-)	73
Cyclo(-ΔApb-L-Ala-)	70
Cyclo(-ΔApp-L-Ala-)	73
Cyclo(-ΔAph-L-Ala-)	70
Cyclo(-L-Phe-L-Ala-)	57
Cyclo(-D-Phe-L-Ala-)	67
Cyclo(-L-Apb-L-Ala-)	63
Cyclo(-D-Apb-L-Ala-)	66
Cyclo(-L-App-L-Ala-)	63
Cyclo(-D-App-L-Ala-)	66
Cyclo(-L-Aph-L-Ala-)	66
Cyclo(-D-Aph-L-Ala-)	68

Abbreviations: Apb, 2-amino-4-phenylbutanoic acid; App, 2-amino-5-phenylpentanoic acid; Aph, 2-amino-6-phenylhexanoic acid; Δ, dehydro.

Layer:	L1 = Silica gel g (Merck).
Solvent:	S1 = Chloroform-methanol, 5:1 v/v.

REFERENCE

1. **Hashimoto, Y., Aoyagi, H., Waki, M., Kato, T., and Izumiya, N.,** *Int. J. Pept. Prot. Res.,* 21, 11, 1983.

Table TLC 24
SOMATOSTATIN ANALOGUES

	L1	L1
Layer		
Solvent	S1	S2
X[a]	**$R_F \times 100$**	
Arginine	6	59
Histidine	63	61
Lysine	21	55
Thialysine	55	59
γ-Fluoro-L-lysine	31	57
δ-Fluoro-L-lysine	57	58
Ornithine	26	54
p-Aminophenyl	82	86

[a] X in the cyclic hexapeptide cyclo (Pro-Phe-D-Trp-X-Thr-Phe).

Layer:	L1	=	Silica gel (Analtech).
Solvent:	S1	=	Chloroform-methanol-concentrated ammonium hydroxide, 8:20:2 v/v.
	S2	=	Ethyl acetate-pyridine-acetic acid-water, 10:5:1:3 v/v.

REFERENCE

1. **Nutt, R. F., Verber, D. F., Curley, P. E., Saperstein, R., and Hirschmann, R.,** *Int. J. Pept. Prot. Res.,* 21, 66, 1983.

Table TLC 25
CYCLIC HEXAPEPTIDE ANALOGUES OF SOMATOSTATIN

	L1	L1
Layer		
Solvent	S1	S2
X[a]	**$R_F \times 100$**	
Δ^Z-Phe	57	39
Ala	55	46
D-Phe	59	43

Abbreviations: Δ^Z-Phe, trans-2,3-dehydrophenylalanyl.

[a] X in the analogue cyclo-(Pro-X-Trp-Lys-Thr-Phe).

Layer:	L1	=	Silica gel 60.
Solvent:	S1	=	Ethyl acetate-pyridine-acetic acid-water, 15:5:1:3 v/v.
	S2	=	Chloroform-methanol-concentrated ammonium hydroxide, 70:30:3 v/v.

REFERENCE

1. **Brady, S. F., Cochran, D. W., Nutt, R. F., Holly, F. W., Bennett, C. D., Paleveda, W. J., Curley, P. E., Arison, B. H., Saperstein, R., and Veber, D. F.,** *Int. J. Pept. Prot. Res.,* 23, 212, 1984.

Table TLC 26
ENKEPHALIN ANALOGUES

Layer	L1	L1	L1
Solvent	S1	S2	S3
Detection	D1	D1	D1

Peptide	$R_F \times 100$		
(Phe⁵)-EK-NH₂	52	41	
(Trp⁵)-EK-NH₂	53	38	
Met-EK-Cys(Cam)-OH	40		71
N,N'-bis(Met-EK)-cystine	26		56

Abbreviations: EK, enkephalin; Met-EK, methionine enkephalin; Cys(Cam), *S*-carbamidomethylcysteine residue.

Layer:	L1	= Silica gel.
Solvent:	S1	= 1-Butanol-acetic acid-water, 4:1:1 v/v.
	S2	= Ethyl acetate-pyridine-acetic acid-water, 60:20:6:11 v/v.
	S3	= Ethyl acetate-pyridine-acetic acid-water, 30:20:6:11 v/v.
Detection:	D1	= Ninhydrin or chlorine-tolidine.

REFERENCE

1. **Nadasdi, L., Yamashiro, D., Li, C. H., and Huidobro-Toro, P.**, *Int. J. Pept. Prot. Res.*, 21, 344, 1983.

Reproduced by permission of Munksgaard International Publishers Ltd.

Table TLC 27
ANALOGUES OF ENKEPHALIN CONTAINING
β-CYCLOPROPYLALANINE (Cpr)

Layer	L1	L1	L1
Solvent	S1	S2	S3
Detection	D1	D1	D1

Compound	$R_F \times 100$		
Boc-Phe-Pen	93	90	78
Boc-Gly-Phe-Pen	91	92	75
Boc-(D)-Ala-Gly-Phe-Pen	77	89	71
TFA-H-Tyr-(BzlCl₂)-(D)-Ala-Gly-Phe-Pen	79	63	60
Boc-(L)-Cpr-Tyr-(BzlCl₂)-(D)-Ala-Gly-Phe-Pen	93	76	82
Boc-(L)-Cpr-Tyr-(D)-Ala-Gly-Phe-Pen	94	68	80
AcOH-(L)-Cpr-Tyr-(D)-Ala-Gly-Phe-Pen	57	38	58
Boc-(D)-Cpr-(BzlCl₂)-(D)-Ala-Gly-Phe-Pen	95	76	78
Boc-(D)-Cpr-Tyr-(D)-Ala-Gly-Phe-Pen	89	80	78
Boc-(L)-Cpr-Gly-Phe-Pen	80	84	80
Boc-Tyr-(L)-Cpr-Gly-Phe-Pen	92	78	81
Tyr-(L)-Cpr-Gly-Phe-Pen	60	54	61
Boc-(D)-Cpr-Gly-Phe-Pen	90	81	75
Boc-Tyr-(D)-Cpr-Gly-Phe-Pen	89	71	82
Tyr-(D)-Cpr-Gly-Phe-Pen	67	49	75
TFA-(DL)-Cpr	65	78	68

Table TLC 27 (continued)
ANALOGUES OF ENKEPHALIN CONTAINING
β-CYCLOPROPYLALANINE (Cpr)

Abbreviations: Cpr, β-cyclopropylalanine; Pen, *n*-pentylamine.

Layer: L1   = Merck pre-coated silica gel 60 or cellulose plates.
Solvent: S1   = $CHCl_3$-MeOH-AcOH-H_2O, 60:30:4:1 v/v.
            S2   = Benzene-AcOH-H_2O, 9:9:1 v/v.
            S3   = *n*-BuOH-AcOH-H_2O, 4:1:1 v/v.
Detection: D1   = UV, ninhydrin, and chlorine/*o*-tolidine.

REFERENCE

1. **Muthukumaraswamy, N., Day, A. R., Pinon, D., Liao, C. S., and Freer, R. J.**, *Int. J. Pept. Prot. Res.*, 22, 305, 1983.

Reproduced by permission of Munksgaard International Publishers, Ltd.

Table TLC 28
LINEAR AND CYCLIC PEPTIDES RELATED
TO Met- AND Leu-ENKEPHALIN AND
PROTECTED INTERMEDIATES

Peptide	Solvent	$R_F \times 100$
Boc-Gly-Gly-Phe-Met-OMe	S1	44
Boc-Gly-Gly-Phe-Met-OH	S7	73
Boc-Phe-Met-Gly-Gly-OMe	S1	43
Boc-Phe-Met-Gly-Gly-OH	S3	86
Boc-D-Phe-Met-OH	S10	70
Boc-D-Phe-Met-εAhx-OMe	S1	50
Boc-D-Phe-Met-εAhx-OH	S1	30
Boc-Phe-D-Met-OH	S10	67
Boc-Phe-D-Met-εAhx-OMe	S1	52
Boc-Phe-D-Met-εAhx-OH	S1	32
Boc-Phe-Leu-OMe	S1	70
Boc-Phe-Leu-OH	S1	35
Boc-Phe-Leu-εAhx-OMe	S1	50
Boc-Phe-Leu-εAhx-OH	S3	46
Boc-Phe-Leu-εAhx-ONp	S1	56
cyclo-(-Phe-Met-)	S3	76
H-Gly-Gly-Phe-Met-OH	S7	30
H-Phe-Met-Gly-Gly-OH	S6	41
cyclo(-Phe-Met-Gly-Gly-)	S4	77
H-Phe-Met(O)-Gly-Gly-OH	S6	17
cyclo(-Phe-Met(O)-Gly-Gly)	S3	54
H-D-Phe-Met-εAhx-OH	S7	61
cyclo(-D-Phe-Met-εAhx-)	S5	83
H-Phe-D-Met-εAhx-OH	S7	61
cyclo(-Phe-D-Met-εAhx-)	S5	84
cyclo(-Phe-Leu-εAhx-)	S2	88

Abbreviations: εAhx, ε-aminohexanoic acid.

Layer: Merck silica gel plates (F_{254}, 0.25 mm).
Solvent: S1   = Toluene-ethanol, 4:1 v/v.
            S2   = Chloroform-methanol, 4:1 v/v.

Table TLC 28 (continued)
LINEAR AND CYCLIC PEPTIDES RELATED TO Met- AND Leu-ENKEPHALIN AND PROTECTED INTERMEDIATES

S3 = Chloroform-methanol-water, 70:30:5 v/v.
S4 = *n*-Butanol-acetic acid-water, 4:1:1 v/v.
S5 = *n*-Butanol-acetic acid-water, 8:1:1 v/v.
S6 = *n*-Butanol-pyridine-acetic acid-water, 8:3:1:4 v/v.
S7 = *n*-Butanol-pyridine-acetic acid-water, 16:3:1:4 v/v.
S8 = Chloroform-methanol, 9:1 v/v.
S9 = Chloroform-trifluoroethanol, 4:1 v/v.
S10 = Methylene chloride-methanol, 4:1 v/v.

Detection: UV (254 nm), fluorescamine and chlorine/*o*-tolidine.

REFERENCE

1. **van Nispen, J. W. and Greven, H. M.,** *Rec. Trav. Chim. Pays-Bas,* 101, 451, 1982.

Reproduced by permission of the Koninklijke Nederlandse Chemische Vereniging.

Table TLC 29
PROTECTED PENTAPEPTIDE PHOSPHONIC ANALOGUES OF ENKEPHALINS

Layer	L1	L1
Solvent	S1	S2
Detection	D1	D1
XP	$R_F \times 100$	
GlyP	23	81
L-LeuP	24	87
D-LeuP	23	80
L-MetP	20	79
D-MetP	26	84
L-PheP	15	83
D-PheP	28	92
D,L-ProP	25	84

Note: The pentapeptides have the formula Boc(Bzl)Tyr-Gly-Gly-Phe-XP(OR)$_2$ where XP stands for aminoalkanephosphonic acid residues corresponding to glycine, leucine, methionine, phenylalanine, and proline. R, ethyl (GlyP, PheP, MetP, ProP), or R, methyl (LeuP).

Layer: L1 = Silica gel plates, Merck 60F$_{254}$.
Solvent: S1 = Benzene-methanol-acetic acid, 15:2:1 v/v.
S2 = Chloroform-methanol, 5:1 v/v.
Detection: D1 = Ninhydrin and iodine.

Table TLC 29 (continued)
PROTECTED PENTAPEPTIDE PHOSPHONIC
ANALOGUES OF ENKEPHALINS

REFERENCE

1. **Kupczyk-Subotkowska, L. and Mastalerz, P.,** *Int. J. Pept. Prot. Res.,* 21, 485, 1983.

Reproduced by permission of Munksgaard International Publishers, Ltd.

Table TLC 30
PENTAPEPTIDE PHOSPHONIC
ANALOGUES OF ENKEPHALINS

Layer	L1	L1
Solvent	S1	S2
Detection	D1	D1
XP	**$R_F \times 100$**	
GlyP	38	72
L-LeuP	47	85
D-LeuP	42	79
L-MetP	45	75
D-MetP	34	68
L-PheP	49	79
D-PheP	44	80
D,L-ProP	39	82

Note: The pentapeptides have the formula Tyr-Gly-Gly-Phe-XP where XP stands for aminoalkanephosphonic acid residues corresponding to glycine, leucine, methionine, phenylalanine, and proline.

Layer: L1 = Silica gel plates, Merck 60F$_{254}$.
Solvent: S1 = Butanol-acetic acid-water, 12:3:5 v/v.
S2 = Butanol-acetic acid-pyridine-water, 6:6:3:5 v/v.
Detection: D1 = Ninhydrin and iodine.

REFERENCE

1. **Kupczyk-Subotkowska, L. and Mastalerz, P.,** *Int. J. Pept. Prot. Res.,* 21, 485, 1983.

Reproduced by permission of Munksgaard International Publishers, Ltd.

Table TLC 31
HUMAN β-ENDORPHIN ANALOGUES

Layer		L1	L1	L1
Solvent		S1	S2	S3
Detection		D1	D1	D1

Analogue	$R_F \times 100$		
(D-Phe[18],Phe[27],Gly[31])-β$_h$-EP[a]	63	49	13
(D-Lys[9],Phe[27],Gly[31])-β$_h$-EP	64	49	9
(D-Thr[2],D-Phe[18],Phe[27],Gly[31])-β$_h$-EP	64	51	13
(D-Thr[2],D-Lys[9],Phe[27],Gly[31])-β$_h$-EP	63	51	9

[a]   β$_h$-EP, human β-endorphin.

Layer: L1 = Silica gel.
Solvent: S1 = 1-Butanol-pyridine-acetic acid-water, 5:5:1:4 v/v.
           S2 = 1-Butanol-pyridine-acetic acid-water, 30:20:6:24 v/v.
           S3 = 1-Butanol-acetic acid-water, 4:1:5 v/v, upper phase.
Detection: D1 = Ninhydrin and Cl$_2$-tolidine reagent.

REFERENCE

1. **Yeung, H.-W., Yamashiro, D., Tseng, L.-F., Chang, W.-C., and Li, C. H.,** *Int. J. Pept. Prot. Res.,* 17, 235, 1981.

Table TLC 32
ALANINE-CONTAINING
ANALOGUES OF
DYNORPHIN-(1-13)

Layer		L1	L1
Solvent		S1	S2
Detection		D1	D1

Peptide	$R_F \times 100$	
Dyn-(1-13)[a]	67	74
(Ala[1])-Dyn-(1-13)	52	56
(Ala[2])-Dyn-(1-13)	61	62
(Ala[3])-Dyn-(1-13)	63	64
(Ala[4])-Dyn-(1-13)	55	69
(Ala[5])-Dyn-(1-13)	56	60
(Ala[6])-Dyn-(1-13)	67	72
(Ala[7])-Dyn-(1-13)	66	59
(Ala[8])-Dyn-(1-13)	54	56
(Ala[9])-Dyn-(1-13)	61	58
(Ala[10])-Dyn-(1-13)	57	51
(Ala[11])-Dyn-(1-13)	61	65

[a]   Dyn-(-13), Dynorphin-(1-13).

Layer: L1 = Silica gel plates; layer thick-
                 ness, 0.25 mm.
Solvent: S1 = 1-Butanol-pyridine-acetic
                 acid-water, 6:6:1.2:4.8 v/v.
           S2 = 1-Butanol-acetic acid-water,
                 4:3:3 v/v.
Detection: D1 = Ninhydrin reagent.

Table TLC 32 (continued)
ALANINE-CONTAINING
ANALOGUES OF
DYNORPHIN-(1-13)

REFERENCE

1. **Turcotte, A., Lalonde, J.-M., St. Pierre, S., and Lemaire, S.,** *Int. J. Pept. Prot. Res.,* 23, 361, 1984.

Reproduced by permission of Munksgaard International Publishers, Ltd.

Table TLC 33
NH₂-TERMINAL PEPTIDE FRAGMENTS OF DERMORPHINS

Peptide	Solvent	$R_F \times 100$
Boc-Phe-Gly-OEt	S1	79
H-Phe-Gly-OEt-HCl	S4	61
Boc-D-Ala-Phe-Gly-OEt	S1	51
Boc-Ala-Phe-Gly-OEt	S1	43
H-D-Ala-Phe-Gly-OEt·HCl	S4	57
H-Ala-Phe-Gly-OEt·HCl	S4	52
Boc-Tyr-D-Ala-Phe-Gly-OEt	S2	53
Boc-Tyr-Ala-Phe-Gly-OEt	S2	59
Boc-Tyr-D-Ala-Phe-Gly-OH	S3	35
	S4	85
Boc-Tyr-Ala-Phe-Gly-OH	S3	30
	S4	85
Boc-Phe-Gly-NH-NH-Z	S2	62
H-Phe-Gly-NH-NH-Z-HCl	S4	77
Boc-D-Ala-Phe-Gly-NH-NH-Z	S2	51
H-D-Ala-Phe-Gly-NH-NH-Z·HCl	S4	76
Boc-Tyr-D-Ala-Phe-Gly-NH-NH-Z	S2	38
Boc-Tyr-D-Ala-Phe-Gly-NH-NH₂	S3	15
	S4	84
Boc-Tyr-Ala-Phe-Gly-NH-NH₂	S3	13
	S4	81

Abbreviations: Z, benzyloxycarbonyl. Boc, tert-butyloxycarbonyl.

Layer:	L1	= Silica gel pre-coated plates 60F₂₅₄ (Merck), 0.25-mm layer thickness, 20-cm length.
Solvent:	S1	= Benzene-petroleum, b.p. 60—80°C-ethyl acetate, 25:5:70 v/v.
	S2	= Benzene-ethyl acetate-acetic acid-water, 100:100:20:10 v/v (upper phase).
	S3	= Benzene-ethyl acetate-acetic acid-water, 100:100:40:15 v/v (upper phase).
	S4	= *n*-Butanol-acetic acid-water, 40:10:10 v/v.
Detection:	D1	= Ninhydrin, chlorine or 1-nitroso-2-naphthol.

REFERENCE

1. **De Castiglione, R., Faoro, F., Perseo, G., and Piani, S.,** *Int. J. Pept. Prot. Res.,* 17, 263, 1981.

Reproduced by permission of Munksgaard International Publishers, Ltd.

Table TLC 34
COOH-TERMINAL PEPTIDE FRAGMENTS OF
DERMORPHINS

Peptide	Solvent	$R_F \times 100$
Z-Ser-NH$_2$	S5	71
	S6	39
Boc-Ser(Bzl)-NH$_2$	S3	82
	S4	91
H-Ser-NH$_2$·HCl	S5	59
	S4	20
H-Ser(Bzl)-NH$_2$·TFA	S4	44
Boc-Pro-Ser-NH$_2$	S3	32
Boc-Hyp-Ser(Bzl)-NH$_2$	S2	13
H-Pro-Ser-NH$_2$·TFA	S4	12
H-Hyp-Ser(Bzl)-NH$_2$·TFA	S4	32
Boc-Tyr(Bzl)-Pro-OH	S2	70
Boc-Tyr-Pro-Ser-NH$_2$	S3	16
	S4	71
Boc-Tyr(Bzl)-Pro-Ser-NH$_2$	S2	17
Boc-Tyr(Bzl)-Hyp-Ser(Bzl)-NH$_2$	S3	46
H-Tyr-Pro-Ser-NH$_2$·TFA	S4	23
	S5	83
H-Tyr(Bzl)-Pro-Ser-NH$_2$·TFA	S4	35
H-Tyr(Bzl)-Pro-Ser-NH$_2$·HCl	S4	35
H-Tyr(Bzl)-Hyp-Ser(Bzl)-NH$_2$·TFA	S4	52

Abbreviations: Z, benzyloxycarbonyl; Boc, tert-butyloxycarbonyl; TFA, trifluoroacetic acid; Bzl, benzyl; Hyp, 4-hydroxyproline.

Layer: L1 = Silica gel pre-coated plates 60F$_{254}$ (Merck), 0.25-mm layer thickness, 20-cm length.

Solvent: S1 = Benzene-light petroleum, b.p. 60—80°C-ethyl acetate, 25:5:70 v/v.

            S2 = Benzene-ethyl acetate-acetic acid-water, 100:100:20:10 v/v (upper phase).

            S3 = Benzene-ethyl acetate-acetic acid-water, 100:100:40:15 v/v (upper phase).

            S4 = *n*-Butanol-acetic acid-water, 40:10:10 v/v.

            S5 = Chloroform-methanol-32% ammonium hydroxide, 65:45:20 v/v.

            S6 = Chloroform-methanol, 70:10 v/v.

Detection: D1 = Ninhydrin, chlorine, or 1-nitroso-2-naphthol.

REFERENCE

1. **De Castiglione, R., Faoro, F., Perseo, G., and Piani, S.,** *Int. J. Pept. Prot. Res.,* 17, 263, 1981.

Reproduced by permission of Munksgaard International Publishers, Ltd.

Table TLC 35
PEPTIDE FRAGMENTS OF DERMORPHINS

Peptide	Solvent	$R_F \times 100$
Boc-Tyr-D-Ala-Phe-Gly-Tyr(Bzl)-Pro-Ser-NH$_2$	S3	15
	S4	81
Boc-Tyr-D-Ala-Phe-Gly-Tyr(Bzl)-Hyp-Ser(Bzl)-NH$_2$	S3	33
	S4	94
Boc-Tyr-Ala-Phe-Gly-Tyr(Bzl)-Pro-Ser-NH$_2$	S3	28
Boc-Tyr-D-Ala-Phe-Gly-Tyr-Pro-Ser-NH$_2$	S4	77
Boc-Tyr-D-Ala-Phe-Gly-Tyr-Hyp-Ser-NH$_2$	S4	75
Boc-Tyr-Ala-Phe-Gly-Tyr-Pro-Ser-NH$_2$	S4	75
H-Tyr-D-Ala-Phe-Gly-Tyr-Pro-Ser-NH$_2$·TFA	S4	51
	S5	77
H-Tyr-D-Ala-Phe-Gly-Tyr-Hyp-Ser-NH$_2$·HCl	S4	44
	S5	66
H-Tyr-Ala-Phe-Gly-Tyr-Pro-Ser-NH$_2$·TFA	S4	39

Abbreviations: Boc, tert-butyloxycarbonyl; TFA, trifluoroacetic acid.

Layer: L1 = Silica gel pre-coated plates 60F$_{254}$ (Merck), 0.25-mm thickness, 20-cm length.
Solvent: S1 = Benzene-light petroleum, b.p. 60—80°C-ethyl acetate, 25:5:70 v/v.
S2 = Benzene-ethyl acetate-acetic acid-water, 100:100:20:10 v/v (upper phase).
S3 = Benzene-ethyl acetate-acetic acid-water, 100:100:40:15 v/v (upper phase).
S4 = *n*-Butanol-acetic acid-water, 40:10:10 v/v.
S5 = Chloroform-methanol-32% ammonium hydroxide, 65:45:20 v/v.
Detection: D1 = Ninhydrin, chlorine, or 1-nitroso-2-naphthol.

REFERENCE

1. **De Castiglione, R., Faoro, F., Perseo, G., and Piani, S.,** *Int. J. Pept. Prot. Res.*, 17, 263, 1981.

Reproduced by permission of Munksgaard International Publishers, Ltd.

Table TLC 36
PEPTIDES RELATED TO DERMORPHIN

Layer	L1	L1	L2
Solvent	S1	S2	S3
Detection	D1	D1	D1
Peptide	**R_F[c]**		
Dermorphin[a]	1.23 Tyr		
(L-Ala)^2-dermorphin	1.08 Tyr		
Tyr-Ala-Phe	1.17 Tyr		
Tyr-D-Ala-Phe[b]	1.31 Tyr		
Ala-Phe	1.00 Tyr	1.66 Tyr	1.39 Tyr
D-Ala-Phe[b]	0.91 Tyr	1.33 Tyr	0.89 Tyr

[a] The natural and synthetic peptides had the same R_F value.
[b] Samples of this peptide obtained by synthesis or by enzymatic degradation of dermorphin had the same R_F value.
[c] R_F relative to tyrosine.

Note: Ala-Phe and D-Ala-Phe were eluted at volumes of 208 and 195 mℓ, respectively, on the amino acid analyzer.

Table TLC 36 (continued)
PEPTIDES RELATED TO DERMORPHIN

Layer: L1   = TLC silica gel plates 60 and TLC plates silica gel
                  60F$_{254}$ (layer thickness 0.25 mm) (Merck).
           L2   = Eastman Chromagram Sheets 13255 cellulose with-
                  out fluorescent indicator.
Solvent: S1   = *n*-Butanol-glacial acetic acid-water, 4:1:1 v/v.
           S2   = Methanol-water, 99:1 v/v.
           S3   = Pyridine-water, 82:18 v/v.
Detection: L1   = 0.1% Ninhydrin in 95% ethanol containing 2% sym-
                  collidine, chlorine, α-nitroso-β-naphthol reagent for
                  tyrosine and isatin for proline.

REFERENCE

1. **Montecucchi, P. C., De Castiglione, R. D., Piani, S., Gozzini,
    L., and Erspamer, V.,** *Int. J. Pept. Prot. Res.,* 17, 275, 1981.

Table TLC 37
OPIOID PEPTIDES

	L1	L1	L1	L1	L1	L1	L1	L1	L1	L1
	S1	S2	S3	S4	S5	S6	S7	S8	S9	S10
	D1	D1	D1	D1	D1	D1	D1	D1	D1	D1
	T1	T1	T1	T1	T1	T1	T1	T1	T1	T1
Peptide						$R_F \times 100$				
β-Lipotropin	13	70	0	15 ± 3	0	0	6 ± 2	0	0	0
β-Endorphin	31	61	10	26 ± 3	40 ± 2	37 ± 6	34 ± 2	0	0	0
Leu-Enkephalin	57 ± 3		52	39 ± 3	52 ± 1	44 ± 1	42 ± 3	80	49	21
Met-Enkephalin	66 ± 2		64	48 ± 3	64 ± 2	57 ± 2	57 ± 2	83	60	34

Layer: L1   = Reversed-phase KC$_{18}$ TLC plates (Whatman Inc.).

Solvent: S1   = Propanol-1-pH 4.1 phosphate buffer, 30:70 v/v.
S2   = Propanol-1-pH 1.5 phosphate buffer, 30:70 v/v.
S3   = Propanol-2-pH 4.1 phosphate buffer, 30:70 v/v.
S4   = Acetonitrile-pH 2.4 formic acid-ammonium hydroxide buffer, 1:2 v/v.
S5   = Propanol-1-pH 4.0 formic acid-pyridine buffer, 30:70 w/v.
S6   = Propanol-1-pH 3.1 formic acid-pyridine buffer, 30:70 w/v.
S7   = Propanol-1-pH 3.1 formic acid-phosphate buffer, 30:70 w/v.
S8   = Methanol + 0.1% TFA-water + 0.1% TFA, 75:10 v/v.
S9   = Methanol + 0.1% TFA-water + 0.1% TFA, 60:40 v/v.
S10   = Methanol + 0.1% TFA-water + 0.1% TFA, 30:70 v/v. The phosphate buffer was prepared using 0.1 M NaH$_2$PO$_4$ adjusted to the desired pH with H$_3$PO$_4$. The formic acid-ammonium hydroxide buffer was prepared using 50 mℓ 88% formic acid and 1.5 mℓ 58% ammonium hydroxide in 1 ℓ deionized water. The formic acid-pyridine buffer was prepared with 0.5 M formic acid titrated to the desired pH with pyridine. The formic acid-phosphate buffer was prepared as 0.1 M NaH$_2$PO$_4$ adjusted to the desired pH with 88% formic acid.

Detection: D1   = Plates were thoroughly dried under cool air following removal from the tank. Peptides were visualized by spraying with 0.01% fluorescamine in acetone followed immediately by spraying with 0.5% pyridine in acetone. The fluorescent spots were viewed under 360 nm UV light through a long-wave UV filter.

Technique: T1   = Plates were developed in tanks lined with filter paper. After equilibration for more than 30 min the plates were developed in solvent to about 2.5 cm from the top of the plate. Run times averaged 3 to 4 hr.

REFERENCE

1. **Ziring, B., Shepperd, S., and Kreek, M. J.,** *Int. J. Pept. Prot. Res.,* 22, 32, 1983.

Reproduced by permission of Munksgaard International Publishers, Ltd.

Table TLC 38
BRADYKININ DERIVATIVES

Layer	L1	L1	L1
Solvent	S1	S2	S3
Detection	D1	D1	D1

Peptide	$R_F \times 100$		
Gly-Leu-Met-Lys-BK[a]	40	14	84
Gly-Leu-Nle-Lys-BK	54	30	85
Gly-Leu-Met-Nle-BK	58	71	80
Gly-Leu-Met(O)-Lys-BK	35	6	84

[a] BK, bradykinin.

Layer: L1 = Silica gel plates (Eastman Chromagram).
Solvent: S1 = 1-Butanol-acetic acid-pyridine-water, 30:6:20:24 v/v.
          S2 = 1-Butanol-acetid acid-ethyl acetate-water, 1:1:1:1 v/v.
          S3 = Chloroform-methanol-ammonia, 2:2:1 v/v.
Detection: D1 = Ninhydrin and chloroplatinic acid.

REFERENCE

1. **Araujo-Viel, M., Juliano, L., and Prado, E. S.,** *Hoppe Seyler's Z. Physiol. Chem.,* 362, 337, 1981.

Reproduced by permission of Walter de Gruyter & Co.

Table TLC 39
CYCLIC DERIVATIVES OF KALLIDIN AND INTERMEDIATES

Compound	L1 S1 D1	L1 S2 D1	L1 S3 D1	L1 S4 D1	L1 S5 D1
	\multicolumn{5}{c}{$R_F \times 100$}				
Z-Arg(Z₂)-Lys-Pro-Pro-Gly-Phe-Gly-Pro-Phe-Arg(NO₂)					
Boc-Ser(Bzl)-Pro-Phe-ONb	90	67			
H-Ser(Bzl)-Pro-Phe-ONb·TFA					
Boc-Phe-Ser(Bzl)-Pro-Phe-ONb	6	46	76	58	
Boc-Phe-Ser(Bzl)-Pro-Phe-N₂H₃	42	92	93	86	80
Boc-Arg(NO₂)-Pro-OH	30	81	90	73	
H-Arg(NO₂)-Pro-OH·TFA	14	50	78	84	
Boc-Arg(NO₂) Z-Lys-Arg(NO₂)-Pro-OH	8	20	22		
H-Arg(NO₂) Z-Lys-Arg(NO₂)-Pro-OH·TFA	2	50	78	68	40
Boc-Phe-Ser(Bzl)-Pro-Phe-Arg(NO₂) Z-Lys-Arg(NO₂)Pro-OH		16	52	54	
Boc-Phe-Ser(Bzl)-Pro-Phe-Arg(NO₂) Z-Lys-Arg(NO₂)-Pro-OH·TFA	6	62	87	80	
Boc-Phe-Ser(Bzl)-Pro-Phe-Arg(NO₂) Z-Lys-Arg(NO₂)-Pro-Pro-Gly-O'Bu	3	82	91	80	
H-Phe-Ser(Bzl)-Pro-Phe-Arg(NO₂) Z-Lys-Arg(NO₂)-Pro-Pro-Gly-OH·HCl		21	76	58	
Z-Lys-Arg(NO₂)-Pro-Pro-Gly-Phe-Ser(Bzl)-Pro-Phe-Arg(NO₂)	3	69	84	62	

Abbreviations: Z, benzyloxycarbonyl; Nb, *p*-nitrophenyl.

Note: Cyclo-6-kallidin, Lys-Arg-Pro-Pro-Gly-Phe-Ser-Pro-Phe-Arg has an R_F of 0.60 in MeOH-H₂O-NH₄OAc, 9.5 mℓ:0.5 mℓ:0.15 g.

Layer: L1 =   Silufol UV 254 plates or silica gel 60 F₂₅₄ plates.
Solvent: S1 =   Chloroform-EtOH-AcOEt-AcOH-H₂O, 85:5:8:2:0.25 v/v.
           S2 =   Chloroform-EtOH-nBuOH-AcOEt-H₂O, 10:6:4:3:1 v/v.
           S3 =   Chloroform-MeOH-H₂O, 40:30:5 v/v.
           S4 =   BuOH-AcOH-H₂O, 4:1:1 v/v.
           S5 =   AcOEt-pyridine-AcOH-H₂O, 5:5:1:3 v/v. (EtOH, ethanol; AcOEt, ethyl acetate; AcOH, acetic acid; nBuOH, *n*-butanol).
Detection: D1 =   UV irradiation, ninhydrin, incubation in iodine vapor, chlorine-benzidine reagent, and Sakaguchi reagent.

REFERENCE

1. Mutulis, F. D., Chipens, G. I., Lando, O. E., and Mutule, I. E., *Int. J. Pept. Prot. Res.*, 23, 235, 1984.

Table TLC 40
DERIVATIVES OF MELANOTROPIN INHIBITING FACTOR (MIF) AND ITS METABOLITES

$R_F \times 100$

Compound	Dns					Ethansyl					Propansyl					Bns					Monoisopropansyl				
Layer	L1	L1	L1	L1	L1	L1	L1	L1	L1	L1	L1	L1	L1	L1	L1	L1	L1	L1	L1	L1	L1	L1	L1	L1	L1
Solvent	S1	S2	S3	S4	S5	S1	S2	S3	S4	S5	S1	S2	S3	S4	S5	S1	S2	S3	S4	S5	S1	S2	S3	S4	S5
Detection	D1	D1	D1	D1	D1	D1	D1	D1	D1	D1	D1	D1	D1	D1	D1	D1	D1	D1	D1	D1	D1	D1	D1	D1	D1
Technique	T1	T1	T1	T1	T1	T1	T1	T1	T1	T1	T1	T1	T1	T1	T1	T1	T1	T1	T1	T1	T1	T1	T1	T1	T1
MIF	49	71	89	25	40	21	80	87	14	23	12	77	91	5	10	8	83	95	4	7	10	58	87	5	29
Pro-Leu	17	69	95	11	57	12	70	85	7	15	7	75	87	4	8	6	81	93	5	7	7	49	79	5	23
Leu-Gly	20	8	89	11	52	16	23	83	5	23	5	25	86	5	10	4	38	83	2	11	9	14	79	3	18
Pro	38	60	65	19	74	19	71	84	7	23	9	80	86	5	11	7	76	93	7	9	7	43	79	7	28
Leu	17	41	82	11	48	10	50	86	6	14	7	61	84	2	7	5	69	93	4	5	5	36	86	3	18
Gly	51	18	53	15	57	29	28	77	6	17	10	36	74	7	7	5	43	84	4	7	10	11	56	6	20
Gly-NH$_2$	56	47	89	47	51	28	58	91	17	26	18	68	89	9	11	9	76	97	4	9	13	32	86	6	24
OH	40	0	0	0	55	40	0	0	12	6	23	0	2	0	0	7	0	0	0	0	28	3	0	0	9

Note: Dns, ethansyl, propansyl, Bns, and monoisopropansyl refer to the *N,N*-dimethyl, diethyl, dipropyl, dibutyl, and *N*-monoisopropyl-aminonaphthylenesulfonyl derivatives, respectively.

Layer: L1 = TLC polyamide precoated microplates, 5 × 5 cm, 25-μm thick.

Solvent: S1 = Formic acid-water, 3:97 v/v.
S2 = Benzene-acetic acid, 9:1 v/v.
S3 = Ethyl acetate-methanol-acetic acid, 20:1:1 v/v.
S4 = Isopropanol-1% acetic acid, 1:9 v/v.
S5 = Acetonitrile-0.01 *M* sodium sulfate buffer, pH 7, 41.5:58.5 v/v.

Detection: D1 = UV at 365 nm.

Technique: T1 = A 1-μℓ volume of the alkylaminonaphthylenesulfonyl derivative solution is spotted on the TLC plate. The spot, not larger than 3 mm in diameter is dried in a stream of warm air and the plate developed with the solvent system at room temperature (20°C). The plate is then air-dried and observed under UV light.

REFERENCE

1. **Hui, K.-S., Salschutz, M., Davis, B. A., and Lajtha, A.,** *J. Chromatogr.*, 192, 341, 1980.

Reproduced by permission of Elsevier Science Publishers B.V.

Table TLC 41
ANALOGUES OF α-MELANOTROPIN

Layer	L1	L1	L1	L1
Solvent	S1	S2	S3	S4
Detection	D1	D1	D1	D1
Peptide		$R_F \times 100$		
α-MSH	18	67	63	90
(Nle^4)-α-MSH	24	68	68	80
(Nle^4,D-Phe^7)-α-MSH	24	79	74	72
Ac-α-MSH$_{4-10}$-NH$_2$	30	61	64	97
Ac-(Nle$_4$)-α-MSH$_{4-10}$-NH$_2$	30	72	70	86
Ac-(Nle^4,D-Phe^7)-α-MSH$_{4-10}$-NH$_2$	30	68	69	87
Ac-(Nle^4)-α-MSH$_{4-11}$-NH$_2$	14	50	56	45
Ac-(Nle^4,D-Phe^7)-α-MSH$_{4-11}$-NH$_2$	14	50	58	53
Ac-α-MSH$_{5-11}$-NH$_2$	3	34	45	58
Ac-(D-Phe^7)-α-MSH$_{5-11}$-NH$_2$	3	43	50	43

Abbreviations: α-MSH, α-melanotropin; Nle, norleucine.

Layer: L1 = Silica gel G plates.
Solvent: S1 = 1-Butanol-AcOH-H$_2$O, 4:1:5 v/v, upper phase only.
S2 = 1-Butanol-AcOH-pyridine-H$_2$O, 15:3:10:12 v/v.
S3 = 1-Butanol-AcOH-pyridine-H$_2$O, 6:6:1.2:4.8 v/v.
S4 = 2-Propanol-25% aqueous ammonia-H$_2$O, 3:1:1 v/v.

Detection: D1 = UV, ninhydrin, or iodine vapor.

REFERENCE

1. **Wilkes, W. C., Sawyer, T. K., Hruby, V. J., and Hadley, M. E.,** *Int. J. Pept. Prot. Res.*, 22, 313, 1983.

Reproduced by permission of Munksgaard International Publishers, Ltd.

Table TLC 42
GASTRIN-LIKE PEPTIDES AND DERIVATIVES

Layer	L1	L1	L1	L1
Solvent	S1	S2	S3	S4
Detection	D1	D1	D1	D1
Peptide		$R_F \times 100$		
H-Trp-Leu-Asp-Phe-NH$_2$	38	48	34	28
Deoxyfructosyl-Trp-Leu-Asp-Phe-NH$_2$	40	52	27	25
Glucosyl-deoxyfructosyl-Trp-Leu-Asp-Phe-NH$_2$	45	41	17	12
H-Thr-Gly-Trp-Leu-Asp-Phe-NH$_2$	28	39	29	15
Deoxyfructosyl-Thr-Gly-Trp-Leu-Asp-Phe-NH$_2$	34	47	24	10
Glucosyl-deoxyfructosyl-Thr-Gly-Trp-Leu-Asp-Phe-NH$_2$	40	37	13	6
H-Ala-Tyr-Gly-Trp-Leu-Asp-Phe-NH$_2$	30	44	30	22
Deoxyfructosyl-Ala-Tyr-Gly-Trp-Leu-Asp-Phe-NH$_2$	31	48	25	7
Glucosyl-deoxyfructosyl-Ala-Tyr-Gly-Trp-Leu-Asp-Phe-NH$_2$	38	40	15	4
H-Glu-Ala-Tyr-Gly-Trp-Nle-Asp-Phe-NH$_2$	8	32	22	4
Deoxyfructosyl-Glu-Ala-Tyr-Gly-Trp-Nle-Asp-Phe-NH$_2$	15	39	20	
Glucosyl-deoxyfructosyl-Glu-Ala-Tyr-Gly-Trp-Nle-Asp-Phe-NH$_2$	27	36	10	

Layer: L1 = Silica gel-60 plates.
Solvent: S1 = Dioxane-water, 9:1 v/v.
S2 = Propanol-34% NH$_4$OH, 7:3 v/v.
S3 = Butanol saturated with 25% NH$_4$OH v/v.
S4 = Chloroform-methanol-acetic acid 75:35:5 v/v.
Detection: D1 = p-Dimethylaminobenzaldehyde reagent.

REFERENCE

1. **Previero, A., Mourier, G., Bali, J.-P., Lignon, M. F., and Moroder, L.,** *Hoppe-Seyler's Z. Physiol. Chem.,* 363, 813, 1982.

Reproduced by permission of Walter de Gruyter & Co.

Table TLC 43
PEPTIDES RELATED TO THE NH$_2$-TERMINAL SEQUENCE OF HUMAN BIG GASTRIN

Peptide	Solvent	$R_F \times 100$
Z-Pro-Gln-Gly-OPh	S3	75
	S1	62
Z-Gly-Pro-Gln-Gly-OPh	S1	55
	S2	80
Z-Leu-Gly-Pro-Gln-Gly-OPh	S2	77
	S6	47
Glu-Leu-Gly-Pro-Gln-Gly-OPh	S2	65
	S6	36
Glu-Leu-Gly-Pro-Gln-Gly-OH	S10	10
	S6	59
Z-Bu'Ser-Leu-Val-Ala-OPh	S6	85
	S7	87
Z-Pro-Bu'Ser-Leu-Val-Ala-OPh	S6	54
	S7	73
Z-His-Pro-Bu'Ser-Leu-Val-Ala-OPh	S2	65
	S7	75
Z-Bu'Ser-Pro	S9	48
	S11	62
Z-Bu'Ser-Pro-His-Leu-Val-Ala-OPh	S12	37
	S8	32
Z-Pro-Bu'Ser-His-Leu-Val-Ala-OPh	S13	40
	S8	45
Z-Bu'Ser-AdocLys-AdocLys-Gln-Gly-OPh	S4	64
	S2	88
Z-Pro-Bu'Ser-AdocLys-AdocLys-Gln-Gly-OPh	S2	47
	S4	94
Glu-Leu-Gly-Pro-Gln-Gly-Pro-Pro-His-Leu-Val-Ala-OH	S2	21
Glu-Leu-Gly-Pro-Gln-Gly-Ser-Pro-His-Leu-Val-Ala-OH	S14	32
Glu-Leu-Gly-Pro-Gln-Gly-Pro-Ser-His-Leu-Val-Ala-OH	S14	26
	S2	27
Human big gastrin	S5	20
	S7	58

Abbreviations: Z, benzyloxycarbonyl; Adoc, adamantyloxycarbonyl.

Layer: L1 = Silica gel (Merck).
Solvent: S1 = Acetonitrile-water, 9:1 v/v.
S2 = Butanol-pyridine-acetic acid-water, 60:20:6:24 v/v.
S3 = Chloroform-2-propanol, 6:1 v/v.
S4 = Butanol-acetic acid-water, 3:1:1 v/v.
S5 = Ethyl acetate-pyridine-acetic acid-water, 120:20:60:11 v/v.
S6 = Chloroform-methanol-33% ammonia, 19:17:3 v/v.
S7 = Chloroform-methanol-acetic acid-water, 60:18:2:3 v/v.
S8 = Chloroform-methanol, 2:1 v/v.
S9 = Chloroform-methanol, 4:1 v/v.
S10 = 2-Butanol-3% ammonia, 3:1 v/v.
S11 = Chloroform-2-propanol, 7:3 v/v.
S12 = Chloroform-2-propanol, 7:1 v/v.
S13 = Chloroform-2-propanol, 4:1 v/v.
S14 = Butanol-pyridine-formic acid-water, 44:24:2:20 v/v.

REFERENCE

1. Choudhury, A. M., Kenner, G. W., Moore, S., Ramachandran, L., Thorpe, W. D., Ramage, R., Dockray, G. J., Gregory, R. A., Hood, L., and Hunka-piller, M., *Hoppe-Seyler's Z. Physiol. Chem.*, 361, 1719, 1980.

Table TLC 44
SERUM THYMIC FACTOR (FTS) ANALOGUES

Layer	L1	L1
Solvent	S1	S2
Peptide	$R_F \times 100$	
<Glu-Ala-Lys-Ser-Gln-Gly-Gly-Ser-Asn-OH	24	8
<Glu-Ala-Lys-Ser-Gln-Gly-Gly-Ser-OH	27	13
<Glu-Ala-Lys-Ser-Gln-Gly-Gly-OH	29	16
<Glu-Ala-Lys-Ser-Gln-Gly-OH	28	20
<Glu-Ala-Lys-Ser-Gln-OH	34	21
<Glu-Ala-Lys-Ser-OH	34	30
H-Ala-Lys-Ser-Gln-Gly-Gly-Ser-Asn-OH	2	6
H-Lys-Ser-Gln-Gly-Gly-Ser-Asn-OH	2	7
H-Ser-Gln-Gly-Gly-Ser-Asn-OH	24	15
H-Gln-Ala-Lys-Ser-Gln-Gly-Gly-Ser-Asn-OH	3	4
H-Gln-Ala-Lys-Ser-Gln-Gly-Gly-Ser-OH	2	7
H-Gln-Ala-Lys-Ser-Gln-Gly-Gly-OH	3	8
H-Gln-Ala-Lys-Ser-Gln-Gly-OH	2	10
H-Gln-Ala-Lys-Ser-Gln-OH	3	10
H-Gln-Ala-Lys-Ser-OH	3	13

Abbreviation: <Glu, pyroglutamic acid.

Layer: L1 = Merck plastic sheets pre-coated with silica gel $60F_{254}$.
Solvent: S1 = 1-Propanol-water, 2:1 v/v.
          S2 = 2-Propanol-formic acid-water, 20:1:5 v/v.

REFERENCE

1. **Gyotoku, J., Imaizumi, A., Terada, S., and Kimoto, E.,** *Int. J. Pept. Prot. Res.*, 21, 135, 1983.

Reproduced by permission of Munksgaard International Publishers, Ltd.

Table TLC 45
ANALOGUES OF THE SERUM THYMIC NONAPEPTIDE FTS

Layer	L1	L1	L1	L1
Solvent	S1	S2	S3	S4
Compound	$R_F \times 100$			
FTS	33	9	26	40
(Gln¹)-FTS	20	1	22	37
(Z-Gln¹)-FTS	52	33	58	73
(des-<Glu¹)-FTS	17	0	22	35
(des-<Glu¹,Ala²)-FTS	17	0	17	30
(des-<Glu¹,Ala²,Lys³)-FTS	31	20	38	63
(Gln⁹)-FTS	38	11	31	50
(D-Asn⁹)-FTS	34	11	30	41
(Asn-NH₂⁹)-FTS	40	1	14	19
(Asp⁹)-FTS	31	27	26	49
(β-Ala-NH₂⁹)-FTS	40	3	21	28
(des-Asn⁹)-FTS	39	12	31	50
(Ala⁸)-FTS	36	11	34	50

Table TLC 45 (continued)
ANALOGUES OF THE SERUM THYMIC NONAPEPTIDE FTS

Compound	$R_F \times 100$			
(D-Ala²)-FTS	32	8	29	45
(des-<Glu¹)(D-Ala²)-FTS	20	0	24	37
(D-<Glu¹)-FTS	31	8	24	39
(D-Gln¹)-FTS	22	1	21	34
(Pro¹)-FTS	18	1	33	60
(Glu¹)-FTS	21	4	25	55
(Nᵉ-Ac-Lys³)-FTS	49	36	58	78
(D-Lys³)-FTS	30	6	23	39
(Orn³)-FTS	31	8	24	40
(Hep³)-FTS	69	60	60	60
(Arg³)-FTS	36	10	15	20
(Har³)-FTS	36	12	18	25
(Ala⁴)-FTS	35	11	31	52
(D-Ser⁴)-FTS	35	10	26	42
(D-Gln⁵)-FTS	31	8	24	39
(Glu⁵)-FTS	28	35	27	52
(Asn⁵)-FTS	30	7	23	37
(Nva⁵)-FTS	46	23	39	62
(D-Ala⁶)-FTS	36	12	32	47
(Nᵉ-Ac-Lys³,D-Ala⁶)-FTS	56	38	63	82
(des-<Glu¹,Ala²)(N-Ac-Lys³)-FTS	38	23	55	68
(des-<Glu¹,Ala²)(D-Lys³)-FTS	18	0	16	30
(des-<Glu¹,Ala²)(Orn³)-FTS	15	0	14	29
(des-<Glu¹,Ala²)(Hep³)-FTS	54	45	60	81
(des-<Glu¹,Ala²)(D-Ala⁶)-FTS	20	0	22	35

Abbreviations: <Glu, L-pyroglutamic acid; FTS, facteur thymique sérique, a circulating thymic factor; Har, homoarginine; Hep, heptyline (2-aminoheptanoic acid).

Layer: L1 = Merck HPTLC plates of silica gel 60F₂₅₄.
Solvent: S1 = Ethanol-*M* ammonium acetate, 7:3 v/v.
         S2 = 1-propanol-water, 2:1 v/v.
         S3 = Chloroform-methanol-concentrated ammonium hydroxide, 2:2:1 v/v.
         S4 = Chloroform-methanol-concentrated ammonium hydroxide, 1:2:1 v/v.

REFERENCES

1. **Blanot, D., Martinez, J., Auger, G., and Bricas, E.,** *Int. J. Pept. Prot. Res.,* 14, 41, 1979.
2. **Martinez, J., Blanot, D., Auger, G., Sasaki, A., and Bricas, E.,** *Int. J. Pept. Prot. Res.,* 16, 267, 1980.

Reproduced by permission of Munksgaard International Publishers, Ltd.

Table TLC 46
FLUORESCAMINE DERIVATIVES OF PEPTIDES OF BIOLOGICAL INTEREST

Layer	L1	L1	L2	L1	L1
Solvent	S1	S2	S2	S3	S4
Detection	D1	D1	D1	D1	D1
Technique	T1	T1	T1	T1	T1

Peptide	$R_F \times 100$				
L-Carnosine	15	60	68	35	31
L-Homocarnosine sulfate	13	58	68	41	31
L-Anserine nitrate	5	38	69	20	29
Glutathione, oxidized	0	59	71	1	4
Bradykinin	3	16	19	3	25
Lysyl-bradykinin	12	61	76	27	99
Methionyl-lysyl-bradykinin	9	63	79	27	99
Gramicidin J	95	97	98	99	94
Bacitracin	28	89	88	65	99
(Asp^{1}Ile^5) Angiotensin I	4	49	72	29	98
(Asp^{1}Ile^5) Angiotensin II	9	58	76	47	98
Oxytocin	46	89	90	89	99
(Arg^8) Vasopressin	15	56	61	22	99
Angiotensin II hexapeptide	35	90	87	89	99
Angiotensin II heptapeptide	19	60	78	55	98
Tridecapeptide renin substrate	19	68	86	57	98
Tetradecapeptide renin substrate	15	67	87	60	99
ACTH (porcine)	0	0	1	0	99
Insulin (bovine pancreas)	0	0	1	1	1
Lysine-vasopressin	31	89	89	88	3
Somatostatin tetradecapeptide	34	89	88	89	99
Caerulein	32	88	88	75	99
Secretin 21-27	54	88	87	89	98
Eledoisin-related peptide[a]	79	94	94	94	98

[a] Lys-Phe-Ile-Gly-Leu-Met.NH_2.2 HCl.

Layer:	L1	= Precoated silica gel 60 TLC plates (Merck).
	L2	= Precoated silica gel plates Q5 (Quantum Industries).
Solvent:	S1	= Chloroform-isopropanol-water, 2:8:1 v/v.
	S2	= Acetone-ethyl acetate-methanol-water, 3:2:2:1 v/v.
	S3	= *n*-Butanol-ethanol-methanol-water, 6:4:2:1 v/v.
	S4	= Dioxane-triethanolamine-methanol-water, 6:1:1:1 v/v.

Detection: D1 = Observation of fluorescence under a long wave (366 nm) UV lamp after air drying.

Technique: T1 = The peptide was derivatized with fluorescamine on the TLC plate by the predevelopment method (see Section IV).

REFERENCE

1. **Nakamura, H. and Pisano, J. J.,** *J. Chromatogr.*, 121, 33, 1976.

Reproduced by permission of Elsevier Science Publishers B.V.

Table TLC 47
DINITROPHENYL DERIVATIVES
OF NEUTRAL DIPEPTIDES

Layer	L1	L1
Solvent	S1	S2
Detection	D1	D1
Technique	T1	T1

DNP derivative of	$R_F \times 100$	
Gly	14	43
Gly-Gly	5	20
Gly-Gly-Gly	2	3
Gly-Gly-Gly-Gly	0	0
Val	20	43
Val-Val	21	39
Ala	7	59
Ala-Ala	4	39
Ala-Gly	2	27
Gly-Ala	2	32
Gly-Val	31	31
Val-Gly	20	27

Layer: L1 = Pre-coated silica gel thin-layer sheets (J. T. Baker).

Solvent: S1 = Benzene-pyridine-acetic acid, 80:20:2 v/v.

           S2 = Chloroform-methanol-acetic acid, 95:5:1 v/v.

Detection: D1 = The intense yellow color of the derivatives.

Technique: T1 = The DNP derivative was applied 3 cm from the bottom of the plate and allowed to run 17 cm.

REFERENCE

1. **Martel, C. and Phelps, D. J.,** *J. Chromatogr.,* 115, 633, 1975.

Table TLC 48
A CYCLOPENTAPEPTIDE AND ITS
INTERMEDIATES

Peptide	$R_F \times 100$
Boc-Leu-Ile-OBzl	94(S1);64(S3)
H-Leu-Ile-OBzl.Tsa salt	54(S1);82(S2)
Boc-D-Val-Leu-Ile-OBzl	85(S1);43(S3)
Boc-Ala-D-Val-Leu-Ile-OBzl	75(S1);11(S3)
Boc-Gly-Ala-D-Val-Leu-Ile-OBzl	61(S1);31(S4)
Boc-Gly-Ala-D-Val-Leu-Ile-N_2H_3	44(S1);93(S2)
Z-Gly-Ala-D-Val-Leu-Ile-OBzl	61(S1);29(S4)
Fmoc-Gly-Ala-D-Val-Leu-Ile-Ozl	73(S1);25(S4)
H-Gly-Ala-D-Val-Leu-Ile-OH-H_2O	52(S2);24(S5)
Cyclo(Gly-Ala-D-Val-Leu-Ile-)	$^{54}_{50}$(S1);$^{89}_{85}$(S2)

Table TLC 48 (continued)
A CYCLOPENTAPEPTIDE AND ITS INTERMEDIATES

Abbreviations: Fmoc, 9-fluorenylmethyloxycarbonyl; Tsa, *p*-toluene-sulfonic acid.

Layer: Silica gel.
Solvent: S1 = Chloroform-methanol, 9:1 v/v.
           S2 = 1-Butanol-acetic acid-water, 4:1:1 v/v.
           S3 = Ethyl acetate-hexane, 3:7 v/v.
           S4 = Ethyl acetate-hexane, 8:2 v/v.
           S5 = Ethyl acetate-pyridine-acetic acid-water, 60:20:6:1 v/v.
Detection: D1 = UV absorption, ninhydrin, fluorescamine, and charring.

REFERENCE

1. **Ji, A.-X. and Bodanszky, M.,** *Int. J. Pept. Prot. Res.,* 22, 590, 1983.

Table TLC 49
INTERMEDIATES FOR THE SYNTHESIS OF SUBSTANCE P

Layer	L1	L1	L1	L1
Solvent	S1	S2	S3	S4
Peptide		$R_F \times 100$		
Boc-Leu-Met(O)-NH$_2$	45		53	
Boc-Gly-Leu-Met(O)-NH$_2$	39	59	23	83
Boc-Phe-Gly-Leu-Met(O)-NH$_2$	48	64	28	87
Boc-Phe-Phe-Gly-Leu-Met(O)-NH$_2$	59	65	27	88
Boc-Gln-Phe-Phe-Gly-Leu-Met(O)-NH$_2$	34	60	10	88
Boc-Gln-Gln-Phe-Phe-Gly-Leu-Met(O)-NH$_2$	11	62	2	89
Boc-Lys(Z)-Pro-Gln-Gln-Phe-Phe-Gly-Leu-Met(O)-NH$_2$	50		15	81
Z-Arg(NO$_2$)-Pro-Lys(Z)-Pro-Gln-Gln-Phe-Phe-Gly-Leu-Met(O)-NH$_2$	15			

Layer: L1 = Silica plates (Merck F$_{254}$).
Solvent: S1 = Chloroform-methanol, 4:1 v/v.
          S2 = *n*-Butanol-ethanol-water, 2:2:1 v/v.
          S3 = Chloroform-methanol-acetic acid, 85:10:5 v/v.
          S4 = Chloroform-methanol-25% ammonia, 2:2:1 v/v.

REFERENCE

1. **Izeboud, E. and Beyerman, H. C.,** *Rec. Trav. Chim. Pays-Bas,* 97, 1, 1978.

Reproduced by permission of the Koninklijke Nederlandse Chemische Vereniging.

Table TLC 50
INTERMEDIATES FOR THE SYNTHESIS OF
S-PEPTIDE ANALOGUES

Layer	L1	L1
Solvent	S1	S2
Detection	D1	D1

Compound	$R_F \times 100$	
H-Cha-OH.HCl	50	60
H-pF-Phe-OH.HCl	80	70
H-Cha-OMe.HCl	60	65
H-pF-Phe-OMe.HCl	80	80
Z-(Cha^8)7-8 OMe	90	95
Z-(pF-Phe^8) 7-8 OMe	90	80
Z-(Cha^8) 7-8 NHNH$_2$	100	100
Z-(pF-Phe^8) 7-8 NHNH$_2$	65	75
Z-(Cha^8,Orn^{10}) 7-12 OMe	80	90
Z-(pF-Phe^8,Orn^{10}) 7-12 OMe	75	90
Z-(Cha^8,Orn^{10}) 1-12 OMe	70	95
Z-(pF-Phe^8,Orn^{10}) 1-12 OMe	60	90
Z-(Cha^8,Orn^{10}) 1-12 NHNH$_2$	85	90
Z-(pF-Phe^8,Orn^{10}) 1-12 NHNH$_2$	80	95

Abbreviations: Z, benzyloxycarbonyl, Cha, cyclohexylalanine; pF-Phe, *p*-fluorophenylalanine.

Layer: L1 = Silica gel (Merck).
Solvent: S1 = *n*-butanol-acetic acid-water, 60:20:20 v/v.
S2 = Ethyl acetate-pyridine-acetic acid-water, 60:20:6:14 v/v.
Detection: D1 = The ninhydrin reagent for free amino groups, the Sakaguchi reagent for guanido groups, the hydrazide reagent for C-terminal hydrazide groups, the Pauly reagent for histidine-containing peptides and a chlorine reagent for all peptide derivatives.

REFERENCE

1. **Borin, G., Filippi, B., Moroder, L., Santoni, C., and Marchiori, F.,** *Int. J. Pept. Prot. Res.,* 10, 27, 1977.

Reproduced by permission of Munksgaard International Publishers, Ltd.

Table TLC 51
INTERMEDIATES IN THE SYNTHESIS OF OF FRAGMENT 335-344 OF HUMAN IgG

Layer	L1	L1
Solvent	S1	S2
Detection	D1	D1
Peptide	**$R_F \times 100$**	
Boc-Gln-Pro-ArgNO₂-OBzl	33	77
Boc-Gly-Gln-Pro-ArgNO₂-OBzl	30	70
Boc-ZLys-Gly-Gln-Pro-ArgNO₂-OBzl	30	70
Boc-Ala-ZLys-Gly-Gln-Pro-ArgNO₂-OBzl	28	70
Boc-ZLys-Ala-ZLys-Gly-Gln-Pro-ArgNO₂-OBzl	30	75
Boc-Ser-ZLys-Ala-ZLys-Gly-Gln-Pro-ArgNO₂-OBzl	28	70
Boc-Ile-Ser-ZLys-Ala-ZLys-Gly-Gln-Pro-ArgNO₂-OBzl	25	70
Z-Thr-Ile-Ser-Lys-Ala-ZLys-ZLys-Gly-Gln-Pro-ArgNO₂-OBzl	23	70

Layer: L1 = Merck silica gel plates GF₂₅₄.
Solvent: S1 = Dichloromethane-methanol, 9:1 v/v.
            S2 = Ethylacetate-pyridine-acetic acid-water, 80:20:5:10 v/v.
Detection: D1 = UV absorption, charring, and ninhydrin.

REFERENCE

1.   **Martinez, J., Laur, J., and Winternitz, F.,** *Int. J. Pept. Prot. Res.,* 22, 119, 1983.

Table TLC 52
Nε-Boc-PROTECTED OLIGO-L-LYSINE PEPTIDES

Layer	L1	L1	L1	L1
Solvent	S1	S2	S3	S4
Detection	D1	D1	D1	D1
Peptides	**$R_F \times 100$**			
H-Lys(Boc)-OSuco	23			
H-Lys(Boc)-(Lys(Boc))₃-Lys(Boc)-OSuco		14	35	
H-Lys(Boc)-(Lys(Boc))₄-Lys(Boc)-OSuco		15	33	
H-Lys(Boc)-(Lys(Boc))₅-Lys(Boc)-OSuco		17	26	
H-Lys(Boc)-(Lys(Boc))₆-Lys(Boc)-OSuco		18	30	
Boc-Lys(Boc)-(Lys(Boc))₈-Lys(Boc)-OSuco		53	65	81
Boc-Lys(Boc)-(Lys(Boc))₈-Lys(Boc)-OH			9	20
Boc-Lys(Boc)-(Lys(Boc))₁₈-Lys(Boc)-OSuco			54	70

Abbreviations: Boc, *tert*-Butyloxycarbonyl; OSuco, *p*-oxymethylbenzylcholestan-3-β-yl succinate.

Layer: L1 = 5 × 10 cm silica gel thin-layer plates (60F₂₅₄,Merck).
Solvent: S1 = Methanol in chloroform, 5:95 v/v.
            S2 = Methanol in chloroform, 7.5:92.5 v/v.
            S3 = Methanol in chloroform, 10:90 v/v.
            S4 = Methanol in chloroform, 15:85 v/v.
Detection: D1 = Treatment with ninhydrin and heating at 120°C for 10 min.

REFERENCE

1.   **Toniolo, C., Bonora, G. M., Lüscher, I. F., and Schneider, C. H.,** *Int. J. Pept. Prot. Res.,* 23, 47, 1984.

Table TLC 53
PROTECTED DI-, TRI-, AND TETRAPEPTIDES

Layer	L1	L1	L1
Solvent	S1	S2	S3
Peptide		$R_F \times 100$	
Boc-Ala-Ala-OMe	85	83	87
Boc-Pro-Ala-Ala-OBzl	92	81	90
Boc-Ala-Pro-Ala-OBzl	87	80	91
Z-Pro-Ala-Ala-Ala-OBzl	89	90	93
Z-Ala-Pro-Ala-Ala-OBzl	78	76	77
Z-Ala-Ala-Pro-Ala-OBzl	77	89	93
Z-Ala-Ala-Ala-Pro-OBzl	54	60	97

Abbreviations: Z, benzyloxycarbonyl; Boc, *tert*-butoxycarbonyl; Bzl, benzyl.

Layer: L1 = Silica gel plates (Silufol).
Solvent: S1 = Butanol-acetic acid-water, 4:1:1 v/v.
S2 = Isopropanol-water, 4:1 v/v.
S3 = Propanol-water, 7:3 v/v.

REFERENCE

1. **Siemion, I. Z., Sobczyk, K., and Nawrocka, E.,** *Int. J. Pept. Prot. Res.,* 19, 439, 1982.

Reproduced by permission of Munksgaard International Publishers, Ltd.

Table TLC 54
DIASTEREOMERS OF PROTECTED DIPEPTIDES

Layer			L1	L1	L1	L1	L1	L1	L1
Solvent			S1	S2	S3	S4	S5	S6	S7
Detection			D1	D1	D1	D1	D1	D1	D1
Technique			T1	T1	T1	T1	T1	T1	T1
	Group on								
Peptide	$-NH_2$	$-CO_2H$				$R_F \times 100$			
Ala-Ala	HCO	OBzl	26	54	31			19	44
	Ac	OBzl	25			13	0	15	42
	ClAc	OBzl	55					47	77
								43	72
	Tfa	OBzl				54	22	77	
							18	73	
	Bz	OBzl				55		59	77
								54	72
	Boc	OBzl				60	20	75	
	Z	OBzl	85		59			51	
	Pht	OBzl				71	11	66	
Ala-Leu	HCO	OMe						26[a]	52[a]
								19	44
	HCO	OBzl	40	68	55			39	67
								34	

Table TLC 54 (continued)
DIASTEREOMERS OF PROTECTED DIPEPTIDES

Peptide	Group on −NH₂	−CO₂H				$R_F \times 100$			
	Ac	OMe	26			13	0	19	50
								15	42
	ClAc	OMe	59					50	78
								45	
	Tfa	OMe				55	25	72	
						51	19	69	
	Bz	OBzl				64		69	85
	Boc	OMe				62	18	76	
	Z	OMe	69				23	50	
	Z	OBzl	97		69			76	
	Pht	OBzl				77	30	75	
Ala-Phe	HCO	OMe	27	65	36			19	48
								15	42
	Ac	OMe	25			13	0	13	43
									37
	ClAc	OMe	57					42	73
	Tfa	OMe				54	21	71	
						51	16	69	
	Bz	OMe				55		52	73
									69
	Boc	OMe				63	16	73	
	Z	OMe	85		59			52	
	Pht	OMe				68	13	64	
Leu-Ala	HCO	OBzl	51	87				51	91
			43	81				44	
	Ac	OBzl	48			31	0	40	42
			39			23		36ᵃ	37
	Bz	OBzl				61	16	73	
			58			58	11		
	Boc	OBzl				75	54	86	
							49ᵃ		
	Z	OBzl				73	41	75	
							35ᵃ		
	Tos	OBzl				64	23	74	94
							17		
	Nps	OBzl				81	45	77	
	Pht	OBzl				75	27	74	
Leu-Leu	HCO	OMe	46					47	
			37					36	
	HCO	OBzl	60	93				68	91
	Ac	ONBzl	55			39		53ᵃ	57
								46	51
	Bz	OBzl				71	38	84	
	Bz	OMe				71	39	76	
						69	32	71	
	Boc	ONBzl				80	58ᵃ	92	
							50		
	Z	OMe	84ᵃ					79	57ᵃ
			83						50
	Z	OBzl				80	42	75	
							37	70	
	Tos	OMe				65	23	69	91
							17		
	Nps	OBzl				84	37	81	
	Pht	OBzl				87	58	87	

Table TLC 54 (continued)
DIASTEREOMERS OF PROTECTED DIPEPTIDES

Peptide	$-NH_2$	$-CO_2H$	$R_F \times 100$						
Leu-Phe	HCO	OMe	52					52	85
			44					34	70
	Ac	OMe	47[a]				31	41	45
			38				23	28	27
	Bz	OMe				61	16	73	
						58	10	66	
	Boc	OMe				72	49[a]	88	
							42		
	Z	OMe				73	46[a]	70[a]	
							37	62	
	Tos	OMe				63	15	68	91
	Nps	OMe				78	35	73	
	Pht	OMe				79	28	78	
Phe-Ala	HCO	OBzl	46	86	66			45	
			42[a]	83[a]	61[a]				
	Ac	OBzl	51	90				44	47
			46					39[a]	41[a]
	Tfa	OBzl				78	48	74	
							43	70	
	Bz	OBzl				67		65	80
						63			
	Boc	OBzl				75	36	88	
	Z	OBzl	82		85		27	62	
							22[a]	57[a]	
	Tos	OBzl				62	12	68	
	Pht	OBzl				77	19	71	
Phe-Leu	HCO	OMe						55[a]	74[a]
								34	69
	HCO	OtBu						64[a]	93[a]
								59	91
	HCO	OBzl	58[a]	93	83[a]			65[a]	91[a]
			54		79			57	87
	HCO	ONBzl			58[a]			58[a]	91[a]
					53			49	86
	Ac	OBzl	66	93			58	62	
			62[a]				54[a]	56[a]	
	Tfa	OMe				78	55	82	
	Bz	OMe				71	32	72	
						65	25	66	
	Bz	OBzl				81		79	77
	Boc	OMe				90	48[a]	91	
							41		
	Z	OMe	82			79	33[a]		
							27		
	Z	OBzl	99		80	82		83	
	Tos	OMe				61	17	74	
							12		
	Pht	OBzl				84	50	83	
Phe-Phe	HCO	OMe	48[a]	90[a]	68[a]			49[a]	
			40	84	58			32	
	Ac	OMe	54	88			48[a]	48[a]	
			47				39	29	
	Tfa	OMe				74	42	76	
							33		

Table TLC 54 (continued)
DIASTEREOMERS OF PROTECTED DIPEPTIDES

Peptide	Group on		$R_F \times 100$			
	$-NH_2$	$-CO_2H$				
	Bz	OMe	70	22	66	82
			66	14	59	77
	Boc	OMe	89	36[a]	85[a]	
			28	82		
	Z	OMe	95	70	25[a]	60[a]
			18	53		
	Tos	OMe	60	10	68	
	Pht	OMe	75	19	71	

[a]   The configuration was identified as L-L by comparison with a standard.

Abbreviations: HCO, formyl; Ac, acetyl; Bz, benzoyl; ClAc, chloroacetyl; Tfa, trifluoroacetyl; Z, benzyloxycarbonyl; Boc, *t*-butoxycarbonyl; Nps, *o*-nitrophenylthio; Tos, tosyl; Pht, phthalyl; OMe, methyl ester; OBzl, benzyl ester; ONBzl, *p*-nitrobenzyl ester; Ot·Bu, *t*-butyl ester.

Layer: L1 = Benzene-acetone, 2:1 v/v.
          S1 = Ethyl acetate.
          S3 = Diethyl ether-acetone, 4:1 v/v.
          S4 = Benzene-acetone, 3:1 v/v.
          S5 = Diisopropyl ether.
          S6 = Diisopropyl ether-isopropanol, 10:1 v/v.
          S7 = Diisopropyl ether-isopropanol, 4:1 v/v.
Detection: D1 = The plates are sprayed with 40% ammonium bisulfate solution and heated at 80°C.
Technique: T1 = 5 to 15 µg of sample in 1 to 3 mm³ of solvent is applied to the plates. After 10 min drying the chromatograms are developed in unsaturated chambers at room temperature.

REFERENCE

1.   **Arendt, A., Kolodziejezyk, A., and Sokolowska, T.,** *Chromatographia,* 9, 123, 1976.

Reproduced by permission of Friedr. Vieweg & Sohn.

Table TLC 55
N-PROTECTED PEPTIDES, AMINO ACIDS, AND AMINOCYCLITOL ANTIBIOTICS

Layer	L1	L1	L1	L1	
Solvent	S1	S2	S3	S4	
Compound	$R_F \times 100$				Color of spot
N-Acetyl-L-Leu	73	79	72	66	Deep orange
N-Acetyl-L-Gln	42	52	48	67	Violet
N-Acetyl-L-Glu	66	63	73	88	Deep orange
N-Acetyl-L-Hyp-Aib-OMe	60	77	57	59	Violet
Nonapeptide from antiamoebin I	62	91	73	90	Deep orange
Antiamoebin I	40	85	15	83	Violet
Antiamoebin I triacetate	50	91	81	85	Violet

Table TLC 55 (continued)
N-PROTECTED PEPTIDES, AMINO ACIDS, AND AMINOCYCLITOL ANTIBIOTICS

Compound	$R_F \times 100$				Color of spot
Emerimicin IV	46	86	19	79	Brown
Alamethicins I and II	37	57	25	73	Violet
	23			46	
Valine- and isoleucine-gramicidins A	74[a]	41	b	82	Violet to dark brown
Gramicidin J	48	4	c	56	Orange to purple
Nocardamin	49	80	21	90	Deep yellow
Valinomycin	89	86	100	92	Yellow to purple
N,N'-Diacetyldeoxystreptamine	14	13	a	a	Faint brown
N,N'-Diacetylstreptamine	5	4	c	c	Brown
N-Hexaacetylneomycin B	4[a]	9	3	a	Brown
Streptidine	c	c	c	c	Faint brown

[a] Elongated spot.
[b] Streak.
[c] Unmoved from origin.

Abbreviations: Aib, aminoisobutyric acid; Hyp, hydroxyproline.

Layer:	L1	= Precoated silica gel G TLC plates, 250-μm thick.
Solvent:	S1	= 1-Butanol-acetic acid-water, 4:1:5 v/v upper layer.
	S2	= Pyridine-ethyl methyl ketone, 3:7 v/v.
	S3	= Chloroform-acetic acid, 1:2 v/v.
	S4	= Chloroform-methanol-water, 65:24:4 v/v.

Detection: The dried TLC plate is thoroughly sprayed with solution a, concentrated HCl, and left at room temperature for 10 to 15 min. The plate is then covered completely with a clear glass plate and heated at about 120°C for 10 to 15 min. After cooling to room temperature the clear glass plate is removed and the TLC plate is heated again at about 120°C for 10 min. The plate is then sprayed with solution b, a 0.5% solution of ninhydrin in 1-butanol, and kept in an oven at about 120°C for 30 to 40 min. Sometimes it is necessary to spray more than once with solution b. Different compounds develop spots of different colors during heating. The principle involved in detecting the N-protected compounds is that the compound is first hydrolyzed on the TLC plate by solution a and is then made visible using solution b.

Technique: After spotting the compounds the plate is developed using an appropriate solvent and the developed plate is dried at room temperature for 10 to 15 min.

REFERENCE

1. **Pandey, R. C., Misra, R., and Rinehart, K. L., Jr.,** *J. Chromatogr.,* 170, 498, 1979.

Reproduced by permission of Elsevier Science Publishers B.V.

Table TLC 56
SYNTHETIC PEPTIDES

Layer	L1	L1	L1
Solvent	S1	S2	S3
Detection	D1	D1	D1

Peptide	$R_F \times 100$		
Z-(Orn(Boc))$_2$-Pro-Thr-Pro-Ala-OEt	80	90	80
Z-(Orn(Boc))$_2$-Ala-Thr-Pro-Ala-OEt		30	
Z-(Orn(Boc))$_2$-Pro-Thr-Val-Ala-OEt	90	95	70
Z-(Orn(Boc))$_2$-Pro-Thr-Pro-Ala-OEt	85	95	70
Z-(Orn(Boc))$_2$-Pro-Thr-Pro-Ala-Thr-Val-Ala-OEt	60	90	70
Z-(Orn(Boc))$_2$-Pro-Ser-(Bzl)-Pro-Ala-OEt	90	95	90
Z-(Orn(Boc))$_2$-Ala-Ser-Val-Ala-OEt	90	90	75

Abbreviations: Z, benzyloxycarbonyl; Boc, *tert*-butyloxycarbonyl; OEt, ethyl ester; Bzl, benzyl.

Layer:	L1	= Silica gel plates F$_{254}$ (Merck).
Solvent:	S1	= Butanol-acetic acid-water, 3:1:1 v/v.
	S2	= Butanol-pyridine-acetic acid-water, 60:40:12:48 v/v.
	S3	= Methanol-chloroform, 15:85 v/v.
Detection:	D1	= Modified chlorine reagent.

REFERENCE

1. **Chessa, G., Borin, G., Marchiori, F., Meggio, F., Brunati, A. M., and Pinna, L. A.,** *Eur. J. Biochem.*, 135, 609, 1983.

Reproduced with permission.

Table TLC 57
PEPTIDE INTERMEDIATES

Compound	Solvent	$R_F \times 100$
Nps-Lys(Z)-ONp	S1	67
Nps-Val-Lys(Z)-ONp	S2	85
Nps-Leu-Lys(Z)-ONp	S1	72
Nps-Leu-Glu(OBzl)-ONp	S3	50
Nps-Leu-OPop	S4	89
HCl,H-Leu-OPop	S4	82
Nps-Lys(Cl$_2$Z)-OH,DCHA	S5	63
Boc-Lys(Cl$_2$Z)-OH,DCHA	S5	58
Nps-Lys(Cl$_2$Z)-Leu-OPop	S5	83
HCl,H-Lys(Cl$_2$Z)-Leu-OPop	S4	63
Nps-Leu-Lys(Cl$_2$Z)-Leu-OPop	S2	80
HCl,H-Leu-Lys(Cl$_2$Z)-Leu-OPop	S2	70
Boc-(Lys(Cl$_2$Z)-Leu)$_2$-OPop	S5	72
Boc(Lys(Cl$_2$Z)-Leu)$_2$-OC$_6$H$_4$OH-o	S5	68
HCl,H-(Lys(Cl$_2$Z)-Leu)$_2$-OC$_6$H$_4$OH-o	S4	73
HCl,H-(Lys(Cl$_2$Z)-Leu)$_2$-OPop	S2	66
Boc-Gly-(Lys(Cl$_2$Z)-Leu)$_2$-OPop	S5	71
Boc-Gly-(Lys(Cl$_2$Z)-Leu)$_2$-OC$_6$H$_4$OH-o	S5	59
HCl,H-Gly-(Lys(Cl$_2$Z)-Leu)$_2$-OC$_6$H$_4$OH-o	S4	54
Nps-Leu-Glu(OBzl)-OPop	S2	74

Table TLC 57 (continued)
PEPTIDE INTERMEDIATES

Compound	Solvent	$R_F \times 100$
HCl,H-Leu-Glu(OBzl)-OPop	S2	69
Nps-Glu(OBzl)-Leu-Glu(OBzl)-OPop	S2	76
HCl,H-Glu(OBzl)-Leu-Glu(OBzl)-OPop	S2	71
Boc-(Leu-Glu(OBzl))$_2$-OPop	S3	40
Boc-(Leu-Glu(OBzl))$_2$-OC$_6$H$_4$OH-o	S6	83
HCl,H-(Leu-Glu(OBzl))$_2$-OC$_6$H$_4$OH-o	S4	72

Abbreviations: Z, benzyloxycarbonyl; ONp, *p*-nitrophenoxy (*p*-nitrophenyl ester). Nps, *o*-nitrophenylthio; Bzl, benzyl; Pop, 2-phenacyloxyphenyl; Cl$_2$Z, 2,4-dichlorobenzyloxycarbonyl; Boc, *tert*-butoxycarbonyl; DCHA, dicyclohexylamine.

Layer: Silica gel (Merck precoated plates 60F$_{254}$).
Solvent: S1 = Acetone-acetic acid, 98:2 v/v.
S2 = Chloroform-methanol, 4:1 v/v.
S3 = Chloroform-acetic acid, 95:5 v/v.
S4 = Ethyl acetate-pyridine-water, 20:10:11 v/v.
S5 = Chloroform-acetic acid-butanol, 85:5:10 v/v.
S6 = Butanol-acetic acid-water, 4:1:1 v/v.

REFERENCE

1. **Brack, A. and Caille, A.,** *Int. J. Pept. Prot. Res.,* 11, 128, 1978.

Reproduced by permission of Munksgaard International Publishers, Ltd.

Table TLC 58
PEPTIDE INTERMEDIATES

Layer	L1
Solvent	S1

Peptide	$R_F \times 100$
Boc-Ser(Bzl)-Asn-OBzl	78
H-Ser(Bzl)-Asn-OBzl·HCl	57
Boc-Gly-Ser(Bzl)-Asn-OBzl	63
H-Gly-Ser(Bzl)-Asn-OBzl·HCl	34
Boc-Gln-Gly-OBzl	58
Boc-Gln-Gly-OH	7
H-Gln-Gly-OH·HCl	0
Boc-Ser(Bzl)-Gln-Gly-OH	18
Boc-Ser(Bzl)-Gln-Gly-Gly-Ser(Bzl)-Asn-OBzl	51
H-Ser(Bzl)-Gln-Gly-Gly-Ser(Bzl)-Asn-OBzl·HCl	6
H-Ala-Lys(Z)-OEt·HCl	95
Z-Gln-Ala-Lys(Z)-OEt	60
Z-Gln-Ala-Lys(Z)-N$_2$H$_3$	36
Z-Gln-Ala-Lys(Z)-Ser(Bzl)-Gln-Gly-Gly-Ser(Bzl)-Asn-OBzl	54
Boc-Gly-Ser(Bzl)-OBzl	82
H-Gly-Ser(Bzl)-OBzl·HCl	68
H-Ser-(Bzl)-Gln-Gly-OH·HCl	0
Z-Gln-Ala-Lys(Z)-Ser(Bzl)-Gln-Gly-OH	0
Z-Gln-Ala-Lys(Z)-Ser-(Bzl)-Gln-Gly-Gly-Ser(Bzl)OBzl	55
Z-Gln-Ala-Lys(Z)-Ser(Bzl)-Gln-Gly-Gly-OBzl	54

Table TLC 58 (continued)
PEPTIDE INTERMEDIATES

Peptide	$R_F \times 100$
Boc-Ser(Bzl)-Gln-OH	17
H-Ser(Bzl)-Gln-OH·HCl	0
Z-Gln-Ala-Lys(Z)-Ser(Bzl)-Gln-OH	74
Z-Gln-Ala-Lys(Z)-Ser(Bzl)-OBzl	63
Boc-Ala-Lys(Z)-N$_2$H$_3$	78
Boc-Ala-Lys(Z)-Ser(Bzl)-Gln-Gly-Gly-Ser(Bzl)-Asn-OBzl	62
Z-Lys(Z)-Ser-(Bzl)-Gln-Gly-Gly-Ser(Bzl)-Asn-OBzl	65

Abbreviations: Z, benzyloxycarbonyl; Bzl, benzyl; Boc, *tert*-butoxycarbonyl.

Layer: L1 = Silica gel G (Merck).
Solvent: S1 = Chloroform-methanol, 5:1 v/v.

REFERENCE

1. **Gyotoku, J., Imaizumi, A., Terada, S., and Kimoto, E.,** *Int. J. Pept. Prot. Res.,* 21, 135, 1983.

Reproduced by permission of Munksgaard International Publishers, Ltd.

Table TLC 59
INTERMEDIATES FOR POLYPEPTIDE SYNTHESIS

Peptide	$R_F \times 100$	Solvent
Z-Ala-Pro-OH,DCHA	49	A
	47	F
H-Ala-Pro-OH	20	G
Z-His(AdOC)-Ala-Pro-OH	68	A
H-His(AdOC)-Ala-Pro-OH	54	H
Z-Asp(O-tBu)-His(AdOC)-Ala-Pro-OH	55	A
	59	F
H-Asp(O-tBu)-His(AdOC)-Ala-Pro-OH	10	A
	62	H
Z-Ser(tBu)-Asp(O-tBu)-His(AdOC)-Ala-Pro-OH	68	A
H-Ser(tBu)-Asp(O-tBu)-His(AdOC)-Ala-Pro-OH	16	A
	17	H
Z-Gly-Ser(tBu)-Asp(O-tBu)-His(AdOC)-Ala-Pro-OH	48	A
Z-Gly-Ser(tBu)-Asp(O-tBu)-His(AdOC)-Ala-Pro-OPcp	73	A
H-Gly-Ser(tBu)-Asp(O-tBu)-His(AdOC)-Ala-Pro-OPcp,HCl	16	A
	55	H
NPS-Ser(Bzl)-Leu-ONP	34	B
	82	C
HCl,H-Ser(Bzl)-Leu-ONP	24	A
NPS-Leu-Ser(Bzl)-Leu-ONP	80	A
HCl,H-Leu-Ser(Bzl)-Leu-ONP	23	A
	41	C
	54	E
NPS-His(Tos)-Leu-Ser(Bzl)-Leu-ONP	82	A
	83	C
	85	E
HCl,H-His-Leu-Ser(Bzl)-Leu-ONP	53	H
HCl,H-Asp(O-tBu)-Leu-ONP	23	A
	58	E
NPS-Leu-Asp(O-tBu)-Leu-ONP	84	D

Table TLC 59 (continued)
INTERMEDIATES FOR POLYPEPTIDE SYNTHESIS

Peptide	$R_F \times 100$	Solvent
HCl,H-Leu-Asp(O-tBu)-Leu-ONP	29	A
	62	E
NPS-Asp(O-tBu)-Leu-Asp(O-tBu)-Leu-ONP	86	D
HCl,H-(Asp (O-tBu)-Leu)$_2$-ONP	48	A
	74	E

Abbreviations: Z, benzyloxycarbonyl; DCHA, dicyclohexylamine; AdOC, adamantyloxycarbonyl; Bzl, benzyl; NPS, *o*-nitrophenylthio; ONP, *p*-nitrophenoxy (*p*-nitrophenyl ester); OPcP, pentachlorophenyl ester; Tos, tosyl (*p*-tolylsulfonyl).

Layer: Merck pre-coated plates, 60F$_{254}$ (silica gel).
Solvents: A = Chloroform-methanol-acetic acid, 85:10:5 v/v.
　　　　 B = Chloroform-methanol, 95:5 v/v.
　　　　 C = Acetone-acetic acid, 98:2 v/v.
　　　　 D = Chloroform-methanol, 4:1 v/v.
　　　　 E = 2-Butanol-ethyl acetate-water, 14:12:5 v/v.
　　　　 F = Ethyl acetate-pyridine-water, 20:10:11 v/v.
　　　　 G = Butanol-acetic acid-water pyridine, 60:6:24:20 v/v.
　　　　 H = Butanol-acetic acid-water, 4:1:1 v/v.

REFERENCE

1. **Trudelle, Y.,** *Int. J. Pept. Prot. Res.,* 19, 528, 1982.

Reproduced by permission of Munksgaard International Publisher, Ltd.

Table TLC 60
INTERMEDIATES OF SYNTHETIC PEPTIDES

	L1	L1	L1	L1
Layer	S1	S2	S3	S4
Solvent	D1	D1	D1	D1
Detection				
Peptide		$R_F \times 100$		
Boc-Pro-Ala-OEt	90	90	70	90
Boc-Thr-Pro-Ala-OEt	80	80	25	90
Fmoc-Ser-(Bzl)-Pro-Ala-OEt	—	90	65	
Z-Thr-Val-Ala-OEt	90	90	20	70
Z-Orn(Boc)-Pro-OH	85	80	40	35
H-Orn(Boc)-Pro-OH	35	60		
Z-Orn(Boc)-Orn(Boc)-Pro-OH	90	80	10	30
Z-Orn(Boc)-Ala-OEt			60	80
Z-Orn(Boc)-Orn(Boc)-Ala-OEt			50	80
Z-Orn(Boc)-Orn(Boc)-Ala-NH-NH$_2$	80	80		60

Abbreviations: Z, benzyloxycarbonyl; Boc, *tert*-butyloxycarbonyl; OEt, ethyl ester; Fmoc, 9-fluorenylmethyloxycarbonyl; Bzl, benzyl.

Layer:　　 L1 = Silica gel plates F$_{254}$ (Merck).
Solvent:　 S1 = Butanol-acetic acid-water, 3:1:1 v/v.
　　　　　 S2 = Butanol-pyridine-acetic acid-water, 60:40:12:48 v/v.
　　　　　 S3 = Chloroform-acetic acid-benzene, 85:10:5 v/v.
　　　　　 S4 = Methanol-chloroform, 15:85 v/v.
Detection: D1 = Modified chlorine reagent.

Table TLC 60 (continued)
INTERMEDIATES OF SYNTHETIC PEPTIDES

REFERENCE

1. **Chessa, G., Borin, G., Marchiori, F., Meggio, F., Brunati, A. M., and Pinna, L. A.,** *Eur. J. Biochem.,* 135, 609, 1983.

Reproduced with permission.

Table TLC 61
PEPTIDE DERIVATIVES

	L1	L1	L1	L1
Layer	S1	S2	S3	S4
Solvent	D1	D1	D1	D1
Detection				
Compound		$R_F \times 100$		
t-BOC-Glu(OBzl)- 0-2-PAOP	45	90	81	91
t-BOC-Ala-Glu(OBzl)- O-2-PAOP	27	79	70	91
t-BOC-Ala-Ala-Glu(OBzl)- O-2-PAOP	10	55	58	88
t-BOC-Glu(OBzl)-Ala-Ala- Glu(OBzl)-O-2-PAOP	3	36	54	88
t-BOC-Lys(Cbz)-Glu(OBzl)-Ala- Ala-Glu(OBzl)-O-2-PAOP	1	24	52	90
t-BOC-Leu-Lys(Cbz)-Glu(OBzl)- Ala-Ala-Glu(OBzl)-O-2-PAOP	0	18	50	90
t-BOC-Ala-Leu-Lys(Cbz)-Glu(OBzl)- Ala-Ala-Glu(OBzl)-O-2-PAOP	0	6	38	86
t-BOC-Ala-Leu-Lys(Cbz)-Glu(OBzl)- Ala-Ala-Glu(OBzl)-O-2-HP	0	9	36	84

Abbreviations: *t*-BOC, *tert*-butyloxycarbonyl; 2-PAOP, 2-phenacyloxyphenyl; Cbz, benzyloxycarbonyl; Bzl, benzyl; 2-HP, 2-hydroxyphenyl.

Layer: L1 = Pre-coated silica gel plates, silica gel 60 F$_{254}$ (Merck).
Solvent: S1 = Benzene-acetone-acetic acid 9:1:0.5 v/v.
    S2 = Chloroform-dioxane-acetic acid 9:1:0.5 v/v.
    S3 = Chloroform-ethanol-acetic acid 9:0.5:0.5 v/v.
    S4 = Chloroform-ethanol-acetic acid 9:1:0.5 v/v.
Detection: D1 = Ninhydrin reagent following light spraying with 3 *N* HCl and heating to 120°C for 10 min (free, Cbz, or *t*-BOC protected amino groups). Phenol esters and phenols (*o*-phenacyloxy-, *o*-hydroxy, and *p*-nitrophenol) are located by the UV absorbing properties quenching the fluorescence of the TLC plate, or their characteristic change of color in NH$_3$ gas (catechol derivatives show dark brown, *p*-nitrophenol derivatives a yellow color).

REFERENCE

1. **Treiber, L. R., Mai Wong, W., Shen, M. E., and Walton, A. G.,** *Int. J. Pept. Prot. Res.,* 10, 349, 1977.

Reproduced by permission of Munksgaard International Publishers, Ltd.

Section II
Techniques

Section II.I.

HIGH-PERFORMANCE LIQUID CHROMATOGRAPHY OF PEPTIDES

INTRODUCTION

During recent years high-performance liquid chromatography (HPLC) has developed as a very rapid and sensitive means of separating many types of peptides and proteins. This development, which has depended largely on the use of normal and reversed phases chemically bonded to microparticles of silica, has occurred at a time when peptide and protein chemistry is undergoing rapid change, particularly in terms of its description of biological phenomena at the molecular level. Much of our present understanding of the structure and function of biological substances derived from amino acids, especially biomolecules available only in small quantities, e.g., neuroendocrine peptides, depends on the ability to optimize a high resolution method such as HPLC so that an inherent property of the peptide, e.g., size, charge, hydrophobicity, can be exploited. The resolution and subsequent analysis of a group of peptides from a synthetic or natural source is a formidable task when a desired component may be available only in microgram or smaller amounts and requires purification 1000-fold or more to reach homogeneity. This section describes the various factors that have to be considered and the steps that may have to be taken to achieve successful peptide separations using HPLC. Peptides considered are principally those below 50 residues long.

ALKYLSILANE BONDED PHASES

Today reversed-phase chromatography is one of the most frequently employed techniques for the separation of a wide variety of mixtures by HPLC; some 60 to 80% of analytical separations may be performed in this way. Surface-modified silica supports used for reversed-phase chromatography are very often octyl (C_8)- or octadecyl (C_{18})-bonded phases. Early alkylsilane-bonded phases were prepared by reacting a trichloroalkylsilane with silica suspended in toluene; most commercially available bonded phases today are prepared using bi- or trifunctional silanes. For steric reasons, residual surface silanols and halosilanes are left after the reaction. When the support is placed in water, a large number of these residual surface silanols become exposed. To overcome this problem the phases are treated further with chlorotrimethylsilane to remove (''cap'') residual silanol groups.[1] Another solution to the problem is to originally treat the silica with monochlorodimethylalkylsilanes.

$$
\begin{array}{ccc}
& CH_3 & CH_3 \\
& | & | \\
silica{\equiv}Si{-}OH + Cl{-}Si{-}R \rightarrow & silica{\equiv}Si{-}O{-}Si{-}R + HCl \\
& | & | \\
& CH_3 & CH_3
\end{array}
\qquad (1)
$$

This bonding chemistry results in heavier surface coverage and fewer residual silanols. Some manufacturers are now beginning to prepare phases using monofunctional silanes.

Supports

Pearson et al.[2] have examined the contribution of several large pore-diameter silica matrices to resolution and recovery of proteins. Evaluation of different silicas cannot be accomplished by studying packed columns from a series of manufacturers, since coating and packing conditions vary to such an extent that definitive statements about silica contribution to support selectivity and resolution are impossible. To overcome this problem silicas were coated with octyl trichlorosilane, $SiCl_3C_8$, using a standard procedure. The silicas were then packed and tested under uniform conditions. Smaller pore silicas, less than 200 Å, were inferior to large pore-diameter supports for resolution and recovery of peptides and proteins. Pore sizes of 300 Å appeared to offer the best resolution. Pores of this size, however, did not necessarily imply that a silica would be successful in every case. Although Spherosil XOBO75 was n-alkylated as completely as other silicas studied, it was not a satisfactory support for reversed-phase separations. Spherisorb® SG30F and LiChrospher Si 300 were both acceptable silicas, but their main disadvantage was inherent fragility. Both silicas crushed easily when back pressures approached 2000 psi. This is a severe handicap, especially when the mobile phase is propanol whose high viscosity may induce operating pressures of 1500 psi or more. The structural integrity of Vydac TP allowed packing pressures greater than 8000 psi without signs of particle collapse. This study by Pearson et al.[2] indicated that pore volume rather than pore diameter is a critical physical parameter when approximating maximum pressures a silica can tolerate.

Wilson et al.[3] have chromatographed the same peptide mixture, the tryptic peptides from rat muscle parvalbumin, on 100 and 300 Å supports (LiChrosorb® RP-18 and Aquapore RP 300, respectively) at low pH.[3] The more hydrophilic nature of the 300 Å support was indicated by the quicker elution of the peptides. Peak resolution was not identical on the two columns, since two peptides which separated on the 100 Å column were not resolved on the 300 Å column. When the pH of the buffers was increased, the peptides were eluted earlier. These pH changes did not affect all peptides equally, resulting in elution at lower acetonitrile concentrations while retaining the same elution order. Some peptides were greatly affected while others were not affected at all. On the 300 Å column, peak broadening was significantly decreased in comparison with a 100 Å packing.

Pellicular supports like Corasil/C_{18} have been used successfully for the analytical or preparative separation of peptides. However, because of their high selectivity, efficiency, and greater capacity fully porous reversed-phase packings such as μBondapak® C_{18} are often preferred.[4] Pellicular packings, such as C_{18}-Corasil or phenyl-Corasil generally show lower efficiencies than fully porous packings due to the peaks exhibiting marked tailing. Peak tailing of peptides has been assumed to reflect the presence of residual silanol groups in the C_{18}- or phenyl-Corasils. However, since similar elution profiles are found after extensive silylation of the packing material, effects other than the presence of residual silanol groups must be in operation during the elution of peptides from reversed-phase pellucular packings with phosphate buffers at about pH 3, the conditions used by Hearn et al.[4]

Chain Length

Bonded phase chain length has little effect on the resolution of nonapeptides.[5] With higher molecular weight peptides and proteins, however, the use of a C_8-bonded phase has been favored because it is possible to elute solutes using less organic solvent. There was no difference in the selectivity of LiChrosorb® RP-8 (C_8 carbon chain) and LiChrosorb® RP-18 (C_{18} carbon chain) on chromatographing peptides obtained during the synthesis of oxytocin.[6] Elution differences between LiChrosorb® RP-8 and LiChrosorb® RP-18 supports were minimum with an l-propanol solvent system. Hydrophilic peptides, however, were sometimes affected.[7] C_2-chain-length supports show poor selectivity, presumably due to the polar character of the stationary phase.[8]

Nice et al.[9] have chromatographed proteins on a number of maximum-coverage alkylsilane-bonded silica packings with carbon chain lengths ranging from C_1 to C_{18}. The greatest range of individual compounds that could be successfully chromatographed was obtained with packings of carbon chain length lower than 6. Optimum recoveries and efficiencies were found with Ultrasphere SAC, a C_3 packing. A wide variety of proteins that could not be chromatographed on C_{18} packings was recovered from the short carbon chain materials. Under constant conditions of gradient elution with acetonitrile in acid saline (pH 2.1), most retention times increased with increasing chain length of the packing, but there were no reversals in retention order indicative of pronounced changes in selectivity. Thus by appropriate alteration of the slope of the gradient profile, a virtually identical elution profile was obtained with all the supports.

MECHANISM OF SEPARATION

Neat aqueous eluents may be used for the separation of small, relatively polar, biological molecules on octadecyl silica.[10] In the absence of an organic component in the eluent, interaction between the solute and the hydrocarbonaceous moiety of the stationary phase must be the cause of solute retention. This means that the chromatographic process is governed by the so-called hydrophobic (solvophobic) effect. Plots of the logarithmic retention factor of solutes against the composition of binary hydro-organic eluents are linear. Retention in reversed-phase chromatography, however, does not always show the regular behavior expected on the basis of the solvophobic effect. Bij et al.,[11] for example, studied the retention characteristics of some hydrophobic peptides with the NH_2-terminal blocked with a *tert*-butyloxycarbonyl (Boc) group and the COOH-terminus in the ethyl ester form. The peptides in question were fragments of glycophorin and were designated Boc-G_7 and NH_2-G_7 for heptapeptide fragments (G_7) with blocked and free NH_2-terminus, respectively. G_7 has the amino acid sequence Val-Met-Ala-Gly-Val-Ile-Gly. Upon reversed-phase chromatography on Supelcosil LC-8, an octyldimethylsilyl bonded phase, using aqueous methanol or acetonitrile containing phosphate buffer, pH 2.25, the logarithmic retention factor vs. the composition of eluent plots for NH_2-G_7 passed through a minimum at relatively low water concentrations. Similar results were obtained when the stationary phase was "naked" silica; BocG_7, however, gave a linear plot. The data suggested that the minimum observed for the unprotected peptide arises from a dual retention mechanism involving silanophilic interactions with the free amino group in addition to the solvophobic effect.

Similar results were obtained by Hearn and Grego,[12] who found that over a wide range of volume fractions of the organic modifier, unprotected peptides did not show linear dependencies of the logarithmic capacity factors on the composition of binary hydro-organic solvents when chromatographed on a μBondapak® C_{18} column. Rather, bimodal plots were obtained showing minima characteristic of the peptide and the organic solvent. Masking the surface silanols by increasing the water concentration of the eluent attenuates silanophilic interactions so that regular retention behavior is observed with peptides. On the other hand, silanol groups at the surface of bonded phases may be essential to obtain adequate selectivity. The similarity between retention data obtained with aqueous eluents on naked silica gel and on C_8 and C_{18} alkyl-silicas implies that the relative polarity of the mobile and stationary phase is the principal criterion for classifying various chromatographic systems.

COMPARISON OF ALKYLSILANE WITH OTHER BONDED PHASES

One problem the research worker using chromatographic techniques faces is knowing what can be expected from a particular type of column. In general this type of information is not easily obtained. A comparison of the performance of different columns has, however, been made in a number of studies. Lundanes and Greibrokk[13] have investigated the separation of dipeptides using the following column packing: Phenyl-Sil-X-1, 13-μm porous particles; μBondapak® NH_2, 10-μm porous particles; Nucleosil® 5 CN, 5-μm porous particles, Spherisorb® S5W-ODS, 5-μm porous particles; Spherisorb® S5 phenyl bonded, 5-μm porous particles; and ODS-Hypersil, 5-μm porous particles.

The amino column showed large deviations from the reversed-phase mode. The nitrile column showed surprisingly constant reversed-phase behavior; the column efficiency and selectivity, however, were lower than those of the ODS columns. The two phenyl columns gave approximately the same results, but the 5-μm Spherisorb® showed the higher efficiency and selectivity. For most of the dipeptides the efficiency and selectivity were considerably better than those of the nitrile column, but poorer than those of the Spherisorb® ODS column. Lundanes and Greibrokk concluded that ODS was to be recommended as a reversed phase for most peptide separations. Both the phenyl and the Spherisorb® ODS columns were strongly affected by basic solutes, while the ODS-Hypersil column was not. Peptides with dominating basic functions that otherwise would be totally retained should be purified on a strictly reversed-phase packing, such as ODS-Hypersil. Ammonium acetate is a suitable buffer for use with this packing.

Wilson et al.[3] have studied the chromatographic separation of the tryptic peptides from rat muscle parvalbumin on a 6-μm cyanopropyl packing and a 10-μm C_{18} packing (Li-Chrosorb® RP-18, 100 Å pore size).[3] The more hydrophobic character of the C_{18} carbon ligand resulted in a better resolution of the peptides, ranging in length from 3 to 15 residues, than was afforded by the cyanopropyl moiety. Rivier[14] similarly found that the best resolution of a peptide mixture was obtained on the most hydrophobic columns, μalkylphenyl or μC_{18}, as compared with μcyanopropyl.

Hansen et al.[15] have studied peptide behavior on Bondapak® phenyl-Corasil, Poragel PN (35 to 75 μm) and Poragel PS (35 to 75 μm). Phenyl-Corasil is like C_{18} packings and is normally eluted with polar solvents. The Poragels are cross-linked organic polymers prepared by copolymerization of styrene with different vinyl compounds. Poragel PN also contains bound ester groups, which are partially hydrolyzed to carboxy groups. Poragel PS is a polystyrene type polymer containing pyridine rings as functional groups. For these reversed-phase packings the retention of peptides was also affected by polar interactions. For phenyl-Corasil the cause was assumed to be residual silanol groups in the silica support which interact with basic groups of the peptides. A similar type of interaction that occurred on Poragel PN was attributed to free carboxyl groups on the gel. The superficially porous phenyl-Corasil and the completely porous Poragel PN behaved very similarly with respect to selectivity and retention of peptides and to column efficiency. Hansen et al.[15] considered these two packing materials to be most suitable for peptides that do not have too basic a character, which could lead to excessive retention.

Poragel PS was considered suitable for peptides containing basic amino acids, since small acidic peptides lacking basic groups are too strongly retained. The fact that Poragel PS, which like phenyl-Corasil and Poragel PN is a reversed-phase packing material, shows a completely different retention pattern, indicated to Hansen et al. that the principal factor contributing to the retention of peptides on these columns is polar interactions between the acidic and basic groups on the peptide and packing material, rather than reversed-phase adsorption or partition. None of the columns showed very high efficiency. Peaks from phenyl-Corasil showed pronounced tailing and Poragel PN showed weakly fronting peaks. This asymmetry in peak shape and the dependency of retention on sample load can be regarded as an overloading phenomena, although these effects are not normally encountered at the quite small sample loads used by Hansen et al. The fact that about 50 μg of peptide on a 3 ft × $^1/_8$ in. column led to overloading showed that the conditions used for the chromatography were far from optimal.

Small peptides such as Met and Leu enkephalin are much more retarded and better separated on a μBondapak® C_{18} column than on a μBondapak® phenylalkyl column.[16] The proteins ribonuclease, lysozyme, and chymotrypsinogen A are slightly more retarded by the C_{18} support and the relative elution times for the proteins are different on the two supports.

Many of the columns mentioned in this article are described in greater detail in Section V.

COMPOSITION OF THE MOBILE PHASE

The eluents used principally in reversed-phase chromatography are acetonitrile-water or methanol-water mixtures that have low viscosity and transparency in the UV. Larger peptides require higher concentrations of organic solvent for elution and many polypeptides are not very soluble in solutions containing high percentages of such organic solvents. However, lower concentrations of *n*-propanol act as well as high concentrations of acetonitrile in eluting large polypeptides without the risk of precipitating them.[17] Thus, α-endorphin with a mol wt of 1750 will elute with 4% *n*-propanol or with 20% acetonitrile from the same column. β-Endorphin, mol wt 3500, and β-lipotropine, mol wt 11,500, will elute with 20% propanol. The only disadvantage of *n*-propanol is its high viscosity, which does not allow high flow rates. This does not really constitute a limitation because large polypeptides diffuse relatively slowly and require low flow rates to give maximum resolution.

Hearn and Grego[18] have studied the retention behavior of peptides and polypeptides at low pH on octadecylsilica columns using neat aqueous, aquo-methanol or aquo-acetonitrile mobile phases, containing phosphate buffers, in isocratic, step- or gradient-elution modes. With acetonitrile as the organic solvent modifier, linear gradients gave better resolution than exponential gradients. Dramatic changes in the capacity factors of some peptides were found following relatively small changes in the concentration of acetonitrile or methanol. This was particularly noticeable with more hydrophobic polypeptides, where efficient chromatography under isocratic conditions could only be performed over a limited range of organic solvent concentrations if the capacity factors of individual peptides are to be kept within reasonable limits, e.g., k' values below 10. Solvent dependent selectivity changes were noted using methanol and acetonitrile. Similar selectivity changes have been observed with hormonal peptides, particularly with very hydrophobic polypeptides, which need higher organic solvent concentrations for their elution.[19]

The efficiency of a μBondapak® C_{18} column is strongly influenced by alterations in the mobile phase composition. Broader peaks are generally observed for smaller peptides using the more viscous propan-2-ol mobile phases than with methanol- or acetonitrile-based eluents.[12] Linear gradients for acetonitrile and shallow convex gradients for methanol and 2-propanol were found to be optimal gradient configurations for polypeptide resolution on surface-modified hydrocarbonaceous silicas.

The ability to elute a peptide at a given buffer percentage in comparison with another solvent decreases in the order propan-1-ol > acetonitrile > methanol. Large peptide fragments or strongly hydrophobic mixtures are optimally separated with propan-1-ol. The absorbance of both propanol isomers in the far UV is relatively high but their high resolving power for both large and small peptides, ease of purification, and low toxicity probably makes their use inevitable. The use of propan-2-ol rather than acetonitrile improved the chromatography of more hydrophobic peptides.[20] Where substances differed greatly in molecular weight, it was found that given changes in the percentage of organic solvent produced much larger changes in the capacity ratio for large molecular weight substances than for small molecules.

ION-PAIRING AGENTS

Hancock et al.[21] found that retention times of known peptides on reversed-phase packings of less than 10-μm particle size were very long and reproducibility was difficult to attain. Several reagents, particularly phosphoric acid, when added to the eluent, gave a marked improvement in reproducibility and retention times. There was a marked decrease in retention time of the tetrapeptide Leu-Trp-Met-Arg on C_{18}-Corasil after the addition of 0.1% phosphoric acid to the acetonitrile-water mobile phase, together with a decrease in the pH of the mobile phase from 6.5 to 2.2. Phosphoric acid thus had the dual effect of causing a change in pH and the equilibria of the ionizable groups of the peptide. Studies with a model peptide in which the amino terminus was blocked and all the side chains were protected, the only ionizable group being the C-terminal carboxyl group, led Hancock et al.[21] to conclude that the formation of an ion pair between the peptide $R-\overset{+}{N}H_3$ and the hydrophilic anion $H_2PO_4^-$ is responsible for the large increase in polarity observed for solute molecules in this study. While hydrophilic pairing agents decrease the capacity factor of a peptide in reversed-phase liquid chromatography, hydrophobic agents increase it.

Wilson et al.[20] chromatographed peptides ranging in length from 2 to 65 residues on reversed-phase packings in buffer systems at low pH, e.g., phosphate-acetonitrile and phosphate-propan-2-ol. The addition of sodium perchlorate to the phosphate buffers, as first described by Meek and Rosetti,[22] significantly improved the chromatographic behavior of more hydrophobic peptides, with a sharpening of the peaks and increased retention. Using a triethylammonium phosphate (TEAP) buffer, Rivier[14] found that a better resolution of peptides was obtained at low pH values, i.e., 2.5 to 3.5. This is the range in which most acidic functional groups are not dissociated and all basic groups are protonated. This pH range has the advantage of being compatible with the column packing materials, as shown by the excellent performance of columns used for more than 1000 different runs.

Dunlap et al.[23] chromatographed synthetic β-endorphin on a μBondapak® C_{18} column, using a linear gradient of acetonitrile. Nonspecific adsorption of β-endorphin to the column was eliminated in the presence of TFA, producing improved resolution together with sharper and more symmetrical peaks. Retention of β-endorphin was greatly decreased on elution with an acetonitrile-TFA system in place of an acetonitrile-formic acid system.

Schaaper et al.[24] have investigated the influence of perfluoroalkanoic acids on the capacity factors of secretin and related peptides. With increasing chain length of the counter ion retention of the peptides increased, as did the separation between secretion and its analogues. The influence of the chain length of the counter ion on retention was much greater with the peptides containing 6 amino acid residues than with those containing 27. An increase in the concentration of the counter ion caused an increase in the retention of peptides but had little effect on selectivity. Schaaper et al. concluded that although mainly ion pair formation occurred in the eluent, a dynamic ion exchange mechanism could not be excluded with perfluoroheptanoic acid, perfluorooctanoic acid, and perfluorodecanoic acid in the eluents used for hexapeptides.

When TFA, pentofluoropropionic acid (PFPA), heptafluorobutyric acid (HFBA), and undecafluorocaproic acid (UFCA) were used as hydrophobic counter ions in the reversed-phase HPLC of a mixture of 12 peptides and proteins profound differences in retention times were observed.[25] As one progressed up the carboxylic acid series the 1-18 ACTH analogue with 8 positive charges had the shortest retention time with TFA, eluted 4th with FPFA, 5th with HFBA, and 7th with UCFA. The absolute elution position of all the peptides was altered, the degree of change being related to the number of positive charges available for ion pairing. TFA, through ion pairing with histidine residues makes histidine-containing peptides relatively more hydrophobic and hence increases their relative retention times on reversed-phase columns.[26] Similarly the formation of an ion pair between the imidazole proton of thyrotropin-releasing hormone (TRHs) histidine residue and the sulfonic acid group of 1-heptanesulfonic acid generated a complex which permitted an excellent separation of TRH from its analogues.[27]

The chromatographic properties of peptides have been examined using a μBondapak®-alkylphenyl column in the presence of hydrophilic (phosphoric acid) or moderately hydrophobic (sodium dodecyl sulfate — SDS) ion pairing reagents. Hydrophobic ion pairing reagents gave rise to a much longer retention time for peptides than did a more polar anion such as dihydrogen phosphate. Bovine insulin, trypsin, and acyl-carrier protein were much more strongly retained on reversed-phase columns in the presence of sodium hexane sulfonate.[28] Dramatic changes in the selectivity of the chromatographic system may be achieved by using different ion pairing reagents. Thus, the elution order of the peptides Gly-Gly-Tyr, Gly-Leu-Tyr, and Arg-Phe-Ala can be reversed by replacing phosphoric acid with SDS in the mobile phase.

When di- to pentapeptides were chromatographed on a reversed-phase μBondapak®-alkylphenyl column with a series of tetraalkyl ammonium reagents, there was a small but progressive increase in retention times as the alkyl chain length of the reagent increased. The tetrabutylammonium reagent, however, caused decreased retention. The addition of tetrapropylammonium ions to the mobile phase produced double peaks with several peptides. The presence of two chromatographic species suggests that two distinct mechanisms are operating via different partition effects. A transition from ion-pair formation to an ion-exchange mode probably occurs with the 12-carbon tetrapropyl group. The shorter retention times found with the tetrabutylammonium cation are consistent with the operation of a different separation mode for this reagent compared to the lower homologs.[29]

Schöneshöfer and Fenner have studied the suitability of reversed-phase systems using the ion pairing reagents ammonium formate and TFA for the separation of biologically important peptides, with particular reference to the later immunoassay of the peptides.[30] Experiments under neutral conditions and at an ammonium formate concentration of 0.05 M gave separations equivalent to those obtained under acidic conditions (pH 1.3) and at the same ionic strength. Ionic strength thus appeared to be the only prerequisite for effective peptide separations, and proton concentration may only affect the polarities of amphoteric amino acids and peptide molecules. Both ammonium formate and TFA are suitable for later lyophilization and immunological quantitation. As lyophilization of TFA was more rapid than that of ammonium formate, Schöneshöfer and Fenner gave priority to this system.

The most important property of the counter ion in influencing retention is probably its charge density. Small, polar counter ions such as $H_2PO_4^-$, ClO_4^-, and $(CH_3)_4N^+$ will tend to have little interaction with the stationary phase and operate with peptides by way of hydrophilic ion-pairing mechanisms, reducing retention times. Reagents of intermediate polarity, such as heptanesulfonate, $(C_4H_9)_4N^+$, will act by way of hydrophobic ion-pairing mechanism, increasing retention times. Equations based on the ligand adsorption model for ion pair reversed-phase chromatography were found to accurately describe the dependence of the capacity factor of dipeptides on the counter ion concentration and on the pH.[31] Depending on the hydrophobic nature of the counter ionic reagent, hyperbolic and parabolic dependencies of the capacity factor on the counter ion concentration were observed. Ion pairing formation appeared to govern retention for short-chain alkyl reagents, whereas ion-exchange mechanisms dominated for long-chain reagents. A minimum in capacity factor occurs when the pH of the mobile phase corresponds to the pI of the peptide. Hearn et al. considered that although it is possible to predict elution conditions based on pH from a minimum set of capacity factors and pH measurements for closely related peptides, there is little practical advantage in using pH alterations as the only method to influence selectivity. Large selectivity differences are readily achieved by adding at suitable pH low concentrations of suitable reagents that can either undergo ion pair formation with the peptides or modify the stationary phase to a dynamic ion exchanger.

OTHER FACTORS AFFECTING CHROMATOGRAPHY

Flow Rate

The effect of flow rate on column efficiency has been studied using 10 peptides ranging in size from TRH (thyrotropin-releasing hormone, mol wt 362) to insulin (mol wt 600) and a Bio-Rad® ODS column.[22] The efficiency of the columns for peptides worsened with increasing molecular weight of the solute or with increasing flow rate of the mobile phase. For high resolution in isocratic separations, flow rates should be less than or equal to 1 mℓ/ min, except for very small peptides. The main cause of band spreading of the peptides studied was probably diffusion, since salts and acids were added to the mobile phase to minimize adsorption. While the size of the solute appeared to be the major cause of band spreading, considerable adsorptive or other nonideal factors contributed to it. The conclusion drawn was that the molecular weight of compounds being chromatographed has an influence via diffusion on the efficiency that can be achieved with any given column.

In a study of the ability of a column to resolve complex mixtures of peptides, the average peak capacity, which is a measure of the number of compounds which can be separated with unit resolution in a chromatographic run, improved as the gradient rate decreased from very fast gradients (5%/min) to very slow gradients (0.5%/min).[22] The flow rate had relatively little effect on resolution, but had a pronounced effect on sensitivity due to the dilution occurring with increasing flow rate. Decreasing the gradient rate decreased the average peak height, although this effect could be minimized by decreasing the flow rate. A compromise must be made between the desire for a short analysis time and the desire for maximum resolution. For several polypeptide hormones chromatographed on a μBondapak® C_{18} column, a reduction in flow rate improved the chromatographic efficiencies, with larger changes being evident for the more hydrophobic peptides.[12] Rivier[14] has studied the effect of changes in flow rate on the resolution of peptides in a mixture while all other parameters were kept constant. Elution volume on a μalkylphenyl column remained constant at different flow rates. It was apparent that the slower the flow rate, the more easily the different components of a mixture would be recovered.

Temperature

Using a μalkylphenyl column, Rivier[14] observed that the higher the temperature, the lower the resolution achieved with a peptide mixture. This is the opposite of what is generally observed for ion-exchange chromatography (e.g., amino acid analyses) or adsorption chromatography (e.g., the resolution of PTH amino acids on a Zorbax® column). In both these cases temperatures of approximately 60°C greatly improve resolution. Increasing temperature was found to decrease retention times and increase the incidence of artifacts such as multiple peaks in pure samples and also to decrease the resolution of the analysis.[32] This occurred with both small tryptic peptides and large cyanogen bromide peptides.

RECOVERY OF PEPTIDES

Reversed-phase supports may show poor recovery and irreversible adsorption of peptides. This applies particularly when small amounts of large or hydrophobic peptides are applied on hydrophobic stationary phases. Wehr et al.[33] investigated the recovery of ranatensin, an 11-residue peptide, from an octadecyl column, the amount of peptide applied to the column being varied over a 100-fold range. Using identical chromatographic conditions, the same samples were applied to a MicroPak® TSK 2000SW column (a steric exclusion column with minimal adsorption of proteins), and the peak areas obtained were defined as 100% recovery. Peptide recovery from the octadecyl column was greater than 80% when amounts of 0.5 μg were applied, but decreased with smaller amounts of peptide. Recovery of enkephalin from an RP-18 column, as determined with radioactive peptides, was close to quantitative while recovery of β-endorphin, using amounts of 300 pmol, was greater than 70%.[34] It was possible to load 10 mg of peptides or small proteins without affecting recovery or retention times.

Nice et al.[9] have studied the recovery of proteins after chromatography on alkylsilane packings with different carbon chain lengths (C_1 to C_{18}). The most notable advantage of the short carbon chain packings was the improved recovery of many of the proteins. This was evident from the size of the "ghost" peaks found with many compounds on the longer chain materials including C_8 packings, and their diminution or absence under identical conditions of gradient elution using the shorter chain C_3 columns. Optimal recoveries with the C_3 packing were obtained with a variety of proteins, and this stationary phase had significant advantages in this regard over C_8, and to a lesser extent C_6 packings. Lewis et al.[34] have used a 1- to 2-hr linear gradient from 0 to 20% *n*-propanol and a column of LiChrosorb® C_{18} for the successful separation of opioid peptides.

OTHER STATIONARY PHASES

LiChrosorb® Diol
Rubinstein[35] has used normal-phase chromatography on a LiChrosorb® diol column for the fractionation of hydrophobic proteins. LiChrosorb diol contains a vicinal alcoholic hydroxyl function at the end of an aliphatic hydrocarbon chain. In several purification procedures, selective precipitation may be obtained by adding organic solvents. The material that remains soluble can then be fractionated on a normal-phase column. For example, 4 mℓ of *n*-propanol is added to 1 mℓ of dialyzed calf serum and after centrifuging the supernatant is added to a LiChrosorb® diol column. When the column is eluted with a gradient of decreasing *n*-propanol concentrations, 6 peaks are observed. In all the cases tested, proteins are eluted in the range of 80 to 50% *n*-propanol. Rubinstein concluded that normal-phase chromatography on LiChrosorb® diol can be carried out only with proteins that are soluble in very high concentrations of organic solvents, e.g., 80% propanol.

Anion Exchangers
Dizdaroglu et al.[36-38] have separated underivatized peptides by HPLC on MicroPak® Ax-10, a difunctional weak anion-exchange bonded phase prepared on LiChrosorb® Si-60 silica, 10 μm. Mixtures of triethylammonium acetate buffer and acetonitrile were used as the eluent. Separation of sequence isomeric dipeptides, resolution of diastereomers, and fractionation of oligomers into classes by chain length were undertaken. For peptides containing a number of acidic amino acids without compensating basic residues, a dilute formic acid solution, pH 2.6, was used as the eluent. Peptides up to about 30 residues long, including somatostatin, ribonuclease S-peptide, α-endorphin, glucagon, and bradykinins were successfully chromatographed. An excellent separation of 12 angiotensins and of diastereomers and analogues of neurotensin was possible. Dizdaroglu et al.[38] point out that multicomponent mixtures of peptides or closely related peptides can be successfully resolved by this technique, which has great advantages in terms of sensitivity, peak symmetry, reproducibility, and high recoveries.

Dizdaroglu et al.[36] chose triethylammonium acetate buffer as the eluent because it shows excellent mobility with acetonitrile, buffers well at the pH employed, and is easily prepared from triethylamine and acetic acid. Its principal advantage is that it allows detection of eluted peptides at wavelengths in the 200 to 220 nm range because of its low absorption in this region. This buffer is easily removed from eluted samples by freeze-drying, which facilitates further isolation and analysis of separated compounds. The experimental conditions appear to allow long column life.

Cellulose Phosphate
Sugihara et al.[39] have described an automatic procedure of cellulose-phosphate chromatography for the separation of peptides and used it in the analysis of a human hemoglobin variant, J. Lome. The application of the procedure was limited because several peptides eluted together. However, in combination with molecular-sieve chromatography on microparticulate polyacrylamide gels, the purification of peptides was efficiently achieved. The

separation of peptides was often inferior to that attained using HPLC methods and ion-exchange resin procedures. Sugihara et al.[39] point out, however, that the superb resolution of HPLC must often be offset in part by the disadvantages of limited loading capacities and variable recoveries. Cellulose phosphate chromatography, designed to analyze materials of the order of 4 to 15 mg, may be suitable for both preparative and analytical purposes.

Similar considerations apply to CM-cellulose and DEAE-cellulose, where comparable amounts of proteins may be applied.

Tripeptide Bonded Phases

Bonded stationary phases, such as the widely used reversed-phase columns, have eliminated many of the problems previously encountered in liquid-liquid partition chromatography. Kitka and Grushka,[40] however, considered that there was a need for a simple and selective method of separating amino acids and peptides, and thought that the best method of approach would be to use stationary phases similar in structure to peptides. The optically active tripeptides Gly-L-Val-L-Phe, L-Val-L-Ala-L-Val, and L-Val-L-Ala-L-Ser were bonded to silica gel and used as stationary phases in conjunction with mobile phases such as 1% citric acid-water and 1% sodium citrate-water. The separation of certain isomeric dipeptides was achieved. Although the separations were superior to those obtained using silica gel, the efficiencies of the columns were low. The results indicated that it was the peptide-bonded phase that controlled the retention order of the peptides.

Grushka and co-workers[41,42] later bonded L-Val-L-Phe-L-Val and L-Val-L-Ala-L-Pro to silica gel for use as stationary phases. The capacity ratios of dipeptides, when distilled and deionized water, pH about 5.5, is used as the mobile phase, are larger than those obtained with acidic and basic buffers. This is probably due to the fact that the ionic strength of deionized water is extremely low. The retention times with a basic mobile phase are longer than those obtained with an acidic mobile phase. Whenever the first amino acid in an isomeric pair of peptides is glycine, that peptide elutes last with an acidic buffer and first with a basic buffer. When the first amino acid in a pair is phenylalanine, that dipeptide elutes first with the acidic mobile phase and last with the basic mobile phase. Separation of dipeptides was difficult with a mobile phase of pH 2.5, but easy at a pH of 7.4. The capacity factors of dipeptides in general were larger than those of the amino acids, except with a very acidic buffer where the reverse was true. The presence of protecting groups on a peptide had a major effect on the retention. The retention mechanism appears to be a combination of anion exchange and hydrophobic interactions. For any bonded peptide phase, the retention order of the peptides, particularly those that are hydrophobic, follows the degree of hydrophobicity of the hydrocarbon side chains of the solutes. The elution orders of the dipeptides resemble the elution orders of their amino acid constituents.

CONCLUSIONS

In spite of the marked improvements in techniques of chromatography that have been made during the past few years, the separation of a complex peptide mixture is by no means routine. Böhlen and Kleeman[43] have discussed the separation of complex peptide mixtures by reversed-phase HPLC. Complete resolution of all peptides is difficult to achieve in a single run, particularly when the mixture contains a large number of peptides as tryptic digests of proteins often do. Gradients designed for eluting a wide variety of largely different peptides are not suitable for separating peptides which behave similarly on chromatography. In practice, the design of a gradient for the optimal separation of all peptides would be difficult and time consuming. It is preferable to use a general linear gradient and collect partially resolved peptides for rechromatography. A useful approach is to rechromatograph a zone of partially resolved peptides under isocratic conditions with the organic modifier concentration altered to give optimal separation. If, as is possible, isocratic elution does not resolve certain peptides, an alternative mode of reversed-phase HPLC with different solute selectivity may be used.

The procedures that may have to be adopted in the separation of a complex peptide mixture are exemplified by the studies of Kamp et al.[44] on the ribosomal protein L 29 from *Escherichia coli*. The tryptic digest of the protein contains hydrophobic peptides that are difficult to separate using purification procedures such as Sephadex® gel filtration or ion-exchange chromatography combined with one- or two-dimensional thin-layer chromatography. The peptides were separated on a RP-18 column, 5 μm (Figure 1). System I employed 0.05% TFA in water at pH 2.0 and acetonitrile. Optimal separation was obtained with a shallow gradient. More rapid increases of the organic solvent in the gradient gave inferior separation, and the isolated peptides were not pure enough for direct sequence analysis. Using the shallow gradient the hydrophobic peptides, e.g., T 7, T 14, T 19, and T 22, were purified in sufficient amounts for direct microsequencing. However, the smaller peptides were eluted together in the first fractions from the chromatogram and two peptides, T 2 and T 1, were not detected. In system II, which was prepared from trace amounts of formic acid and ammonia in water at pH 7.8 with methanol as the organic modifier, all but two peptides, T 8 (Ala-Lys) and T 9 (Glu-Lys), were obtained in purified form from the tryptic digest. System II thus proved more suitable for the separation of the tryptic peptides of protein L 29 than system I. The differences observed when using the two systems may be due to differences in the solubilities of the peptides in the digest.

Kamp et al.[45] have separated the 30S proteins of the *Escherichia coli* ribosome using different reversed-phase columns and four buffer-gradient systems. The columns employed were (1) LiChrosorb® RP-2, (2) SynChropak RP-P, (3) Vydac TP-RP, and (4) C8- and C18-coated silica supports with small pore sizes. It was not possible to recover many of the ribosomal proteins from the last class of columns, independent of the gradient system used. On the LiChrosorb® RP-2 column the buffer-gradient systems used were (1) 0.1 *M* ammonium acetate, pH 4.1/80% acetonitrile gradients; (2) 0.1% TFA in water/acetonitrile gradients; and (3) 0.1 *M* ammonium acetate, pH 4.1/80% 2-propanol gradients. Separations with 2-propanol gradients were superior to those using acetonitrile. When the same buffer systems were employed in conjunction with the SynChropak RP-P column lower resolution was obtained than with the other columns used. Additionally, microsequencing showed the protein fractions to be less pure.

Chromatography of the 30S proteins on the Vydac TP-RP column gave excellent resolution of the peaks and good reproducibility between runs. Four of the 30S proteins were found to be pure by microsequence analysis. Most of the other proteins were found highly enriched in their fractions. Recoveries for most of the proteins were in the range of 40 to 80% and were highest for small hydrophilic proteins. On the Vydac TP-RP column, the best separation conditions were obtained with acetonitrile gradients. In general, best resolutions and recoveries were found with supports of large pore size (about 30 nm), which allow a better penetration of proteins. Supports with small pore sizes, e.g., Hypersil® supports, trapped virtually all the proteins, with the exception of the LiChrosorb® coated with a C_2-phase, from which many 30S proteins were eluted. This may have been due to weaker interactions of the short aliphatic groups of the support with the side chains of the proteins.

Since the abilities of the supports to separate ribosomal proteins were greatly different and varied from supplier to supplier and sometimes from batch to batch, the particle size distribution and shapes of the supports were examined under the electron microscope. Uniform and globular-shaped particles gave better resolution of complex protein mixtures than did irregular-shaped particles. The SynChropak support, showing an irregular shape, gave a poorer resolution of the ribosomal protein mixture. It is possible that uneven particles size distribution is often responsible for peak tailing.

Marked differences in behavior between columns were encountered by Rivier et al.[46] in their isolation of corticotropin releasing factor (CRF) from ovine hypothalami. Trialkylammonium buffers and TEAP in particular were found to increase recovery and overall column performance. The role of the added alkylamine to the mobile phase was to competitively inhibit the participation of solute in the ion exchange or adsorption reactions with the nonbonded silanols on the stationary phase. Many columns, including Supelco Sil C_{18} (5 μm), DuPont ODS and CN (5 μm) columns, and several solvent systems were tried without success. In most cases, biological activity was lost or no separation was achieved. It was only when a μBondapak® C_{18} (but not Supelco Sil or DuPont C_{18}) column immersed in ice water was employed that the TEAP-acetonitrile buffer system gave acceptable resolution and recovery of biological activity. Lowering of the temperature to find isocratic conditions on the μBondapak® CN column was necessary to achieve some further purification. Many difficulties in the original isolation scheme were later resolved by using large pore size silica (300 Å) that had been properly derivatized (C_{18}) and end-capped, i.e., Perkin Elmer HC-ODS SIL-X-1 (10 μm) or Vydac (5 μm).

ACKNOWLEDGMENT

I am indebted to Dr. B. Wittmann-Liebold for information and advice during the preparation of this section.

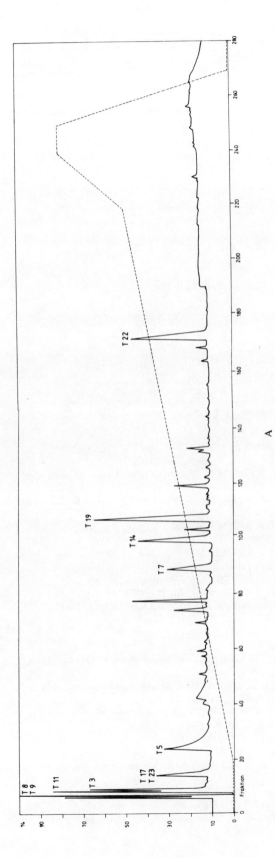

FIGURE 1.   Separation of tryptic peptides of ribosomal protein L 29.   The reversed-phase column support was RP-18, ODS Hypersil®, 5 μm.   Solvent (A) buffer A, 0.05% TFA in water at pH 2.0; buffer B, acetonitrile with 20% of buffer A.   Gradient, 0% B for 10 min, 0% B to 50% B in 100 min, 50% B to 80% B in 10 min, hold at 80% B for 5 min, 80% B to 0% B in 2 min, reconditioning for 30 min with buffer A.   (B) Buffer A, ammonium formate, pH 7.8, 1.6 mℓ 25% ammonia + 0.25 mℓ 98% HCO₂H/2 ℓ of water; buffer B, 80% methanol + 20% buffer A.   Gradient, 0% B for 5 min, 0% B to 90% B in 90 min, hold at 90% B for 5 min, 90% B to 0% B in 2 min, reconditioning for 30 min with buffer A.   The increase of buffer B (%) is shown by the interrupted line.   Detection at 220 nm.   (From Kamp, R. M., Yao, Z.-J., and Wittmann-Liebold, B., *Hoppe-Seyler's Z. Physiol. Chem.*, 364, 141, 1983.   Reproduced by permission of Walter de Gruyter & Co.)

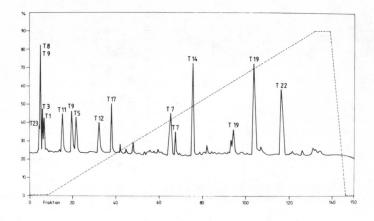

FIGURE 1B.

REFERENCES

1. **Roumeliotis, P. and Unger, K. K.,** *J. Chromatogr.,* 149, 211, 1978.
2. **Pearson, J. D., Lin, N. T., and Regnier, F. E.,** *Anal. Biochem.,* 124, 217, 1982.
3. **Wilson, K. J., Van Wieringen, E., Klauser, F., Berchtold, M., and Hughes, G. J.,** *J. Chromatogr.,* 153, 407, 1982.
4. **Hearn, M. T. W., Bishop, C. A., Hancock, W. S., Harding, D. R. K., and Reynolds, G. D.,** *J. Liq. Chromatogr.,* 2, 1, 1979.
5. **Regnier, F. E. and Gooding, K. M.,** *Anal. Biochem.,* 103, 1, 1980.
6. **Nachtmann, F.,** *J. Chromatogr.,* 176, 391, 1979.
7. **Wilson, K. J., Honegger, A., Stötzel, R. P., and Hughes, G. J.,** *Biochem. J.,* 199, 31, 1981.
8. **Blevins, D. D., Burke, M. F., and Hruby, V. J.,** *Anal. Chem.,* 52, 420, 1980.
9. **Nice, E. C., Capp, M. W., Cooke, N., and O'Hare, M. J.,** *J. Chromatogr.,* 218, 569, 1981.
10. **Horvath, C., Melander, W., and Molnar, I.,** *J. Chromatogr.,* 125, 129, 1976.
11. **Bij, K. E., Horvath, C., Melander, W. R., and Nahum, A.,** *J. Chromatogr.,* 203, 68, 1981.
12. **Hearn, M. T. W. and Grego, B.,** *J. Chromatogr.,* 218, 497, 1981.
13. **Lundanes, E. and Greibrokk, T.,** *J. Chromatogr.,* 149, 241, 1978.
14. **Rivier, J. E.,** *J. Liq. Chromatogr.,* 1, 343, 1978.
15. **Hansen, J. J., Griebrokk, T., Currie, B. L., Johansson, K. N.-G., and Folkers, K.,** *J. Chromatogr.,* 135, 155, 1977.
16. **Henderson, L. E., Sowder, R., and Oroszlan, S.,** *Chemical Synthesis and Sequencing of Peptides and Proteins,* Liu, C. et al., Eds., Elsevier/North-Holland, Amsterdam, 1981, 252.
17. **Rubinstein, M., Chen-Kiang, S., Stein, S., and Udenfriend, S.,** *Anal. Biochem.,* 95, 117, 1979.
18. **Hearn, M. T. W. and Grego, B.,** *J. Chromatogr.,* 203, 349, 1981.
19. **O'Hare, M. J., and Nice, E. C.,** *J. Chromatogr.,* 171, 209, 1979.
20. **Wilson, K. J., Honegger, A., and Hughes, G. J.,** *Biochem. J.,* 199, 43, 1981.
21. **Hancock, W. S., Bishop, C. A., Prestidge, R. L., Harding, D. R. K., and Hearn, M. T. W.,** *J. Chromatogr.,* 153, 391, 1978.
22. **Meek, J. L. and Rossetti, Z. L.,** *J. Chromatogr.,* 211, 15, 1981.
23. **Dunlap, C. E., III, Gentleman, S., and Lowney, L. I.,** *J. Chromatogr.,* 160, 191, 1978.
24. **Schaaper, W. M. M., Voskamp, D., and Olieman, C.,** *J. Chromatogr.,* 195, 181, 1980.
25. **Bennett, H. P. J., Browne, C. A., and Solomon, S.,** *J. Liq. Chromatogr.,* 3, 1353, 1980.

26. **Acharya, A. S., Di Donato, A., Manjula, B. N., Fischetti, V. A., and Manning, J. M.,** *Int. J. Pept. Prot. Res.*, 22, 78, 1983.
27. **Spindel, E. and Wurtman, R. J.,** *J. Chromatogr.*, 175, 198, 1979.
28. **Hancock, W. S., Bishop, C. A., Meyer, L. J., Harding, D. R. K., and Hearn, M. T. W.,** *J. Chromatogr.*, 161, 291, 1978.
29. **Hancock, W. S., Bishop, C. A., Battersby, J. E., Harding, D. R. K., and Hearn, M. T. W.,** *J. Chromatogr.*, 168, 377, 1979.
30. **Schöneshöfer, M. and Fenner, A.,** *J. Chromatogr.*, 224, 472, 1981.
31. **Hearn, M. T. W., Grego, B., and Hancock, W. S.,** *J. Chromatogr.*, 185, 429, 1979.
32. **Mahoney, W. C. and Hermodson, M. A.,** *J. Biol. Chem.*, 255, 11199, 1980.
33. **Wehr, C. T., Correia, L., and Abbott, S. R.,** *J. Chromatogr. Sci.*, 20, 114, 1982.
34. **Lewis, R. V., Stein, S., and Udenfriend, S.,** *Int. J. Pept. Prot. Res.*, 13, 493, 1979.
35. **Rubinstein, M.,** *Anal. Biochem.*, 98, 1, 1979.
36. **Dizdaroglu, M., Krutzsch, H. C., and Simic, M. G.,** *J. Chromatogr.*, 237, 417, 1982.
37. **Dizdaroglu, M., Krutzsch, H. C., and Simic, M. G.,** *Anal. Biochem.*, 123, 190, 1982.
38. **Dizdaroglu, M., Simic, M. G., Rioux, F., and St. Pierre, S.,** *J. Chromatogr.*, 245, 158, 1982.
39. **Sugihara, J., Imamura, T., Yanase, T., Yamada, H., and Imoto, T.,** *J. Chromatogr.*, 229, 193, 1982.
40. **Kitka, E. J., Jr. and Grushka, E.,** *J. Chromatogr.*, 135, 367, 1977.
41. **Fong, G. W.-K. and Grushka, E.,** *J. Chromatogr.*, 142, 299, 1977.
42. **Fong, G. W.-K. and Grushka, E.,** *Anal. Chem.*, 50, 1154, 1978.
43. **Böhlen, P. and Kleeman, G.,** *J. Chromatogr.*, 205, 65, 1981.
44. **Kamp, R. M., Yao, Z.-J., and Wittmann-Liebold, B.,** *Hoppe-Seyler's Z. Physiol. Chem.*, 364, 141, 1983.
45. **Kamp, R. M., Yao, Z.-J., Bosserhoff, A., and Wittmann-Liebold, B.,** *Hoppe-Seyler's Z. Physiol. Chem.*, 364, 1777, 1983.
46. **Rivier, J., Rivier, C., Spiess, J., and Vale, W.,** *Anal. Biochem.*, 127, 258, 1982.

Section II.II.

GEL PERMEATION CHROMATOGRAPHY OF PEPTIDES

INTRODUCTION

The technique of gel permeation chromatography (size exclusion chromatography) distinguishes between different molecules on the basis of size. Certain porous materials have pores so small that even at the level of molecular size, they act as gates permitting smaller molecules to pass easily while excluding larger molecules. For a given particle size this effect is a continuous decrease in accessibility for molecules of increasing size rather than a step function. If a mixture of compounds of different molecular sizes is passed through a column of a porous material of this type, the compounds will leave the column in the order of decreasing molecular size (Figure 1). Fractionation of proteins according to size on cross-linked dextran or polyacrylamide gel was first described by Porath and Flodin.[1] The technique is now widely used for hydrophilic and some hydrophobic macromolecules employing aqueous buffers with or without the addition of organic modifiers.

SUPPORTS

Gel permeation chromatography on Sephadex® has been used extensively for the separation and molecular weight determination of proteins and peptides. However, since Sephadex® is not mechanically stable under pressure, it is impossible to increase the flow rate through the column by applying pressure, and analysis times are thus long. Recently column packings have been developed that can be operated at high speed under pressure, yield high resolution, and do not adsorb proteins. Examples of these column packings are described in Section V.

Globular proteins with molecular weights from 5000 to 250,000 have been separated using 3 TSK-GEL SW columns of different pore sizes, G20000SW, G3000SW, and G4000SW.[2] These 3 columns showed the highest separation efficiencies for molecular weights of less than 30,000, 30,000 to 500,000, and greater than 500,000, respectively. The separation of a mixture of proteins on two gels of different particle sizes is illustrated in Figure 2. Enzymes may be purified by high-speed gel filtration on TSK-GEL G3000SWG columns (SWG columns are preparative columns of large diameter).[3] A 15-fold purification of crude β-galactosidase from bacterial cells and of commercial urease was obtained in a single run. Samples of up to 100 mg could be applied without loss of separation efficiency. Enzymes were eluted from the column within 1 hr, and almost complete recovery of enzymatic activity was obtained.

Apolipoproteins can be resolved on an analytical high-performance steric exclusion column, the MicroPak® TSK 3000SW, with a recovery of greater than 80%, but the separation is not as useful for analytical purposes as that obtained by anion exchange.[4] Using a preparative MicroPak® 3000SW column, a sample introduction and fraction collection scheme for the isolation of apolipoproteins has been devised, based on a study of the performance of the column as a function of flow rate, sample load, and mobile phase composition. On TSK-GEL 2000SW and 3000SW columns, the eluent being phosphate buffer containing sodium sulfate (pH 6.5), a linear relationship between the logarithm of the molecular weight and retention volume was observed for 16 proteins.[5] Lysozyme eluted slower than the expected retention volume from the 2000SW column. The adsorption of solutes on the column was reduced by adding sodium sulfate or sodium chloride, but adsorption of lysozyme

* Figures and tables follow text of this section.

and smaller aromatic compounds was still seen with salt concentrations greater than 0.2 M. The exclusion limits for the 2000SW and 3000SW columns were about 60,000 daltons and larger than 480,000 daltons, respectively.

The molecular weight of small peptides has been determined using Bio-Sil® TSK 20 and Bio-Gel® TS 125 columns.[6] The TSK 20 column gave a good separation of peptides in the mol wt range 1000 to 10,000, the retention times being inversely correlated with the logarithms of their molecular weights. Two combined TSK 125 columns allowed reliable molecular weight determination in the range 800 to 3500. Shioya et al.[7] studied the chromatography of over 50 peptides with mol wt between 200 and 10,000 using a TSK GEL G2000SW size-exclusion column; post-column detection of peptides was carried out using o-phthalaldehyde. The elution volumes (capacity factors) of the peptides differed from one solvent to another. In the solvent systems used, peptides were eluted from the column in the approximate order of their molecular weights. Two groups of peptides, however, were eluted in unexpected positions. One group adsorbed to the resin and eluted later than expected. It contained Gly-Trp, angiotensins I and II, and insulin A and B chain S-sulfonates. The second group, containing melittin and mastoparan, eluted faster than expected. Shioya et al. concluded that the estimation of the molecular weight of peptides by simple size exclusion chromatography was not promising.

Using model peptides of approximately the same molecular weight but possessing different elution volumes, the relationships between the flow rate and the HETP (height equivalent to a theoretical plate) have been studied.[7] When the molecular weights of the peptides were approximately the same, the flow rates to obtain optimum HETP were the same, even though the optimum HETP value for each peptide was different due to different elution mechanisms. Peptides that adsorbed to the column and were eluted later showed a smaller HETP. When the relationship between molecular weight and the flow rate required to obtain optimum HETP for the model peptides was studied, it was found that the optimum flow rate was inversely and linearly related to log molecular weight. This relationship held even for peptides that were eluted from the column at unexpected positions and also held when the elution solvent was a simple phosphate-saline. When the molecular weights of model peptides were estimated on the basis of these findings, the results were superior to those obtained using the conventional method.

COMPOSITION OF THE MOBILE PHASE

The separation of proteins according to size is only possible by adjusting the pH and ionic strength of the eluent. Relatively high concentrations of nonvolatile buffers must be used to increase ionic strength, which tends to minimize ionic interactions of positively charged proteins with the negatively charged surface, producing a reduced elution volume. At the same time it decreases electrostatic repulsion of negatively charged proteins with the negative surface and results in a larger elution volume than is theoretically predicted. Low ionic strength generally results in poor resolution and recovery of proteins of widely different isoelectric points or hydrophobic character.

Rivier[8] considered that there was a need for a UV-transparent, biologically compatible, and if possible volatile buffer, which would permit good resolution and recovery of a wide range of proteins and peptides.[8] The solvent system then described was believed to largely meet this need. Triethylammonium formate (TEAF) and triethylammonium phosphate buffer (TEAP), which had been effective as solvents for reversed-phase HPLC, were employed. The TEAP and TEAF buffers in the presence of a certain amount of acetonitrile were compatible with the protein analysis column PAC I-125. A linear relationship between log molecular weight vs. retention time was obtained for molecules ranging from acetic acid or thyrotropin releasing factor to globulins and including proteins with quite different isoelectric points.

The separation of larger proteins improved at lower concentration of acetonitrile, whereas a higher concentration of acetonitrile was favorable for smaller peptides. A noteworthy point was the low pH (<3) of the aqueous buffer used for separations. Rivier claimed that the advantages possessed by the technique over other methods of determining molecular weights include high sensitivity, accuracy, and rapidity. The volatility of TEAF or the UV transparency and compatibility of TEAP with most biological systems made both systems more versatile than those containing detergents on high salt concentrations.

An HPLC gel permeation system which allows the rapid separation of proteins in the mol wt range 500 to 90,000 consists of protein analysis columns, one I-60, two I-125s, and one I-1250 connected in series; runs were made isocratically at room temperature.[9] With 0.2 M TEAF, pH 3.0 as the eluting buffer, a very useful fractionation range of 500 to 150,000 daltons in less than 50 min is possible. A very high limit of sensitivity is obtained. Molecular weights of proteins can be accurately determined. The use of TEAF buffer alone, however, leads to aggregation, mainly dimerization, of various peptides and proteins. To overcome this problem, the columns are run using 4 and 6 M guanidine hydrochloride in 0.2 M TEAP, pH 3.0. A linear calibration curve is obtained, together with sharp peaks. No aggregation is observed, although reduction of the fractionation range to 500 to 90,000 daltons is noted. Sample recoveries as determined with radiolabeled proteins always exceeds 70%.

A gel permeation column of TS-G3000 PW equilibrated and developed with 36 or 45% acetonitrile in 0.1% TFA efficiently fractionated mixtures of peptides.[10] The elution volumes of 11 peptides and proteins were linearly related to the logarithms of their molecular weights at high and low flow rates. A mobile phase consisting of pH 7.0 phosphate buffer containing NaCl and 20% v/v ethanol used in conjunction with an I-125 column minimizes the ionic and hydrophobic interactions of proteins with the stationary phase, and can be used to determine the molecular weights of proteins between 10,000 and 70,000.[11]

In studies of the gel permeation chromatography of proteins, nondenaturing solvents have generally been employed in order to maintain the native configuration of the protein conformation and its antigenic or enzymatic properties.[12] However, the resolving power of columns can be greatly enhanced by the use of denaturants such as sodium dodecyl sulfate (SDS) or guanidine hydrochloride. The addition of denaturants increases the accuracy of molecular weight determination, presumably because proteins assume a more uniform conformation and because protein interactions with the gel matrix are minimized. Apparent molecular weights determined in this way showed an average deviation of 5 to 7% from reported values.

Schneider et al.[13] encountered great difficulties in attempting to resolve peptide mixtures obtained by enzymic digestion subunit e of the hemocyanin from the tarantula, *Eurypelma californicum*. On gel filtration the peptides tended to aggregate very strongly unless they were very short. Consequently, 8 M urea had to be added to all the solvents.

Gel permeation chromatography separates molecules primarily on the basis of hydrodynamic volume. For this to be achieved the chromatographic support must be neutral and the polarity must be nearly equal to that of the mobile phase. If this is not the case, the support may play a part in the separation process. As the magnitude of surface contributions becomes greater, the deviation from ideal behavior increases. Kopaciewicz and Regnier[14] chromatographed a number of proteins on SynChrom GPC-100, TSK-G2000SW, and TSK-G3000SW columns at low ionic strength. A protein could be selectively adsorbed, ion excluded, or chromatographed in an ideal size-exclusion mode by varying the pH of the mobile phase relative to the isoelectric point of the protein. These effects were considered to be the result of electrostatic interactions between proteins and silanols on the surface of the support.

Hansen et al.[15] have studied the chromatography of four different peptides on Hydrogel IV. The Hydrogel series is based on a polar organic polymer compatible with aqueous solvents. This packing is designed for a molecular weight range of about 500 to 40,000, so that all the peptides used should have been well retained and eluted slightly before or in the total solvent volume of the column. This situation was best approached by 40% acetonitrile. This solvent exerted a major influence on elution for at least two peptides, and 40% acetonitrile was a stronger eluent than water in these two cases, indicating some degree of reversed-phase character. The excessive retention and the very broad peaks obtained showed the influence of factors other than gel filtration in determining retention. The most strongly retained peptides were the larger ones containing the most basic functions.

COMPARISON OF GEL PERFORMANCES

With the object of achieving gel chromatographic separation of proteins similar to that possible under HPLC conditions, Rutschmann et al.[16] studied the chromatography of a protein mixture on three different gels of similar fractionation range, namely Sephadex® G-75 superfine, Ultrogel AcA 54, and Bio-Gel® P-100 minus 400 mesh. Proteins with mol wt between 13,000 and 68,000 were resolved with different efficiencies. Of the three gels, Bio-Gel® P-100 was found to have optimal resolving power. As judged by polyacrylamide gel electrophoresis, fractionation of the protein mixture on this gel resolved all the components, including carbonic anhydrase and chymotrypsinogen A, two proteins which differ in molecular weight by as little as 5000.

Jenik and Porter[17] have evaluated two gel permeation columns, an I-125 protein column and a TSK gel 4000 SW column, reported to have molecular weight exclusion limits of 2000 to 80,000 and 5000 to 1,000,000, respectively. The columns did not function in the size exclusion mode for all the proteins studied, some kind of solute-column interaction occurring for several proteins. Most gel permeation media contain negatively charged groups which give rise to an ion-exclusion effect, whereby the diffusion of anionic solutes into the gel interior is restricted by electrostatic interaction between the solute and gel matrix. On the I-125 protein column, negatively charged molecules such as ATP or acyl carrier protein are repelled by the silanol groups present on the silica support, and their molecular weights are overestimated. Basic molecules such as cytochrome *c* or chymotrypsinogen A are retarded or bound by electrostatic adsorption to the silanol groups, resulting in underestimation of their molecular weights and/or poor recovery from the column.

The ion-exclusion effect can be suppressed by increasing the ionic strength of the mobile phase. Acetate buffer is much poorer in suppressing ionic effects than the same concentration of phosphate buffer, and ionic effects are thus more pronounced with the former. The deviation from a linear relationship between the retention time and molecular weight for certain proteins is thus decreased by using a phosphate mobile phase. Since the retention time of some proteins on gel permeation columns is affected by the composition and concentration of the mobile phase, the molecular weights for unknown proteins obtained on these columns should be checked in other ways. Recovery of proteins varies from 60 to 95%.

The amount of protein that can be applied to gel permeation columns is limited and the technique is therefore generally used in analytical (microgram) separations. These columns afford the best resolution when samples are applied in volumes smaller than 50 $\mu\ell$. As the sample volume is increased peak spreading occurs, and above a sample volume of about 300 $\mu\ell$ spreading interferes with the resolution and separation of protein mixtures. Separations of HPLC with ion-exchange or reversed-phase supports is better suited for semipreparative work on peptides and proteins. A disadvantage of reversed-phase HPLC is the instability of most enzymes in the organic solvents employed for elution.

Kamp et al.[18] have applied the protein mixture extracted from 30S subunits of *Escherichia coli* ribosomes (TP30) to a number of gel filtration columns. Optimal separation was achieved with 0.05 to 0.1 *M* ammonium acetate on a TSK 2000 SW column, a similar elution profile being obtained on a Bio-Sil® TSK-125 column. A slow flow rate gave the best results. An I-125 column did not resolve the protein mixture and low resolution was obtained with LiChrospher-100 Diol or Si-200-Polyol columns.

Roumeliotis and Unger[19] showed that LiChrosorb® diol was a column packing well suited for the separation of proteins and enzymes over a mol wt range of 10,000 and 100,000. Seventy-five milligrams of a mixture of three proteins which differ in molecular weight by a factor of 2 could be separated in 1 run. At a flow rate of about 20 mℓ/min, solutes were eluted within 6 min. As the large bore LiChrosorb® diol column contains some 50 g of packing, 30 mg of a protein can be chromatographed without overloading the column. The performance of this column was contrasted with that of a Sephacryl® S-200 superfine column, a polysaccharide-base organic packing for high performance gel filtration. Under optimal conditions 30 mg of protein can be separated on the LiChrosorb® diol column in 6 min, whereas the same separation on Sephacryl® requires several hours. Moreover, on the former column solute is eluted in about 20 mℓ, while the latter requires large volumes.

PROTEIN RECOVERY

Kamp et al.[18] have commented on protein recoveries after chromatographic separation of *Escherichia coli* 30S ribosomal proteins. According to amino acid analysis, the yields of the proteins depend on the support selected for chromatography and decrease with the hydrophobicity and mass of the individual proteins. Size exclusion chromatography gave the best recoveries, with values of up to 90% for some proteins. Yields after reversed-phase chromatography were lower, many being in the range of 40 to 60%. The reversed-phase columns retained acidic proteins and trapped hydrophobic peptides more than hydrophilic peptides. Lowest yields were obtained for large hydrophobic proteins.

PROTECTED PEPTIDES

Gel permeation chromatography may be used to purify protected peptides.[20,21] The G-series of Sephadex® gels is effective over a wide range of molecular weights, but is generally used in conjunction with aqueous buffers in which fully protected peptides are usually insoluble. Sephadex® LH-20, a hydroxypropylated derivative of G-25, has been successfully employed for the chromatography of protected peptides using dimethylformamide or other organic solvents as eluent. The exclusion limit is between mol wt 2500 and 3000. G-50 and G-75 Sephadex® gels with 5% water in hexamethyl phosphoramide have been used for large protected peptides outside the range of the Sephadex® LH-20-dimethylformamide system. Enzakryl K2, together with dimethylformamide or *N*-methylpyrrolidone, has been used to purify protected peptides of mol wt up to 10,000, but this gel proved to have variable resolving power, particularly with sparingly soluble peptides. Sephadex® LH-60, a higher homolog of LH-20, has both hydrophilic and lipophilic properties. Dimethylformamide and *N*-methylpyrrolidone proved generally applicable for use with this gel, which performed well at relatively high flow rates and showed good resolution. Although the majority of protected peptides chromatographed satisfactorily, in some cases there was evidence of aggregation, which is not normally observed.

THIN-LAYER GEL FILTRATION

An alternative to column gel filtration for the determination of molecular weights, which is in some respects superior to the column technique, is thin-layer gel filtration. It requires small amounts of samples and affords the possibility of separating, in a relatively short time, several samples applied side by side. The migration distance of the peptide or protein, or the migration distance relative to a standard protein, is plotted against the logarithm of the molecular weight.

Klaus et al.[22] used Sephadex® G-200 (superfine grade) for the estimation of molecular weights of polypeptide chains. Proteins were dissolved in guanidine hydrochloride in buffer at pH 8.5; disulfide bonds were reduced with dithiothreitol, followed by alkylation of the thiol groups with iodoacetamide. Excess reagents were removed by precipitating the protein with ethanol, and the precipitates were dissolved in guanidine hydrochloride for thin-layer gel filtration. At the end of the run, the protein spots were transferred to Whatman 3MM paper; the paper was removed, dried, and immersed in 10% trichloroacetic acid for 15 to 30 min. After scraping residual gel from the back, the print was again dried and stained with Coomassie Brilliant Blue R250, followed by destaining in tap water and then in methanol-acetic acid-water, 50:10:50 v/v.

The polypeptides investigated exhibited a straight line relationship of distribution coefficient K_t vs. log molecular weight. This method of data treatment was derived in an effort to relate thin-layer gel filtration to the known fundamental transport and flow phenomena of liquid chromatography and to correlate the data of thin layer and column experiments. The fact that K_t gave excellent linear regressions when plotted against log molecular weight lent support to its usefulness in practical terms.

Heinz and Prosch[23] estimated the molecular weight of polypeptides by thin layer gel chromatography in 6 M guanidine hydrochloride. Chromatography of chains with a molecular weight of less than 100,000 was carried out with Sephadex® G-75; at higher molecular weights G-100 was used. After chromatography, a dry filter paper was pressed on the wet gel layer for several minutes. The spots corresponding to proteins were then easily observed as transparent patches on the filter paper or by spraying the filter with diazotized sulfanilic acid. For proteins with a molecular weight of less than 20,000 it was preferable to estimate the molecular weight using Sephadex® G-50. The use of proteins with intact disulfide bonds by Heinz and Prosch extended the fractionation range of Sephadex® gels, since the Stokes' radii of unreduced proteins are smaller than those of the corresponding reduced proteins.

Thin-layer gel filtration has been used to estimate the molecular weight of peptides and glycopeptides by Hung et al.[24] (Table 1). Samples are dissolved in 6 M guanidine hydrochloride-0.1 M phosphate buffer prior to application to thin-layer plates coated with Sephadex® G-50 Superfine. On completion of the chromatography, the peptides are transferred to a sheet of Whatman 3MM paper which is removed and dried. The paper print is sprayed with a solution of fluorescamine in acetone, is heated at 60°C in an oven for 5 min, and peptides are then located by viewing under a UV lamp. Excellent resolution is obtained with both peptides and glycopeptides. There is a linear relationship between R_F and log molecular weight for peptides ranging in molecular weight from 1700 to 10,500. The behavior of glycopeptides in the molecular weight range of 1900 to 3300 does not differ significantly from that of peptides.

Wasyl et al.[25] have used thin-layer gel filtration of reduced polypeptide chains in 8 M urea and 0.1% SDS solution on Sephadex® G-150 to evaluate their molecular weight.[25] Using the relationship they proposed between the migration distance of protein zones and the molecular weight, a good agreement was found with a number of proteins.

Table 1
ESTIMATION OF THE MOLECULAR WEIGHT OF PEPTIDES AND GLYCOPEPTIDES BY THIN-LAYER GEL FILTRATION

Layer	L1
Solvent	S1
Detection	D1

Peptide or glycopeptide	$R_F \times 100^a$	Molecular weight	
		Actual	Observed
Rattlesnake insulin B-chain tryptic peptide	53.7	841	
Avian pancreatic polypeptide (APP) tryptic peptide	57.7	1,759	1,841
Horse cytochrome *c* CNBr peptide II	61.5	1,763	2,163
Bovine insulin A chain	64.2	2,516	2,427
Cytochrome *c* CNBr peptide III	68.9	2,764	2,958
Glucagon	69.5	3,411	3,040
Insulin B chain	70.5	3,481	3,162
Bovine α-lactalbumin CNBr peptide II	75.9	4,018	3,981
APP	75.5	4,240	3,917
Horse myoglobin CNBr peptide I	85.4	6,220	5,958
Cytochrome *c* CNBr peptide I	92.4	7,654	8,017
Myoglobin CNBr peptide II	90.9	8,166	7,516
α-Lactalbumin CNBr peptide I	100.0	10,500	11,066
Glucosylgalactosylhydroxylysine	50.3	481	
Ovalbumin glycopeptide (GP) IV	56.2	1,497	
Ovalbumin GP III	57.8	1,918	1,849
Human transferrin GP	65.0	2,621	2,506
Fetuin GP (asparagine linked)	68.5	3,325	2,911

[a] Relative to α-lactalbumin CNBr peptide I.

Layer: L1 = Sephadex® G-50 Superfine on glass plates, 20 × 40 cm.
Solvent: S1 = 6 *M* guanidine hydrochloride-0.1 *M* phosphate buffer, pH 7.0.
Detection: D1 = By transferring to a Whatman 3 MM paper sheet which was then stained with fluorescamine.

REFERENCE

1. **Hung, C.-H., Strickland, D. K., and Hudson, B. G.,** *Anal. Biochem.,* 80, 91, 1977.

Reproduced by permission of Academic Press, Inc.

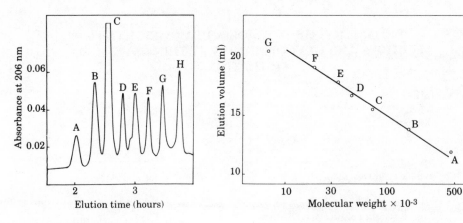

A: Apoferritin (480,000)
B: γ-globulin (160,000)
C: Human albumin (68,000)
D: Ovalbumin (45,000)

E: Carbonic anhydrase (31,000)
F: Myoglobin (17,000)
G: Aprotinin (6,000)
H: Tyrosine (180)

FIGURE 1.   Molecular weight determination by high-performance gel permeation chromatography. Column: UltroPac® TSK-G 3000 SW, 7.5 × 600 mm. Eluent: 0.1 M Na$_2$HPO$_4$/NaH$_2$PO$_4$, 0.1 M NaCl, pH 6.8. Flow rate: 100 μℓ/min. Temperature: 22°C. Detection: UV at 206 nm. Reproduced by permission of LKB-Produkter AB.

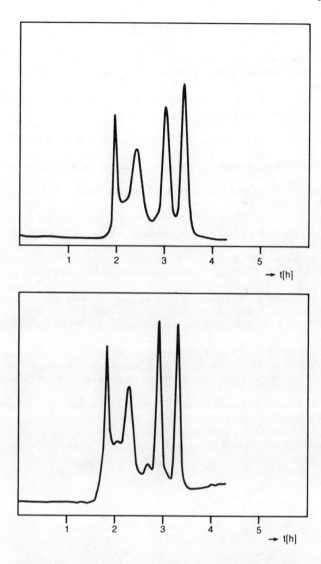

FIGURE 2.   Gel permeation chromatography using gels of different particle size. The separation of thyroglobulin, bovine serum albumin, and myoglobin using (A) Fractogel TSK HW-55 (F), particle size (moist with water), 0.032 to 0.063 mm and (B) Fractogel TS HW-55 (S), particle size (moist with water), 25 to 40 μm. Gel bed 640 × 26 mm. Eluent: Phosphate buffer, 1/15 mol/ℓ, pH 7: KCl, 0.2 mol/ℓ. Flow rate: 1.1 mℓ/min. Pressure (A) 0.8 bar; (B) 1.5 bar. Detection: UV. Thyroglobulin appears as two fractions and is eluted in the first two peaks. Reproduced by permission of E. Merck, Darmstadt, West Germany.

REFERENCES

1. **Porath, J. and Flodin, P.,** *Nature (London),* 183, 1657, 1959.
2. **Kato, Y., Komiya, K., Sasaki, H., and Hashimoto, T.,** *J. Chromatogr.,* 190, 297, 1980.
3. **Kato, Y., Komiya, K., Sawada, Y., Sasaki, H., and Hashimoto, T.,** *J. Chromatogr.,* 190, 305, 1980.
4. **Wehr, C. T., Cunico, R. L., Ott, G. S., and Shore, V. G.,** *Anal. Biochem.,* 125, 386, 1982.
5. **Rokushika, S., Ohkawa, T., and Hatano, H.,** *J. Chromatogr.,* 176, 456, 1979.
6. **Richter, W. O., Jacob, B., and Sohwandt, P.,** *Anal. Biochem.,* 133, 288, 1983.
7. **Shioya, Y., Yoshida, H., and Nakajima, T.,** *J. Chromatogr.,* 240, 341, 1982.
8. **Rivier, J. E.,** *J. Chromatogr.,* 202, 211, 1980.
9. **Lazure, C., Dennis, M., Rochement, J., Seidah, N. G., and Chretien, M.,** *Anal. Biochem.,* 125, 406, 1982.
10. **Swergold, G. D. and Rubin, C. S.,** *Anal. Biochem.,* 131, 295, 1983.
11. **Hefti, F.,** *Anal. Biochem.,* 121, 378, 1982.
12. **Montelaro, R. C., West, M., and Issel, C. J.,** *Anal. Biochem.,* 114, 398, 1981.
13. **Schneider, H.-J., Drexel, R., Feldmaier, G., Linzen, B., Lottspeich, F., and Henschen, A.,** *Hopppe-Seyler's Z. Physiol. Chem.,* 364, 1357, 1983.
14. **Kopaciewicz, W. and Regnier, F. E.,** *Anal. Biochem.,* 126, 8, 1982.
15. **Hansen, J. J., Greibrokk, T., Currie, B. L., Johansson, K. N.-G., and Folkers, K.,** *J. Chromatogr.,* 135, 155, 1977.
16. **Rutschmann, M., Kuehn, L., Dahlmann, B., and Reinauer, H.,** *Anal. Biochem.,* 124, 134, 1982.
17. **Jenik, R. A. and Porter, J. W.,** *Anal. Biochem.,* 111, 184, 1981.
18. **Kamp, R. M., Yao, Z.-J., Bosserhoff, A., and Wittmann-Liebold, B.,** *Hoppe-Seyler's Z. Physiol. Chem.,* 364, 1777, 1983.
19. **Roumeliotis, P. and Unger, K. K.,** *J. Chromatogr.,* 185, 445, 1979.
20. **Galpin, I. J., Kenner, G. W., Ohlsen, S. R., and Ramage, R.,** *J. Chromatogr.,* 106, 125, 1975.
21. **Galpin, I. J., Jackson, A. G., Kenner, G. W., Noble, P., and Ramage, R.,** *J. Chromatogr.,* 147, 424, 1978.
22. **Klaus, G. G. B., Nitecki, D. E., and Goodman, J. W.,** *Anal. Biochem.,* 45, 286, 1972.
23. **Heinz, F. and Prosch, W.,** *Anal. Biochem.,* 40, 327, 1971.
24. **Hung, C.-H., Strickland, D. K., and Hudson, B. G.,** *Anal. Biochem.,* 80, 91, 1972.
25. **Wasyl, Z., Luchter-Wasyl, E., and Bielanski, W., Jr.,** *Biochim. Biophys. Acta,* 285, 279, 1972.

Section II.III.

TECHNIQUES OF PEPTIDE MAPPING

INTRODUCTION

The techniques of peptide mapping have produced valuable information on the primary structure of proteins. In sequence studies, peptide maps are used to estimate the number of peptides and the completeness of digestion and to establish purification conditions. A major problem is the separation and detection of progressively smaller quantities of peptides, as many proteins of importance in molecular biology can only be obtained in microgram or lesser amounts. Thin-layer and HPLC procedures have both been used to obtain peptide maps; these are described in the present section.

THIN-LAYER TECHNIQUES

Selective Cleavage of Aspartyl-Prolyl Bonds

Sonderegger et al.[1] have described a technique for peptide mapping of protein bands obtained by polyacrylamide gel electrophoresis (PAGE). The method uses selective acid hydrolysis of aspartyl-prolyl bonds, which are found in proteins at an average frequency of 1/400 amino acid residues. PAGE in the presence of sodium dodecyl sulfate (SDS-gel electrophoresis) is performed in a slab gel apparatus. In a first step the proteins are run on a polyacrylamide (10%) gel and the protein bands generated are visualized by staining with 0.25% Coomassie brilliant blue G-250 in 30% methanol, 7% acetic acid for 5 min, and are destained with the same solvent for 5 min.

In an alternative method for visualizing the protein bands, the gel is immersed in 4 *M* sodium acetate for 10 to 30 min in order to precipitate SDS not bound to protein. During several minutes the protein bands can be observed as transparent cut-outs on an opaque background when the gel is illuminated obliquely from below. After visualization the band to be examined is cut out and the gel piece is subjected to *in situ* chemical cleavage. The gel piece is placed in a 10-mℓ test tube and equilibrated with 5 mℓ of 75% formic acid during 4 hr at room temperature. The formic acid solution is then sucked off and the gel piece covered with 2 mℓ highly liquid paraffin in order to prevent evaporation of formic acid and leakage of the protein and peptides from the gel piece. The gel piece is incubated at 37°C for 18 to 24 hr and the paraffin replaced by Tris buffers to remove the formic acid. After soaking in 10 mℓ of 250 m*M* Tris-chloride, 0.5% SDS, pH 9.0, for 2 × 1 hr at room temperature, the gel piece is kept in 125 m*M* Tris-chloride, 0.5% SDS, pH 6.8, 20% glycerol for a further 2 × 1 hr.

In the next step the gel piece is placed directly into the sample well of a second SDS-polyacrylamide gel, and the peptides are separated electrophoretically. After equilibration with 125 m*M* Tris-chloride, pH 6.8, 0.5% SS, the gel pieces are carefully placed into the sample wells and overlaid with 100 to 200 μℓ of 125 m*M* Tris-chloride, 0.5% SDS, pH 6.8, 20% glycerol, 0.001% bromophenol blue. Due to the rarity of aspartyl-prolyl bonds their selective hydrolysis yields usually relatively large peptides that are suitable for electrophoresis. On the other hand, many proteins do not contain aspartyl-prolyl linkages and are therefore not amenable to formic acid cleavage. The *in situ* acidic cleavage is highly reproducible. An advantage of the method is the independence of the chemical cleavage of the amount of substrate.

Selective Cleavage of Tryptophanyl Bonds

Lischwe and Ochs[2] have reported a fast and simple method of using *N*-chlorosuccinimide-urea to obtain partial peptide maps from proteins in gel slices, using standard conditions that apply to almost any protein. *N*-Chlorosuccinimide selectively cleaves the tryptophanyl bonds in proteins. SDS-PAGE is first performed to separate the proteins under study. The proteins are then visualized with 4 *M* sodium acetate or stained with 0.05% Coomassie blue R-250 in 40% methanol-10% acetic acid. Procedures for digestion and washing of the protein-containing gel slices are carried out in small beakers on a shaker at room temperature. The gel slices are washed first with 25 mℓ of water for 20 min with one change and then with 10 mℓ of urea-water-acetic acid, 1 g:1mℓ:1mℓ for 20 min with one change. Cleavage is performed using 0.015 *M N*-chlorosuccinimide in urea-water-acetic acid (5 mℓ) for 30 min. The slices are washed with water as above, followed by equilibration in 10% glycerol, 15% β-mercaptoethanol, 3% SDS, and 0.0625 *M* Tris-HCl, pH 6.8 (10 mℓ) for 1.5 hr with 3 changes and are then loaded on the resolving gel. After electrophoresis, the peptide bands are revealed by a silver-staining technique. The method is applicable to nanogram quantities of unlabeled or labeled proteins.

Analysis of Peptides Separated by Thin-Layer Procedures

Fishbein et al.[3] have described procedures which allow the determination of the amino acid composition and sequence of peptides separated by thin-layer methods. Using silica gel G thin layer plates, peptide-containing areas can be scraped off and hydrolyzed directly for amino acid analysis. Peptides may also be recovered from silica by elution and then subjected to sequence analysis. All solvents are distilled under appropriate conditions and stored under nitrogen at $-20°C$. Glassware is washed with deionized distilled water and baked at 600°C for 8 hr in a furnace. Pasteur pipettes are heat cleaned at 500°C for 8 hr. All glassware is stored in dust-proof containers until needed. Silica gel G thin-layer plates (250 μm Merck silica, 20 × 20 cm plates without fluorescent indicator) are used without activation. One to five nanomoles of, for example, a tryptic digest of a protein, are applied to the plates in spots of about 1 mm diameter. Chromatography is performed in chloroform-ammonium hydroxide (34%)-ethanol, 2:1:2, and the plates are dried for 2 hr at 60°C. Electrophoresis is then carried out perpendicular to the direction of chromatography at 900 V for 60 to 120 min at 8 to 10°C. After drying for 2 hr at 60°C, the plates are sprayed with 1% triethylamine in acetone and again with 0.01% fluorescamine in acetone.

To determine their amino acid composition, the peptides are scraped onto glassine paper using a thin layer chromatography (TLC) scraper which is rinsed in 15% formic acid after use. The silica powder is added to clean hydrolysis tubes, 5 × 50 mm or 10 × 70 mm. Using a Pasteur pipette, an aliquot of constant boiling HCl is added; this aliquot is 50 μℓ for the small tubes and 250 μℓ for the larger tubes. The tubes are gently heated to expel trapped gases, sealed *in vacuo,* and heated for 24 hr at 110°C. After heating, the tubes are centrifuged, opened, and recentrifuged for 3 min. The HCl from over the pelleted silica is removed with a Pasteur pipette and placed in a 2-mℓ conical plastic tube. Using another Pasteur pipette, 250 μℓ of HCl is added to the hydrolysis tube in order to wash the walls and resuspend the silica. The hydrolysis tube is recentrifuged for 5 min and the supernatant transferred to the tube containing the first HCl aliquot. The HCl is evaporated using a Speed-Vac concentrator centrifuge (Savant Institute, New York). The residue is then resuspended in 65 μℓ of citrate buffer, pH 2.2, and the solution is centrifuged to remove insoluble material. The amino acid composition may then be determined using an automatic analyzer. For tryptophan analysis, the peptide is hydrolyzed with 4 *N* methane sulfonic acid reagent.

Peptides to be sequenced are eluted from the silica, peptide spots being scraped from the plates onto separate pieces of glassine powder paper. The silica is transferred to a pipette plugged with one fourth of a glass fiber filter, the filter having been previously washed with deionized distilled water. A 250-$\mu\ell$ aliquot of elution solvent is added to the pipette and the silica powder is suspended for 15 min. The solvent is washed through the filter under nitrogen pressure and collected in a plastic centrifuge tube. The process is repeated 3 times. All the eluates are collected in one tube. The samples are dried in a centrifuge evaporator and stored *in vacuo* over P_2O_5 before sequencing. From a 1.6-cm^2 area of Merck silica gel (0.01 g) approximately 30 pmol of contaminating amino acids, primarily glycine and serine, were released on acid hydrolysis. On a weight basis this is about 10 times lower than that obtained with hydrolyzed cellulose. A rise of only about 20% in background contaminants was observed during the fingerprinting procedure.

When amino acids were hydrolyzed in the presence of silica, a recovery greater than 90% was found, there being no significant difference between fluorescamine treated and untreated samples. The lowest recoveries were found for tyrosine, phenylalanine, histidine, and lysine. When peptides separated by chromatography and electrophoresis on silica gel G were located by spraying with fluorescamine, scraped off the plates, and hydrolyzed, recoveries of 80% or more were obtained. This compares with less than 60% recovery from silica using elution with 6 N HCl. The amino-terminal residues of peptides located with fluorescamine were recovered quantitatively; however, for amino-tryptophanyl peptides recovery was 20%.

HPLC TECHNIQUES

A common procedure in peptide mapping involves subjecting an enzymatic digest of protein to paper chromatography in one dimension, followed by high-voltage electrophoresis in the second dimension. Techniques of this character, however, have disadvantages. Their use for preparative purposes has been hindered by the need for chromogenic or fluorogenic reagents which may irreversibly react with amino groups. Additionally, peptides purified on paper are only obtained in low yield. The sensitivity of peptide mapping can be reduced from the milligram to the microgram level by using fluorescent detection, radioiodination, or the incorporation of radioactively labeled amino acids into the protein. These methods, however, may be limited to the detection of only some of the peptides in a digestion mixture.

Hancock et al.[4] have described the use of hydrophilic ion-pairing reagents, such as phosphoric acid, in combination with reversed-phase columns which allows the rapid analysis of an underivatized digestion mixture; 44 pmol of acyl carrier protein for example was easily analyzed in this way. The use of phosphoric acid in the mobile phase has the advantage that detection can be carried out at 200 nm. Because of the significant end absorption of the amide bond at this wavelength observation of all the peptides is possible. At this low wavelength, however, commonly used solvents such as methanol or acetonitrile, unless highly purified, make gradient elution difficult. Isocratic elution, however, may be applied successfully, particularly when maximum sensitivity is necessary.

The use of three different concentrations of organic solvent usually allows peptides with a wide range of different polarities to be observed. The resolving power of the reversed-phase system is excellent. The extreme sensitivity of the method makes the actual identification of peaks difficult unless an equally rapid and sensitive system of amino acid analysis is available. The use of multiple wavelength detection can give further information on the nature of the peptide. In addition, the use of up to three different concentrations of organic solvent allows a comparison of the retention times, which will usually detect nonpeptide samples because of their anomalous behavior.

Isobe et al.[5] have developed an HPLC system for analytical peptide mapping and preparative peptide separation. A macroreticular cation-exchange resin of the styrene-divinylbenzene type (Hitachi-Gel 3013C) with a relatively small particle size (6 ± 1 μm) and a high cross-linkage (35%) is used. Elution is performed at 70°C with a linear gradient from water to 0.4 M ammonia containing 50% acetonitrile and 25% isopropanol adjusted to pH

6.2 with methanesulfonic acid, the eluted peptides being monitored at 210 nm. Excellent peak resolution together with mechanical and chemical stability was demonstrated. Analysis of tryptic digests of *S*-aminoethyl Bence-Jones proteins was achieved in less than 2 hr using nanomolar to micromolar quantities of sample. The isolated peptides could be subjected to amino acid analysis or sequencing directly or after desalting.

Takahashi et al.[6] have described an HPLC system for analytical and preparative separations of peptides using a macroreticular anion-exchange resin, Diaion CDR-10, with a particle size distribution of 5 to 7 μm. The peptide fraction from the ion-exchange column could be dried *in vacuo* without desalting and hydrolyzed with HCl for amino acid analysis.

Rubinstein et al.[7] have mapped tryptic digests of proteins using a highly sensitive automated fluorescamine monitoring system which allows measurement of peptides in the picomole range. The high resolving power of reversed-phase HPLC permits the analysis of highly complex peptide mixtures in relatively short times with excellent reproducibility. There was no alteration in the elution position of marker peptides even after repeated use over several months. The recoveries in many cases were better than 90%. Rubinstein et al.[7] consider the advantages of column chromatography for peptide mapping to be that it can be automated and is easier to quantify by measuring peak heights rather than the size of spots. Additionally, the resolved peptides are collected in tubes ready for further analysis without extraction procedures required in thin-layer or paper chromatography.

Oray et al.[8] have compared the two-dimensional thin-layer technique and HPLC for the separation of complex mixtures of peptides obtained by tryptic digestion of proteins with respect to reproducibility, recovery, and sensitivity. Peptides isolated by HPLC gave better agreement with the expected amino acid compositions than those isolated by thin-layer methods. Peptides isolated by the latter technique generally showed extra aspartic acid, threonine, serine, glutamic acid, glycine, and histidine. The level of background contamination observed with the thin-layer techniques was markedly reduced when the peptides were isolated by HPLC. All the peptides isolated by HPLC when analyzed for amino acids showed a high level of ammonia due to ammonium acetate buffer used in the chromatographic separation. Although one peptide isolated by the thin-layer method did not contain glutamic acid, glycine, and histidine, contaminating levels of these amino acids were found on analysis. The recovery of peptides by HPLC was uniformly superior to that from the two-dimensional thin-layer method.

Oray et al.[8] concluded that both techniques were applicable in structural studies of proteins available only in limited quantity. The thin-layer two-dimensional techniques have the advantage that they require much less sophisticated and expensive equipment. Moreover, the separation of small numbers of peptides is practically assured on a first attempt, without the need for establishing more complex gradients for peptide resolution by HPLC. On the other hand, HPLC is more reproducible than the thin-layer method. The HPLC method also allows the detection of peptides by, for example, the absorbance at 220 nm and does not require reaction of the peptides with fluorescamine. In general, the HPLC procedures are more satisfactory, particularly if further structural analysis such as amino acid composition and sequencing is to be performed.

REFERENCES

1. **Sonderegger, P., Jaussi, R., Gehring, H., Brunschweiler, K., and Christen, P.,** *Anal. Biochem.*, 122, 298, 1982.
2. **Lischwe, M. A. and Ochs, D.,** *Anal. Biochem.*, 127, 453, 1982.
3. **Fishbein, J. C., Place, A. R., Ropson, I. J., Powers, D. A., and Sofer, W.,** *Anal. Biochem.*, 108, 193, 1980.

4. **Hancock, W. S., Bishop, C. A., Prestidge, R. L., and Hearn, M. T. W.**, *Anal. Biochem.*, 89, 203, 1978.
5. **Isobe, T., Takayasu, T., Takai, N., and Okuyama, T.**, *Anal. Biochem.*, 122, 417, 1982.
6. **Takahashi, N., Isobe, T., Kasai, H., Seta, K., and Okuyama, T.**, *Anal. Biochem.*, 115, 181, 1981.
7. **Rubinstein, M., Chen-Kiang, S., Stein, S., and Udenfriend, S.**, *Anal. Biochem.*, 95, 117, 1979.
8. **Oray, B., Jahani, M., and Gracy, R. W.**, *Anal. Biochem.*, 125, 131, 1982.

Section II.IV.

THE PREDICTION OF PEPTIDE RETENTION TIMES IN HIGH-PERFORMANCE LIQUID CHROMATOGRAPHY (HPLC)

INTRODUCTION

The possibilities for separating peptides were markedly improved on the introduction of reversed-phase HPLC. This technique depends on hydrophobic interactions between a hydrocarbonaceous column and the peptides to be separated — the more hydrophilic compounds being retained more strongly by the column. To elute strongly retained compounds, the mobile phase must contain a high percentage of organic solvent. Choice of the optimum mobile phase and chromatographic conditions for given peptides can be found by trial and error, after examining the ratio of hydrophobic and hydrophilic amino acids in the peptide. Prior knowledge of the retention time of a given peptide would simplify the selection of suitable chromatographic conditions. For mixtures of closely related peptides obtained by synthesis, the ability to predict from the retention times the likely sequence variants, such as desired product, deletions, transaminations, etc., would be particularly useful.

Several attempts have been made to predict peptide retention times, with a considerable degree of success. Many examples can now be cited where the elution order of peptides from a chromatographic column agrees with the predicted order. These predictions have been made in a number of different ways.

HYDROPHOBICITY CONSTANTS

Molnar and Horvath,[1] employing increasing concentrations of acetonitrile to elute peptides from a column of LiChrosorb® RP-18, found that the adjusted retention times of phenylalanine and its oligomers were a linear function of the number of residues. A good correlation was obtained between the retention times and the sum of the side chain "hydrophobicity numbers" of Rekker. Rekker's[2] constants are based on the partition coefficients of free amino acids between water and octanol. O'Hare and Nice[3] similarly predicted the retention order of peptides with less than 15 residues by summing the contributions of all individual hydrophobic residues in terms of their Rekker constants with reasonable accuracy, but the procedure was empirically improved by restricting the calculation to the 5 most strongly hydrophobic residues.

One of the most important attempts to predict the retention times of peptides in chromatography was made by Meek.[4] Since under either isocratic or gradient elution conditions overall hydrophobicity is the principal factor determining retention, it should be possible to predict the elution order for peptides using suitable indexes which take into account the effective contribution of each amino acid side chain to the retention process. At low pH values such as 2.5, glycine and its oligomers show little retention on alkylsilicas with aqueous or aqueous-organic eluents, suggesting that the peptide chain as such makes only a small contribution to the retention process. By treating peptides as oligoglycine derivatives in which the side chain components are introduced in a defined fashion, the net contribution of the side chains to the retention behavior of a particular peptide can be assessed.

Retention times of 25 peptides were determined on a Bio-Rad® ODS column with a gradient from 0.1 M NaClO$_4$ to 60% acetonitrile-0.1 M NaClO$_4$. Retention coefficients for the amino acids were computed using a calculator programed to change the retention coefficients of all the amino acids sequentially to obtain a maximum correlation between actual and predicted retention times. Amino acids with aromatic or aliphatic side chains have a marked positive contribution to retention which changes relatively little with pH. On the other hand, residues with acidic side chains have a marked negative contribution to retention which increases in magnitude as ionization increases. Basic and neutral residues have little effect on retention. A high correlation was seen when the actual retention times of the peptides studies were plotted against the times predicted by summing the appropriate retention coefficients for each peptide.

Retention times for peptides obtained with the Bio-Rad® column were similar to those found with columns from Waters Associates, 10-μm particle size, LiChrosorb® 5 μm, and DuPont, 5 μm. Predicted retention times for several peptides of biological interest agreed reasonably well with values reported in the literature. For all peptides up to 20 amino acid residues long, the average error was 4 min. Meek and Rossetti[5] later chromatographed 100 peptides on a Bio-Rad® ODS column with a linear gradient of acetonitrile containing 0.1% H_3PO_4, starting with 0.1 M $NaClO_4$-0.1% H_3PO_4 or 0.1 M Na_2HPO_4-0.1% H_3PO_4. Based on these results a revised list of retention coefficients was published. The data obtained with mobile phases containing perchlorate and phosphate were generally similar. Basic groups such as lysine, arginine, histidine, and amino terminal groups were found, however, to have a more negative effect on retention with the phosphate mobile phase.

A good correlation exists between the retention times of dipeptides related to the COOH-terminus of cysteinyl-alanyl-glycyl-tyrosine and the average side chain hydrophobicity functions.[6] The relative retention values of three synthetic peptides of the angiotensin type also agreed with relative hydrophobicity functions. Met-,(D-Ala²)-Met- and Leu-enkephalins eluted from a MicroPak MCH-10 column in the order of increasing Rekker sums.[7] Browne et al.[8] chromatographed a series of 25 peptides from the rat neurointermediary lobe using identical acetonitrile gradients on the same μBondapak® C_{18} column, in one case with 0.1% TFA as the hydrophobic counter ion throughout, and in the other with 0.13% heptafluorobutyric acid (HFBA). The retention coefficients had to change on proceeding from the TFA system to the HFBA system in order to account for the changes in the relative elution order of the peptides. For both sets of data a very good correlation was obtained between the observed and predicted retention times. Browne et al.[8] expressed the retention value as an acetonitrile concentration, rather than an elution time, so that the data would be more widely useful. The mean difference between the observed and predicted elution positions was 0.28% acetonitrile or 0.9 min for the TFA system and 0.48% acetonitrile or 1.6 min for the HFBA system. It is difficult to directly compare the retention coefficients of Browne et al. with those of Meek and Meek and Rossetti, partly because the former are calculated as retention times and the latter are calculated in terms of percentage acetonitrile.

Hearn and Grego[9] determined apparent capacity factors for several peptides on a μBondapak® C_{18} column using a linear acetonitrile gradient and compared them with values computed by linear regression for two sets of hydrophobic retention indices derived by Rekker and Meek. Both set of indexes gave reasonable correlation between the observed and predicted elution order, although some discrepancies were noted.

Wilson et al.[10] have studied the chromatographic separation of peptides on columns of LiChrosorb® RP-18 (100 Å pore size) and Aquapore RP 300 (300 Å pore size) at pH 2.2 and 7.0. At the lower pH of 2.2 the peptides eluted as a function of their increasing polarity when the values of Meek and Rossetti were used to calculate peptide hydrophobicity. This was to be expected since the buffer systems used by Wilson et al. and Meek and Rossetti were essentially the same. When the amino acid hydrophobicity values determined by Wilson et al. in a different buffer system, e.g., pyridine-formate-acetate using l-propanol as the organic eluent were used, the estimation of peptide elution order was less satisfactory.

Gazdag and Szepesi[11] examined the applicability of Meek's prediction method for isocratic elution with different eluents. When the predicted capacity factors for ACTH derivatives were plotted against the measured capacity factors using two different eluent systems, a roughly linear relationship was obtained. Predicted capacity ratios were calculated for a number of other polypeptides and compared with measured values. The good correlation between estimated and measured elution order indicated to these authors that data calculated by the Meek method can be transferred to different eluent systems. Calculated data can thus be used to select suitable chromatographic systems for the separation of polypeptides.

DEVIATIONS FROM PREDICTED BEHAVIOR

Although in most chromatograms the order of elution of closely related peptides is explicable in terms of the hydrophobicity of the amino acid residues, in some instances this is not the case.[12] For instance, (Gly^1)-ACTH-(1-18)-NH_2 is eluted faster than $(\beta$-$Ala^1)$-ACTH-(1-18)-NH_2, and this elution order is reversed with ACTH-(1-18)-NH_2 and $(\beta$-$Ala^{10})$-ACTH(1-18)-NH_2. The replacement of one amino acid residue by its enantiomer may alter the conformation of the molecule, thus making a large difference in the partition coefficients of epimers, such as (Ala^{10})-ACTH-(1-18)-NH_2 and $(D$-$Ala^{10})$-ACTH-(1-18)-NH_2, which are separable. The pair of peptides, (Lys^3, Ser^{11})-ACTH-(1-18)-NH_2 and ACTH-(1-18)-NH_2, can also be separated chromatographically. These peptides have the same amino acid composition but different sequences. They probably have different conformations and may therefore have different affinities with the stationary phase.

When the Rekker values for the hydrophobicity of peptides were plotted against the organic modifier concentration required for elution from a reversed-phase C_{18}-diol column with a k' of 2 there was reasonably good agreement.[7] For larger peptides with more than 30 residues, the correlation breaks down, presumably due to the effects of secondary and tertiary structure that tend to remove hydrophobic residues from the interactive peptide surfaces.

Using data obtained with 57 different peptides chromatographed on the same μBondapak® C_{18} column, Su et al.[13] examined assumptions explicit to a solvophobic model using two different methods of numerical analysis, namely multiple regression via an iterative forcing approach and by a mathematical routine for solving linear equations. A high degree of correlation was obtained between the observed and predicted values of k', using both methods of numerical analysis. However, the fit of the data points to a straight line fell short of the expectations based solely on a solvophobic model. Su et al. conclude that several assumptions made in the calculations may be responsible for the observed divergence from the predicted relationships. For instance, no correlation terms were included in the calculations to accommodate differences in specific electrostatic and hydrogen bonding interactions which arise during the distribution of ionized peptides between polar mobile phases and hydrocarbonaceous silicas.

The heterogeneity of the stationary phase surface of alkyl-bonded silicas has been examined in several studies. Even with well "capped" alkylsilicas of high carbon coverage, unprotected peptides show dual retention behavior exemplified by the concave binodal dependence of ln k' on the volume fraction of water in the mobile phase. This dual retention behavior is attributed to solvophobic-silophilic interactions. In addition, specific solvation effects can lead to individual selectivity divergencies for some peptides. It is unlikely that these electrostatic hydrogen bonding, or solvation components in the sorption process remain constant for different peptides. With larger peptides, where secondary and tertiary structural features are likely to be important, greater deviations from an additive effect may be expected.

When Sasagawa et al.[14] reinvestigated the chromatography of over 100 peptides on a μBondapak® C_{18} column with aqueous TFA as the mobile phase, the retention constants they derived were of similar magnitude to those reported previously.[14] Both Meek and O'Hare and Nice reported a linear relationship between the retention times (t_r values) of small peptides and the sum of their amino acid retention constants. The study of Sasagawa et al., however, clearly showed an exponential relationship. This apparent discrepancy may simply be due

Table 1
AMINO ACID HYDROPHOBICITY CONSTANTS

Amino acid	Constants according to reference								
	2	4[a]	4[b]	5[c]	5[d]	15	8[e]	8[f]	14
Alanine	0.53	0.5	−0.1	1.1	1.0	−0.3	7.3	3.9	0.13
Arginine	−0.82	0.8	−4.5	−0.4	−2.0	−1.1	−3.6	3.2	0.26
Asparagine	−1.05	0.8	−1.6	−4.2	−3.0	−0.2	−5.7	−2.8	−0.45
Aspartic acid	−0.02	−8.2	−2.8	−1.6	−0.5	−1.4	−2.9	−2.8	0.10
Cysteine	0.93								
Cystine	1.11	−6.8	−2.2	7.1	4.6	6.3	−9.2	−14.3	
Glutamic acid	−0.07	−16.9	−7.5	0.7	1.1	0	−7.1	−7.5	0.27
Glutamine	−1.09	−4.8	−2.5	−2.9	−2.0	−0.2	−0.3	1.8	0.36
Glycine	0	0	−0.5	−0.2	0.2	1.2	−1.2	−2.3	0.22
Histidine	−0.23	−3.5	0.8	−0.7	−2.2	−1.3	−2.1	2.0	0.34
Isoleucine	1.99	13.9	11.8	8.5	7.0	4.3	6.6	11.0	1.38
Leucine	1.99	8.8	10.0	11.0	9.6	6.6	20.0	15.0	1.34
Lysine	0.52	0.1	−3.2	−1.9	−3.0	−3.6	−3.7	−2.5	0.05
Methionine	1.08	4.8	7.1	5.4	4.0	2.5	5.6	4.1	0.85
Phenylalanine	2.24	13.2	13.9	13.4	12.6	7.5	19.2	14.7	1.71
Proline	1.01	6.1	8.0	4.4	3.1	2.2	5.1	5.6	0.48
Serine	−0.56	1.2	−3.7	−3.2	−2.9	−0.6	−4.1	−3.5	0.18
Threonine	−0.26	2.7	1.5	−1.7	−0.6	−2.2	0.8	1.1	0.12
Tryptophan	2.31	14.9	18.1	17.1	15.1	7.9	16.3	17.8	2.34
Tyrosine	1.70	6.1	8.2	7.4	6.7	7.1	5.9	3.8	1.23
Valine	1.46	2.7	3.3	5.9	4.6	5.9	3.5	2.1	0.38

[a] pH 7.4.
[b] pH 2.1.
[c] NaClO₄ system.
[d] NaH₂PO₄ system.
[e] TFA system.
[f] HFBA system.

to a difference in the range of peptides investigated. Meek only studied small peptides less than 29 residues long and obtained data approximating a linear relation. The data of Sasagawa et al. also gave a linear relation during the first 20 min of elution. The nonlinear equation accurately described the dependence of retention time on amino acid composition over a wider range of peptide size.

Wilson et al.[15] observed that peptides eluted from a RP-8 or RP-18 column as expected, i.e., dependent on their respective hydrophobicities. Deviation of peptide chromatographic behavior was essentially independent of hydrophobicity and of propan-1-ol concentration. Peptides longer than about 18 residues tended to elute more rapidly than predicted from hydrophobic considerations alone. Low correlations which were observed between amino acid hydrophobicity constants and peptide retention times were independent of the source of the constants. Predicting exact peptide retention times, at least for the chromatographic system used by Wilson et al., was difficult.

Nice et al.[16] compared the elution order of several proteins with their net hydrophobicity and found that, as with polypeptides, a general correlation exists but that numerous individual anomalies occur. These are probably due to conformational differences, but net hydrophobicity does give a general guide as the probability of an individual protein being successfully chromatographed.

Grego et al.[17] have commented on the difficulties involved in predicting peptide retention values for a given chromatographic system using retention coefficients derived from a different system (see Table 1). Comparisons between observed and predicted retention for elution systems of different selectivities, e.g., for phosphate and TFA-based mobile phases shows reasonable to good correlation, but correlation is less reliable as the selectivity of the two systems diverge, e.g., in comparing phosphate with pyridine formate systems. Comparisons between buffer effects can only be made reliably when the same stationary phase is employed throughout, the organic solvent modifier is common to all elution systems, and the linear gradient compositional limits are identical.

In order to permit conversion of data between two chromatographic systems, comparability in selectivity must exist. It is possible to assess this requirement for various peptides from plots of observed retention times found with different eluents on the same stationary phase. The form which the interdependency can take for gradient systems of different buffer composition, but with a fixed rate of change of the common organic modifier, has been presented by Grego et al. and used to correlate data in different elution systems. Even when the essential criteria of selectivity comparability are satisfied, however, interconversion of data from one elution system to another still requires caution. Thus, Sasagawa et al.[18] demonstrated that a column of porous spherical polystyrene-divinylbenzene copolymer (PRP-1) was useful for the reversed-phase resolution of a wide range of peptides. Retention constants for amino acid residues were computed from the data they obtained on chromatographing 47 peptides on this column. These constants differed significantly from the corresponding values for an octadecyl silica column. This allows the further resolution of difficult mixtures by successive combinations of columns.

CONCLUSIONS

From the above discussion it will be seen that reasonable prediction of the elution order of peptides of different composition is possible. Browne et al.[19] have given a striking demonstration of the use of amino acid retention coefficients to predict elution times. These authors isolated γ_3-melanotropin (γ_3-MSH) from the neurointermediary lobe of the rat pituitary by HPLC. From the rat γ_3-MSH sequence predicted by DNA sequencing, a retention coefficient of 42 was calculated, which under their experimental conditions gave a retention time between 9 and 12 min. Material eluting in a UV peak at 10 min was rechromatographed to give a single symmetrical peak. Amino acid analysis of material isolated from this peak gave the result expected from the reported DNA sequence, except that there was an unexpected lysine residue. The analysis also indicated that an amino sugar peak was present.

When the retention behavior of positional or stereoisomers is predicted less reliable results are obtained. Pietrzyk et al.[20] have shown dramatic differences in retention for position isomers of small chain peptides when lipophilic groups are moved away from ionic groups. For short-chain peptides, however, peptide shape is probably not a significant factor in determining retention. For larger polypeptides secondary and tertiary structural effects reduce the number of exposed hydrophilic residues and the accuracy of prediction is again reduced. With additional refinements the assignment of sequential features with a specific peptide may be possible.

ACKNOWLEDGMENT

I am indebted to Dr. J. L. Meek for help and advice during the preparation of this chapter.

REFERENCES

1. **Molnar, I. and Horvath, C.,** *J. Chromatogr.,* 142, 623, 1977.
2. **Rekker, R. F.,** *The Hydrophobic Fragmental Constant,* Elsevier, New York, 1977.
3. **O'Hare, M. J. and Nice, E. C.,** *J. Chromatogr.,* 171, 209, 1979.
4. **Meek, J. L.,** *Proc. Natl. Acad. Sci. U.S.A.,* 77, 1632, 1980.
5. **Meek, J. L. and Rossetti, Z. L.,** *J. Chromatogr.,* 211, 15, 1981.
6. **Hearn, M. T. W., Bishop, C. A., Hancock, W. S., Harding, D. R. K., and Reynolds, G. D.,** *J. Liq. Chromatogr.,* 2, 1, 1979.
7. **Wehr, C. T., Correia, L., and Abbott, S. R.,** *J. Chromatogr. Sci.,* 20, 114, 1982.
8. **Browne, C. A., Bennett, H. P. J., and Solomon, S.,** *Anal. Biochem.,* 124, 201, 1982.
9. **Hearn, M. T. W. and Grego, B.,** *J. Chromatogr.,* 203, 349, 1981.
10. **Wilson, K. J., Van Wieringen, E., Klauser, S., Berchtold, M., and Hughes, G. J.,** *J. Chromatogr.,* 241, 407, 1982.
11. **Gazdag, M. and Szepesi, G.,** *J. Chromatogr.,* 218, 603, 1981.
12. **Terabe, S., Konaka, R., and Inouye, K.,** *J. Chromatogr.,* 172, 163, 1979.
13. **Su, S.-J., Grego, B., Niven, B., and Hearn, M. T. W.,** *J. Liq. Chromatogr.,* 4, 1745, 1981.
14. **Sasagawa, T., Okuyama, T., and Teller, D. C.,** *J. Chromatogr.,* 240, 329, 1982.
15. **Wilson, K. J., Honegger, A., Stotzel, R. P., and Hughes, G. J.,** *Biochem. J.,* 199, 31, 1981.
16. **Nice, E. C., Capp, M. W., Cooke, N., and O'Hare, M. J.,** *J. Chromatogr.,* 218, 569, 1981.
17. **Grego, B., Lambrou, F., and Hearn, T. W.,** *J. Chromatogr.,* 266, 89, 1983.
18. **Sasagawa, T., Ericsson, L. H., Teller, D. C., Titani, K., and Walsh, K. A.,** *J. Chromatogr.,* 307, 29, 1984.
19. **Browne, C. A., Bennett, H. P. J., and Solomon, S.,** *Biochem. Biophys. Res. Commun.,* 100, 336, 1981.
20. **Pietrzyk, D. J., Smith, R. L., and Cahill, W. R., Jr.,** *J. Liq. Chromatogr.,* 6, 1645, 1983.

Section II.V.

FLUORESCENCE TECHNIQUES FOR THE DETECTION OF PEPTIDES

INTRODUCTION

To analyze an unknown peptide mixture, peptide identity must be established by sequence analysis. With prior knowledge about the peptide mixture, amino acid analysis or NH_2-terminal identification might suffice. However, all these techniques are off-line and increase the time and expense of the analysis. On-line methods available for detecting key residues directly as peptides are eluted from a column. The determination of absorbance ratios is a simple technique of identification of peptides containing aromatic residues. The chromatogram is monitored at a wavelength where all peptides absorb, such as 220 nm and also at 254 nm, where only peptides with aromatic amino acids are detected. Detection at 254 nm has the disadvantages that sensitivity is much poorer than at lower wavelengths, and it is difficult to establish the identity of the aromatic amino acid. This peak identification could be difficult if 2 or more closely related peptides contain a residue absorbing at 254 nm.

NATIVE FLUORESCENCE

Schlabach and Wehr[1] have used the native fluorescence of tyrosine and tryptophan to detect peptides containing these amino acids. Both tyrosine and tryptophan fluoresce strongly, so detection limits should be lower than when using absorbance measurements at 254 nm. Since tryptophan fluoresces with greater intensity and downfield from tyrosine, it should be possible to select conditions where tryptophan yields peaks of greater intensity and thus distinguish between tryptophan and tyrosine peptides. The sensitivities of detection for tryptophan and tyrosine are as low as 3 ng/mℓ for the former and 5 ng/mℓ for the latter. Schlabach and Wehr chose excitation at 220 nm to provide the greatest sensitivity. By collecting emission at 330 nm, where both tyrosine and tryptophan emission overlap, fluorescence detection should be selective for peptides with these two groups (Figures 1 and 2).

Separation of a peptide mixture showed that the fluorescence chromatogram resembled that observed at 254 nm with some notable exceptions. A peptide containing phenylalanine was not seen in the fluorescence chromatogram. The baseline of the fluorescence profile was flatter and less noisy and tryptophan peptides constituted the predominant peaks. Fluorescence detection was more than 6 times more sensitive than 254 nm detection for peptides containing tryptophan. The ratio of absorbance peaks at 215 nm to fluorescence peaks differed for peptides with tyrosine from those with tryptophan. There was also appreciable variation in these ratios among peptides containing tyrosine only. There was a similar scattering in this ratio for peptides with only one tryptophan residue. Fluorescence-to-absorbance ratios determined from the respective peak heights were uniformly higher for peptides containing tryptophan. The group average for peptides containing a single tyrosine group was 195, while the average for peptides with one tryptophan residue was 707. This ratio should thus be useful in distinguishing peptides with a single tyrosine residue from those with a single tryptophan.

FLUOROGENIC REAGENTS

Detection in Column Eluates

Peptides lacking aromatic residues can only be detected by absorbance measurements at

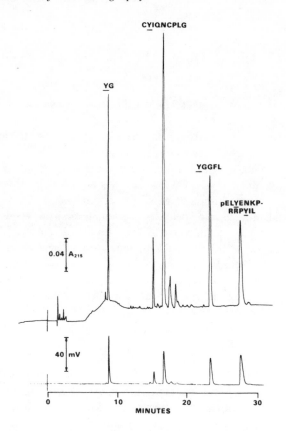

FIGURE 1.   Fluorescence and absorbance detection of peptides containing tyrosine. The upper chromatogram was recorded at 215 nm and fluorescence in the lower. The sample contained tyrosylglycine, oxytocin, Leu-enkephalin, and neurotensin in order of their elution. Oxytocin has a C-terminal amide group. The peptides were chromatographed on a MicroPak® MCH-5 column, 15 × 0.4 cm. Peptide elution began with 0.05% TFA in distilled water. The second solvent was 90% acetonitrile and distilled water containing the same amount of TFA. The gradient ramped linearly to 20% second solvent 5 min after commencement. During the next 25 min the gradient was linearly incremented to 40% second solvent. On the last 10 min 60% second solvent was reached. The flow rate was 0.8 mℓ/min. (From Schlabach, T. D. and Wehr, T. C., *Anal. Biochem.*, 127, 222, 1982. Reproduced by permission of Academic Press, Inc.)

low wavelengths where amide groups begin to absorb. These peptides may be detected with a postcolumn reaction. Ninhydrin has been commonly used in such reactions. In recent years two fluorogenic reagents, which are more sensitive than ninhydrin and are simple in use, have been used for the detection of peptides. These reagents are fluorescamine and *o*-phthalaldehyde.

Creaser and Hughes[2] have described a relatively simple system for the analysis and separation of peptides in the range of 5 to 10 μmol. The system comprises a gradient-generating device which pumps volatile pyridine buffers to a column of cation-exchange resin (Beckman PA 35). Eluate from the column is fed through a proportioning pump to a fluorocolorimeter, the output from which is displayed on a recorder. For analytical runs the eluate is mixed with *o*-phthalaldehyde in borate buffer containing Brij and 2-mercaptoethanol before its passage into the detector. For preparative work the eluate is split, one part reacting

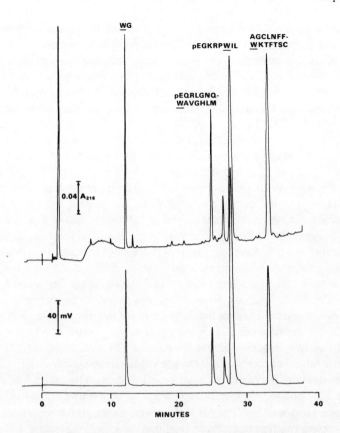

FIGURE 2.   Fluorescence and absorbance detection of peptides containing tryptophan. The sample contained tryptophylglycine, bombesin (an amide), xenopsin, and somatostatin in order of their elution. The chromatographic and detector conditions were as described for Figure 1. (From Schlabach, T. D. and Wehr, T. C., *Anal. Biochem.*, 127, 222, 1982. Reproduced by permission of Academic Press, Inc.)

with *o*-phthalaldehyde and the other being collected. The sensitivity of the *o*-phthalaldehyde reaction and its detection by the fluorocolorimeter were very much greater than necessary.

In early experiments fluorescamine was used as the fluorophore. Its use was discontinued because this compound is soluble in acetone but not in aqueous buffers. Fluorescamine, therefore, required the use of a third stream in the flow system, and it was essential to use silicone tubing, which had a very short life, to pump acetone solutions. In addition, fluorescamine gave less yield of fluorescence. *o*-Phthalaldehyde does not react appreciably with the amino terminus of peptides, but does react with the ε-amino group of lysine residues in peptides. Peptides containing lysine produce more than 50 times the fluorescence of those lacking lysine.[1] Although lysine is important in peptide fluorescence using *o*-phthalaldehyde, it is by far the least reactive of the naturally occurring primary amino acids when these are tested individually.[3] The reaction of the ε-amino group of lysine with *o*-phthalaldehyde, however, is not unique since peptides purified by paper chromatography and stained with ninhydrin were found to contain very little of both the NH$_2$-terminal amino acid and COOH-terminal lysine. This is presumably the result of reaction of ninhydrin with the ε-amino group of lysine residues.

The wide differences in fluorescence produced by peptides indicated to Joys and Kim[3] that the *o*-phthalaldehyde reaction is far from ideal for the detection of peptides. The fluorescence obtained with *o*-phthalaldehyde and lysine-containing peptides was linear with peptide concentration within experimental error. The lack of a linear relationship between fluorescence and peptide concentration with peptides that do not contain lysine probably indicates that the fluorescence produced is nonspecific and due to background contamination. Fluorescamine showed good fluorescence with all the peptides tested and was considered to be a superior reagent for the analysis of small amounts of protein digests.[3]

Frei et al.[4] have used Fluram (fluorescamine) for the postcolumn fluorescence derivatization of the pharmaceutically important nonapeptides oxytocin, lysine[8]-vasopressin, and ornipressin following their chromatographic separation. An extensive investigation of the reaction of these nonapeptides with fluorescamine was undertaken. Fluorescamine reacts with free amino groups, including the amino terminus of peptides to yield a fluorescent product. The fluorescence of fluorescamine derivatives is greatly dependent on the type of organic solvent used in the mobile phase of a chromatographic system or in the reagent solution. A decrease in net fluorescence (sample minus blank) was observed with increasing proportions of organic solvent. Acetonitrile gave the best net fluorescence despite having the largest increase in blank fluorescence. This result was fortunate, as acetonitrile is a very useful solvent for reversed-phase separations of these peptides and for the reagent solution.

The effect of pH on the reaction of oxytocin with fluorescamine was studied in the system water-acetonitrile, 80:20 v/v, a system which was found to be the most promising for chromatographic purposes. These optimal pH range for the reaction was 6 to 8. Frei et al.[4] concluded that one should work at as low an acetonitrile content as possible in the mobile phase and add as little reagent solution as possible. In a study of the effect of reagent concentration, solutions of various concentrations of fluorescamine in acetonitrile were added to aqueous solutions of oxytocin buffered to pH 7 or 8. The optimal concentration of fluorescamine appeared to be about 20 mg/100 mℓ of acetonitrile at both pH 7 and 8. The net fluorescence yield was lower at pH 8, however, owing to the higher blank. The time needed for complete reaction under these conditions at room temperature was about 50 sec.

Mendez and Gavilanes[5] have used fluorogenic reagents to detect peptides in fractions collected during column chromatography. Although reaction with fluorescamine yields more fluorescence with intact peptides than does reaction with o-phthalaldehyde, after alkaline hydrolysis o-phthalaldehyde yields approximately twice as much fluorescence as fluorescamine. In the opinion of these authors, peptides collected by column chromatography should be hydrolyzed with alkali prior to fluorogenic assay, since hydrolysis greatly increases the number of primary amino groups available for reaction. Furthermore, peptides having proline, hydroxyproline, or blocked amino groups at the NH_2-terminal yield no fluorescent products without hydrolysis. The fluorescence yield of dipeptides with o-phthalaldehyde is 3 to 20 times lower than that of amino acids.[6] Extra nitrogen atoms in the fluorescent chromophore seem to lower the fluorescence yield; however, arginine is an exception to this rule.

When using reagents such as fluorescamine and o-phthalaldehyde, the fluorescence of the blank is the chief factor limiting sensitivity. Precautions must be taken to avoid solvent contamination by amines and other reactive compounds. Mendez and Gavilanes[5] used a water purification system producing water with a resistance greater than 4×10^5 ohms. Their distilled water was filtered to eliminate microorganisms, twice deionized, and filtered to eliminate organic compounds. They found, however, that distilled or deionized water from several sources had at times introduced high blanks with o-phthalaldehyde but not with fluorescamine. Borate buffers prepared from boric acid obtained from several commercial sources also gave high fluorescence blanks with o-phthalaldehyde, but not with fluorescamine. Ammonium ions, which give poor fluorescence with fluorescamine, was possibly the principal contaminating species.

A comparison of the ease of use and sensitivity of UV and fluorescence detection systems has been made by Wilson et al.[7] Considerable sample dilution occurred with a fluorescamine-detection system which was not observed when the absorbance of the effluent was directly measured. For this reason the ''fine'' detail of their acetonitrile- or propan-2-ol-NaClO$_4$ systems were more obvious using UV detection. Using fluorescence detection the necessity of employing additional pumps to deliver an *o*-phthalaldehyde-buffer solution and fluorescamine in acetone can cause difficulties due to pump maintenance and performance. The sensitivity of both systems is theoretically high enough to provide for detection at the picomolar level, but both depend on the purity of solvents used. Fluorescence detection requires the removal of primary amine impurities, which is readily achieved by distillation over ninhydrin. UV absorbance detection, however, necessitates the removal of virtually all impurities, which is not easy for solvents such as acetonitrile. Practically speaking, the sensitivity of fluorescence detection is five - to tenfold greater than that of absorbance detection. Additionally, samples that have been through several manipulations acquire extraneous material that also has UV absorbance. In these cases fractions containing peptides are far more readily determined using fluorescence, which virtually measures only primary amines.

Detection on Paper and Thin-Layer Chromatograms

Lindeberg[8] has used *o*-phthalaldehyde to detect amino acids and peptides on thin-layer chromatograms.[8] Thin-layer chromatogram plates are sprayed generously with a solution of 0.1% *o*-phthalaldehyde and 0.1% 2-mercaptoethanol in acetone followed 5 min later by 1% triethylamine in acetone. After 10 min the plates are viewed under a long wave (350 nm) UV lamp. The spray reagent is stable for several days when kept at room temperature in a closed bottle protected from light, but deteriorates rapidly on adding base (1% triethylamine). The procedure allows the detection of as little as 500 to 100 pmol of many amino acids. In most cases the sensitivity is 2 to 5 times higher on silica gel, but aromatic amino acids, and especially cysteine, are more readily detected on cellulose. There is a striking difference in stability on the two sorbents. On silica gel the spots decay within a few minutes at room temperature, while on cellulose they are essentially resistant even to heating.

Amino acids substituted with labile N-protecting groups, e.g., *tert*-butyloxycarbonyl, 2-phenylisopropyloxycarbonyl, and *p*-methoxypenzyloxycarbonyl, give no reaction when the plate is dried at room temperature but can be detected at the 500 pmol level when heated at 100°C for 2 hr before spraying. Decreasing the concentration of *o*-phthalaldehyde in the reagent to 0.01% produces a slight decrease in sensitivity while an increase above 0.1% has no effect. The optimal concentration of 2-mercaptoethanol is one to ten times that of the aldehyde. The composition of the triethylamine spray is not critical but high concentrations of this reagent result in increased background fluorescence. The best results are obtained when the base is applied immediately after the phthalaldehyde-2-mercaptoethanol, or is mixed with it immediately before spraying. The stability is improved by substitution of triethanolamine for triethylamine, with somewhat reduced sensitivity. Lindeberg concluded that the *o*-phthalaldehyde reagent is comparable to fluorescamine in sensitivity and convenience and has the additional advantage of being less expensive.

Mendez and Lai[9] have described the detection of peptides on paper with fluorescamine. After electrophoresis or chromatography the papers are dried in air and treated as follows: (1) heated at 55°C for at least 1 hr, (2) washed with acetone, (3) wet with 1% v/v triethylamine in acetone and allowed to dry for 5 min at room temperature, (4) wet with a solution of 10 mg of fluorescamine in 100 mℓ of acetone and allowed to dry for 5 min, and (5) washed with acetone and dried. The fluorescent spots are viewed under a long wave (336 nm) UV lamp.

For amino acid analysis the fluorescent peptides are eluted from the paper with 0.2 *M* NH$_4$OH and evaporated to dryness on a rotary evaporator. Hydrolysis is carried out at 110°C for 20 hr with 5.7 *N* HCl in evacuated and sealed tubes.

The lowest concentration of fluorescamine solution required for optimal visualization is 10 mg%. The contrast is greatly improved by washing the paper with acetone 5 min after application of fluorescamine. As little as 76 pmol of a peptide obtained from rabbit muscle aldolase was easily detected. On hydrolysis of peptides, all amino acid residues including the NH_2-terminal residue are recovered quantitatively. The sensitivity of detection of peptides with fluorescamine is 4 to 6 times higher than with the ninhydrin staining method. Mendez and Lai suggested that the reaction of fluorescamine with peptides and amino acids on paper proceeds differently from the reaction in aqueous medium, stopping at an intermediate step to yield a fluorescent compound that can revert to the free amino acid on hydrolysis.

On paper chromatograms amino acids are more readily visualized at the 500 pmol level with *o*-phthalaldehyde than with fluorescamine, while peptides are more easily recognized with fluorescamine. The fluorescent spots produced by fluorescamine remain visible for several weeks, while the spots produced by *o*-phthalaldehyde disappear after several hours. Fluorescamine shows greater sensitivity for the detection of peptides on thin-layer chromatograms.

Fujiki and Zurek[10] have used fluorescamine for the detection of tryptic peptides on thin-layer chromatograms. The visualized peptides could be further characterized by determining their amino acid composition using an amino acid analyzer. The recovery of the NH_2-terminal amino acid of a fluorescamine-stained peptide was as good as that of a purified identical peptide which had not been stained. Fluorescence techniques thus play an important role in the detection of peptides and will probably continue to do so.

REFERENCES

1. **Schlabach, T. D. and Wehr, T. C.,** *Anal. Biochem.,* 127, 222, 1982.
2. **Creaser, E. H. and Hughes, G. J.,** *J. Chromatogr.,* 144, 69, 1977.
3. **Joys, T. M. and Kim, H.,** *Anal. Biochem.,* 94, 371, 1979.
4. **Frei, R. W., Michel, L., and Santi, W.,** *J. Chromatogr.,* 126, 665, 1976.
5. **Mendez, E. and Gavilanes, J. G.,** *Anal. Biochem.,* 72, 473, 1976.
6. **Taylor, S. and Tappel, A. L.,** *Anal. Biochem.,* 56, 140, 1973.
7. **Wilson, K. J., Honegger, A., and Hughes, G. J.,** *Biochem. J.,* 199, 43, 1981.
8. **Lindeberg, E. G. G.,** *J. Chromatogr.,* 117, 439, 1976.
9. **Mendez, E. and Lai, C. Y.,** *Anal. Biochem.,* 65, 281, 1975.
10. **Fujiki, H. and Zurek, G.,** *J. Chromatogr.,* 140, 129, 1977.

Section II.VI.

THE ELECTROCHEMICAL DETECTION OF PEPTIDES

Recent considerable advances in column technology have produced a situation where liquid chromatography is beginning to rival gas chromatography in importance. However, one limitation to further advances may be the lack of suitable detector systems possessing high sensitivity and wide applicability. Electrochemical methods of detection fill this gap. HPLC with electrochemical detection has been used for the determination of a wide range of compounds. Fleet and Little[1] employed this method for the detection of a number of amino acids, electroactivity being attributed to the NH_2 group. Peptides with a free terminal amino group should also be electroactive. The phenol, indole, and thiol substituents of tyrosine, tryptophan, and cysteine, respectively, are more electroactive than the free amino group and peptides containing these amino acids are eminently suitable for analysis by HPLC and electrochemical detection.

Mousa and Couri[2] have used a procedure of this type for the separation of enkephalins and β-endorphin. The method is at least 100 times more sensitive than HPLC with UV detection. Neuropeptides in brain tissue can be accurately determined without prior derivatization.[3] White[4] has studied the HPLC determination of oxytocin using HPLC and electrochemical detection. In order to select a suitable operating potential the relationship between detector response and working electrode potential was investigated. An electrode potential of 0.9 V was found to give the optimum signal to noise ratio. The detection limit was then comparable to that obtained with UV detection at 220 nm. At higher electrode potentials an enhanced electrochemical detector response is obtainable, but the increase in background current is greater, reducing the signal-to-noise ratio. A high electrode potential is undesirable as selectivity is lost.

Electrochemical has an advantage over UV detection in that increased peptide selectivity can be obtained without loss of sensitivity. Peptides containing tyrosyl or tryptophyl residues can be detected selectively by UV absorption at 280 nm but electrochemical detection is more sensitive. Although most peptides are electroactive because of the terminal amino group, the response due to this group can be eliminated by selecting a suitable electrode potential. This can be demonstrated using Leu-enkephalin and (des-Tyr[1])-Leu-enkephalin. The electroactivity of the latter peptide is due to the terminal amino group, whereas Leu-enkephalin contains an additional tyrosyl residue which can be detected at a much lower electrode potential. Figure 1 shows the chromatograms obtained with a mixture of these two peptides. At an electrode potential of 0.9 V only the Leu-enkephalin is detected electrochemically, whereas both are detected using UV absorption at 220 nm.

The response of an electrochemical detector depends not only on the potential applied but also on the composition of the mobile phase.[3] When a high potential has to be applied to the buffer pH, the salts and the organics used are critical. Citrate buffers cause a rapid loss of detector sensitivity, which can be prevented by using phosphate buffers. In HPLC the detection limit also depends on the shape of the peaks; long retention times produce broad peaks which eventually cannot be distinguished from baseline. Chromatographic conditions must therefore be chosen so as to obtain short retention times. The conclusion can be drawn that for the selective detection of tyrosine-containing peptides, electrochemical detection affords a more sensitive alternative to UV absorption at 280 nm. Detection limits equivalent to or better than those found using UV absorption at 220 nm are possible, although the actual sensitivity depends on the number and type of amino acid residues present.

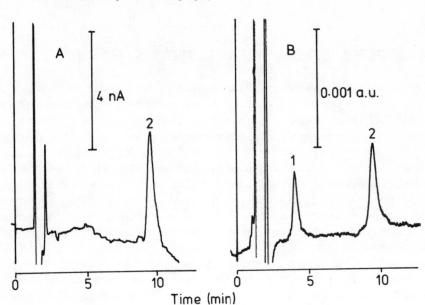

FIGURE 1.   The electrochemical detection of peptides. A mixture of (des-Tyr^1)-Leu-enkephalin and Leu-enke-phalin applied to a column 15 cm × 4.6 mm, I.D. of ODS-modified silica is used. Mobile phase, acetonitrile-0.025 M K$_2$HPO$_4$, 15:85. Flow rate: 1.2 mℓ/min. Detection (A) electrochemical (B) UV at 220 nm. (From White, M. W., *J. Chromatogr.*, 262, 420, 1983. Reproduced by permission of Elsevier Science Publishers B.V.)

REFERENCES

1. **Fleet, B. and Little, C. J.,** *J. Chromatogr. Sci.*, 12, 747, 1974.
2. **Mousa, S. and Couri, D.,** *J. Chromatogr.*, 267, 191, 1983.
3. **Sauter, A. and Frick, W.,** *Anal. Biochem.*, 133, 307, 1983.
4. **White, M. W.,** *J. Chromatogr.*, 262, 420, 1983.

Section III
Detection Reagents

Section III.I.

DETECTION REAGENTS FOR PAPER AND THIN-LAYER CHROMATOGRAPHY

The detection reagents described are principally for paper and/or thin-layer chromatography. Some reagents may detect more than one class of compound.

ALPHABETICAL INDEX OF DETECTION REAGENTS

AMMONIUM BISULFATE

Preparation: 40% ammonium bisulfate solution.
Procedure: Following chromatography, thin layer plates coated with silica gel are sprayed with the solution and heated at 180°C.
Result: Detects protected peptides.
Reference: Arendt, A., Kolodziejczyk, A., and Sokolowska, T., *Chromatographia*, 9, 123, 1976.

COPPER (II)-NINHYDRIN

Preparation: A copper-ninhydrin reagent is prepared by dissolving appropriate amounts of cupric nitrate, to give a final concentration of 25 mmol/ℓ, and ninhydrin, to give a final concentration of 1% in a mixture of 3 mℓ of water and 1 mℓ of glacial acetic acid. The solution is made up to 30 mℓ with acetone.
Procedure: The paper chromatogram is dried, uniformly sprayed with the reagent, then dried in air and heated at 65°C for 30 min.
Results: Small α-peptides, with the exception of those containing NH_2-terminal L-tryptophan, give a yellow chromophore with an absorption maximum at 395 nm. All the protein amino acids, together with asparagine and glutamine, give different shades of color with the reagent, but none gives a yellow color. The chromophore of each amino acid has its own characteristic absorption spectrum and all the chromophores show two or more absorption maxima. Nonprotein amino acids gave different chromophores, but none give a yellow chromophore. Polyamines do not give a colored product.
Reference: Ganapathy, V., Ramachandramurty, B., and Radhakrishnan, A. N., *J. Chromatogr.*, 213, 307, 1981.

N-(7-DIMETHYLAMINO-4-METHYLCOUMARINYL) MALEINIMIDE (DACM)

Preparation: Solution a: 10% pyridine acetate, pH 6.4 + 1% β-mercaptoethanol
Solution b: pyridine acetate, pH 6.4
Solution c: pyridine-isoamyl alcohol-water, 35:35:30 v/v
Solution d: 0.01 m*M* DACM in 90% acetone + 10% solution b
Procedure: A mixture of peptides, such as a tryptic digest of a protein, is dissolved in solution a, and is incubated for 10 min at room temperature to reduce cystine residues to cysteine residues. The reaction mixture is subjected to high voltage paper electrophoresis on Whatman 3 MM paper in solution b, followed by ascending chromatography for 17 hr in solution c. The peptide map is then dipped in solution d and visualized under UV light.
Results: Cysteine-containing peptides are revealed as fluorescent spots.
Comments: DACM is a highly specific and easy to handle fluorescent dye. The spots may be eluted with 50% pyridine in distilled water, purified by high voltage electrophoresis and chromatography on paper, eluted with 6 *N* HCl, and hydrolyzed for amino acid analysis. After staining with DACM, other peptides may be stained with fluorescamine within 1 hr of the DACM staining.
Reference: Klasen, E. C., *Anal. Biochem.*, 121, 230, 1982.

FERRIC FERRICYANIDE

Preparation: Solution a: 0.5 *M* ferric chloride in 0.5 *M* HCl

Solution b: 0.2 *M* potassium ferricyanide in deionized water; both solutions a and b are stored in brown bottles

Solution c: 10 mℓ of concentrated H_2SO_4 are diluted with water to 2 ℓ and the solution saturated with powdered calcium sulfate to give an acidic calcium sulfate solution

Solution d: Water is saturated with calcium sulfate to give a neutral calcium sulfate solution

Procedure: All plates or papers are freed of developing solvent in a current of warm air. For a 50 × 100 mm chromatogram plate, about 10 mℓ each of solutions a and b are mixed. The dark brown ferric ferricyanide solution is immediately poured onto the plate in a shallow dish. When color development is complete in 10 to 30 sec the plate is quickly rinsed with water.

Results: Peptides and proteins containing amino acid residues with at least moderate reducing power, such as tryptophan, tyrosine, and cysteine, give deep blue spots visible against a brown background. Higher sensitivity is obtained by removing unreacted ferric ferricyanide. Paper chromatograms or cellulose-coated plates are freed of excess reagents by rinsing with 6 *M* HCl and then water. Silica gel coated plates are placed face down on a circular glass frame about 90 mm in diameter having three upright projections of sufficient length to accommodate a magnetic stirrer between the plate and the bottom of the dish. The plate is covered with the acidic calcium sulfate solution and the solution is stirred for 2 to 3 hr or until the plate is free from brown color. The plate is then rinsed once with neutral calcium sulfate solution and dried.

Comments: It is stated that ferric ferricyanide may be superior to ninhydrin, chlorine-tolidine, and the Pauly and Ehrlich reagents for the detection of certain peptides and proteins.

Reference: Lutz, W. B. and Folkers, K., *Anal. Biochem.*, 120, 410, 1982.

FLUORESCAMINE

Technique I.

Preparation: Solution a: 10% triethylamine in dichloromethane

Solution b: 0.01% fluorescamine in acetone

Procedure: Spray with solution a and b and solution a again and observe under long wavelength UV (366 nm).

Results: Detects picomole quantities of amino acids and peptides.

Comment: When peptides were recovered, hydrolyzed with acid, and analyzed for constituent amino acids, *o*-phthalaldehyde and ninhydrin as spray reagents resulted in destruction of the amino-terminal residue, while fluorescamine did not.

References: Felix, A. M. and Jiminez, M. H., *J. Chromatogr.*, 89, 361, 1974. Schiltz, E., Schnackerz, K. D., and Gracy, R. W., *Anal. Biochem.*, 79, 33, 1977.

Technique II.

Preparation: Solution a: 1% v/v triethylamine in acetone

Solution b: 10 mg of fluorescamine in 100 mℓ acetone

Solution c: 0.1 *M* acetic acid in acetone

Solution d: 0.1 *M* N-chlorosuccinimide in acetone

Procedure: After electrophoresis or chromatography the papers are dried in air and heated at 55°C for at least 1 hr. They are then washed with acetone, wet with solution a, allowed to dry for 5 min at room temperature, wet with solution b, allowed to dry for 5 min, and washed with acetone and dried. The fluorescent spots are detected by viewing under a long-wave (336 nm) UV lamp.

Comment: It is suggested that the reaction of fluorescamine with peptides and amino acids on paper proceeds differently from the reaction in aqueous medium by stopping at an intermediate step to yield a fluorescent compound that can revert to the free amino acid on hydrolysis.

Reference: Mendez, E. and Lai, C. Y., *Anal. Biochem.*, 65, 281, 1975.

Technique III.

Preparation: Solution a: 0.2 M borate buffer, pH 8.0-acetone, 1:1
             Solution b: 50 mg of fluorescamine in 100 mℓ of acetone

Procedure: Peptides (for example, a tryptic digest) are separated on a silica gel 60 thin-layer plate without fluorescent indicator. Separation in the first dimension is achieved by electrophoresis in 2 M acetic acid-0.6 M formic acid, pH 1.9, and in the second dimension by ascending chromatography in pyridine-acetic acid-n-butanol-water, 40:14:68:25 v/v. Spray with solution a and then with solution b. View under long-wave (366 nm) UV light.

Results: Detects peptides.

Comments: The amino acid composition of the peptides may be determined by scraping the fluorescent spots off the plates during the first 3 to 4 min after the peptides have been visualized, stopping the fluorogenic reaction by adding 5.7 N HCl and then hydrolyzing at 110°C for 24 hr before analysis in an automatic analyzer. The fluorogenic reaction with the NH_2-terminal amino acid does not have an adverse effect on the amino acid composition.

Reference: Fujiki, H. and Zurek, G., *J. Chromatogr.*, 140, 129, 1977.

FLUOROGENIC REAGENTS

Fluorogenic reagents may be used to detect peptides in fractions collected during column chromatography. The reagents may be used directly or after alkaline hydrolysis of the peptides.

Preparation: Solution a: 0.5 M sodium borate, pH 8.5
             Solution b: Fluorescamine in acetone, 0.3 mg/mℓ
             Solution c: 0.5 M or 0.25 M sodium borate, pH 10, containing 0.05% mercaptoethanol
             Solution d: o-Phthalaldehyde in water, 0.3 mg/mℓ
             Solution e: 0.5 M NaOH
             Solution f: 0.5 M HCl

Procedure: Direct assay. Aliquots from the chromatographic column are placed in borosilicate culture tubes, 13 $\times$ 100 mm and dried at 110°C. For assay with fluorescamine, 1.6 mℓ of solution a is added to each tube. While the tube is vigorously shaken on a vortex type mixer, 2 0.2-mℓ aliquots of solution b are added and mixed for 10 to 15 sec. For assay with o-phthalaldehyde, 1.6 mℓ of solution c is added to each tube followed by 0.4 mℓ of solution d, mixing being performed for 2 to 3 sec. Measurements are carried out in the same tubes in a spectrofluorometer with excitation at 390 nm and emission at 475 nm for fluorescamine and excitation at 340 nm and emission at 455 nm for o-phthalaldehyde. Fluorescence is measured after 30 or more minutes. Assay after alkaline hydrolysis. After drying the aliquots from the column, 0.2 mℓ of solution e is added and hydrolysis is carried out at 120°C in an autoclave for 30 min. After cooling, 0.2 mℓ of solution f and 1.2 mℓ of solution a or solution c are added, followed by additions of fluorescamine or *o*-phthalaldehyde as described above.

Results: Detects peptides.

Comments: Peptides collected after column chromatography should preferably be hydrolyzed with alkali prior to the fluorogenic reaction. This is because peptide bond hydrolysis increases several-fold the number of primary amino groups available for reaction. Additionally, some peptides such as those having proline, hydroxyproline, or blocked amino acid at the NH_2-terminal, yield no fluorescent products unless they are hydrolyzed.

Reference: Mendez, E. and Gavilanes, J. G., *Anal. Biochem.*, 72, 473, 1976.

FORMALDEHYDE, FORMALDEHYDE-OZONE, AND FORMALDEHYDE-HYDROCHLORIC ACID

Procedure: Filter papers and thin layers are treated by one of four different procedures.

a. Formaldehyde gas. Formaldehyde condensation of compounds is induced by exposure to formaldehyde gas at 65 to 75°C for 1 hr in a closed jar (1000 mℓ) containing about 5 g of paraformaldehyde that has been equilibrated in air at about 50% relative humidity.

b. Formaldehyde gas + ozone. Ozone is introduced into the reaction vessel by discharge from a Tesla coil for 10 to 15 min, and then formaldehyde condensation is carried out as in procedure a.

c. Formaldehyde gas + HCl. Treatment with formaldehyde is carried out as in procedure a except that concentrated HCl, usually 0.05 mℓ, is present in the reaction vessel. Controls are run without formaldehyde.

d. Prochazka method. The filter papers and silica gel thin layers are sprayed with a mixture of formalin, 6 *N* HCl, and ethanol, 1:2:1, and dried in an oven at 65 to 75°C for 1 hr.

The filter papers and thin layers are viewed under UV light.

Results: The procedures detect tryptophan and tryptophan-containing peptides. Procedure a gives fairly strong fluorescence on silica gel with tryptophan and peptides with NH_2-terminal tryptophan. On filter paper the formaldehyde-induced fluorescence of these compounds is poor. Peptides with tryptophan in the COOH-terminal or intermediate position give no formaldehyde-induced fluorescence on either silica gel or filter paper. On filter paper procedure b induces strong fluorescence with tryptophan and peptides with NH_2-terminal tryptophan. Peptides with tryptophan in intermediate or COOH-terminal positions remain non-fluorescent. Procedures c and d give very intense fluorescence with tryptophan and indoleacetic acid, as well as with all tryptophan-containing peptides, regardless of the position of tryptophan in the molecule, on both filter paper and silica gel.

Reference: Larsson, L.-I., Sundler, F., and Hakanson, R., *J. Chromatogr.*, 117, 355, 1976.

NINHYDRIN

Preparation: A ninhydrin solution consisting of ninhydrin, absolute ethanol, glacial acetic acid, and 2,6-lutidine (1 g, 100 mℓ, 210 mℓ, and 29 mℓ, respectively) is made fresh every week.

Procedure: The thin-layer chromatographic plate coated with cellulose on which the chromatography has been performed is lightly sprayed with the ninhydrin solution. The chromatograms are then incubated at 40 to 45°C for 5 min and then at room temperature for 24 hr.

Results: Detects peptides.

Reference: Wainwright, I. M. and Shapsak, P., *J. Chromatogr. Sci.*, 17, 535, 1979.

NINHYDRIN-CADMIUM ACETATE

Preparation: The reagent consists of cadmium acetate, 0.5 g, water, 50 mℓ, glacial acetic acid, 20 mℓ, and propanone to 500 mℓ. To each portion required for use, ninhydrin is added to give a final concentration of 0.2%.

Procedure: Following chromatography on thin layers of purified cellulose powder, the plates are sprayed with the reagent until they appear translucent. The plates are then heated at 60°C for 30 min and cooled.

Results: The reagent detects peptides and amino acids. The R_F value and initial color of each peptide complex are noted. Estimation of the final color of each complex is made after allowing the sprayed plates to stand for about 18 hr in an ammonia-free atmosphere. The amount of each spot may be determined by quantitative densitometry.

Comments: The NH_2-terminal amino acid has an appreciable influence on the color of the cadmium ninhydrin complex formed by a peptide. For example, all peptides with a glycyl NH_2-terminal residue give a yellow color similar to that given by proline. When the positions of the NH_2-terminal and COOH-terminal residues are reversed in a dipeptide, the observed color is determined by the nature of the new NH_2-terminal amino acid residue. The magnitude of the color yield also seems to be influenced by the NH_2-terminal amino acid. In the case of NH_2-terminal valyl peptides the color yields are so low that such peptides would not be easily detected in routine TLC analysis of biological fluids.

References: 1. Heathcote, J. G., Keogh, B. J., and Washington, R. J., *J. Chromatogr.*, 79, 187, 1973.
2. Heathcote, J. G., Washington, R. J., and Keogh, B. J., *J. Chromatogr.*, 92, 355, 1974.
3. Heathcote, J. G., Washington, R. J., Keogh, B. J., and Glanville, R. W., *J. Chromatogr.*, 65, 397, 1972.

PERCHLORIC ACID

Preparation: Solution a: 70% $HClO_4$

Procedure: After ascending chromatography on precoated silica gel glass plates, the plate is briefly air-dried with a hair dryer. The plate is sprayed with solution a for 5 sec and the fluorescence observed in the dark under long-wave UV light. After development with alkaline solvents, the plate is sprayed for 10 sec.

Results: Detects 3-substituted indoles including tryptophan, tryptamine, and peptides containing tryptophan. The indoles give an intense yellowish fluorescence immediately after spraying with solution a. The fluorescence is stable for at least 30 min. As little as 40 to 850 pmol of 3-substituted indoles can be detected.

Reference: Nakamura, H. and Pisano, J. J., *J. Chromatogr.*, 152, 167, 1978.

o-PHTHALALDEHYDE

Technique I.

Preparation: Solution a: 0.1% *o*-phthalaldehyde and 0.1% 2-mercaptoethanol in acetone
 Solution b: 1% triethylamine in acetone

Procedure: Spray with solution a and 5 min later with solution b. After 10 min observe under long-wave (350 nm) UV.

Results: Detects amino acids or peptides on thin-layer plates. As little as 50 to 100 pmol of many amino acids can be detected.

Reference: Lindeberg, E. G. G., *J. Chromatogr.*, 117, 439, 1976.

Technique II.

Preparation: Solution a: 10% triethylamine in dichloromethane
 Solution b: 0.05% phthalaldehyde in methanol containing 0.2% 2-mercapto-ethanol and 0.09% Brij-35

Procedure: Spray with solution a, solution b, and solution a again. Observe under long-wave UV.

Results: Detects picomole quantities of amino acids and peptides on thin-layer chromatograms.

Reference: Schiltz, E., Schnackerz, K. D., and Gracy, R. W., *Anal. Biochem.*, 79, 33, 1977.

Technique III.

Preparation: Solution a: 1% triethylamine and 0.05% 2-mercaptoethanol in acetone
 Solution b: *o*-phthalaldehyde, 0.3 mg/mℓ in acetone

Procedure: After electrophoresis or chromatography the paper is dried at 50°C for 1 hr, is washed with acetone, and then dipped in a tray containing solution a. After 5 min at room temperature the paper is dipped in a tray containing solution b and allowed to dry for 5 min at room temperature. Finally the paper is washed with acetone and dried. The fluorescent spots are detected under a long-wave (366 nm) UV lamp.

Results: Detects amino acids and peptides. Amino acids are more readily visualized at the 500 pmol level with *o*-phthalaldehyde than with fluorescamine. While the fluorescent spots produced by fluorescamine remain visible for several weeks the spots produced by *o*-phthalaldehyde disappear after several hours.

Reference: Mendez, E. and Gavilanes, J. G., *Anal. Biochem.*, 72, 473, 1976.

4,4'-TETRAMETHYLDIAMINO-DIPHENYLMETHANE (TDM)

Preparation: Solution a: Sodium hypochlorite containing 13 to 14% active chlorine diluted to 6 times its original volume
 Solution b: A mixture of the following amounts of solutions c and d with 1.5 mℓ of solution e added
 Solution c: 2.5 g 4,4'-tetramethyldiamino-diphenylmethane dissolved in 10 mℓ glacial acetic acid and diluted with 50 mℓ water

Solution d: 5 g potassium iodide in 100 mℓ water

Solution e: 0.3 g ninhydrin dissolved in 90 mℓ water and 10 mℓ glacial acetic acid added

Procedure: Thin-layer plates are dried at 100°C for 15 min after chromatography and sprayed with solution a until damp. Silica gel plates are then dried for 10 min in warm air, cellulose plates for 45 min in warm air or for 5 min at 100°C, and are lightly sprayed with solution b.

Results: Amines, amides, peptides, and amino acids produce green spots, which slowly turn blue-green to dark blue. The reaction can be speeded up by a stream of warm air or by UV irradiation.

Reference: Von Arx, E., Faupel, M., and Brugger, M., *J. Chromatogr.*, 120, 224, 1976.

Section III.II.

SUMMARY TABLES FOR DETECTION REAGENTS

The following tables represent summary information on several detection reagents and the sensitivity of detection for peptides and peptide derivatives. Procedures for the preparation of detection reagents are included in the earlier part of this section.

Table 1
DETECTION LIMITS OF PEPTIDES

Detection method	TDM[a]	RH[b]
Peptide	**Detection limit (µg)**	
Z-Trp-Leu-OH	0.05	0.08
Z-Gln-Trp-Leu-OH	0.05	0.05
Z-Val-Gln-Trp-Leu-OH	0.02	0.05
H-Trp-Leu-OH	0.05	0.05
Z-Arg-Val-Gln-Trp-Leu-OH	0.05	0.05
H-Gln-Trp-Leu-OH	0.05	0.10
H-Val-Gln-Trp-Leu-OH	0.06	0.06
Z-Arg-Val-Glu(OC(CH$_3$)$_3$)-Trp-Leu-OH	0.05	0.10
Z-Trp-Leu-OCH$_3$	0.05	0.10
Z-Gln-Trp-Leu-OCH$_3$	0.05	0.10
H-Met-Glu-Arg-Val-Gln-Trp-Leu-Arg-Lys-Lys-Lys-Gln-Leu-Val-Arg-His-Asn-Phe-OH	0.05	0.08
Z-Val-Gln-Trp-Leu-OCH$_3$	0.05	0.10

Note: Peptides were separated by TLC on silica gel.

[a] TDM: Detection using 4,4'-tetramethyldiamino-diphenylmethane (see Section III.I).
[b] RH: Reindel-Hopper method. Chlorine-*o*-tolidine after chlorination by spraying with sodium hypochlorite solution.

REFERENCE

1. **Von Arx, E., Faupel, M., and Brugger, M.,** *J. Chromatogr.*, 120, 224, 1976.

Reproduced by permission of Elsevier Science Publishers B.V.

.

Table 2
THE DETECTION LIMITS OF FREE AND PROTECTED PEPTIDES IN THE UV

Peptide	Detection limit	
	µg/mℓ	ng
Cbo-S-benzyl-Cys-Pro-Leu-Gly-NH$_2$	0.84	42
Cbo-Asn-S-benzyl-Cys-Pro-Leu-Gly-NH$_2$	0.42	21
Cbo-Gln-Asn-S-benzyl-Cys-Pro-Leu-Gly-NH$_2$	0.34	17
Gln-Asn-S-benzyl-Cys-Pro-Leu-Gly-NH$_2$	1.00	50

Abbreviation: Cbo, benzyloxycarbonyl.

REFERENCE

1. **Nachtmann, F.,** *J. Chromatogr.*, 176, 391, 1979.

Table 3
FLUORESCENCE OF AMINO ACIDS AND PEPTIDES ON TREATMENT WITH FLUORESCAMINE[a]

Compound	Relative fluorescence intensity
Glycine	13.0
Valine	11.6
Isoleucine	11.6
Aspartic acid	11.6
Arginine	10.9
Lysine	0.7
Ornithine	0.7
Tryptophan	11.1
Phenylalanine	12.2
Tyrosine	10.3
Threonine	13.7
Sarcosine	0.3
Glycyl-phenylalanine	3.0
Glycyl-leucine	4.1
Arginyl-phenylalanine	0.9
Lysyl-glycine	2.4
Tryptophyl-leucine	0.6
Ammonium sulfate	3.4

[a]   5 nmol of amino acid or peptide in 100 $\mu\ell$ are mixed with 2.9 mℓ of buffered *o*-phthalaldehyde-mercaptoethanol, pH 9.5. Fluorescence intensities are expressed relative to an arbitrary quinine sulfate standard (1 μg/mℓ 0.1 NH_2SO_4) of 70.

REFERENCE

1. **Taylor, S. and Tappel, A. L.**, *Anal. Biochem.*, 56, 140, 1973.

Reproduced by permission of Academic Press, Inc.

Table 4
RELATIVE FLUORESCENCE OF PEPTIDES ON REACTION WITH FLUORESCAMINE AND *o*-PHTHALALDEHYDE

Peptide	Conc. (nmol/mℓ)	Fluorescence OPA[a]		FA[b]	
Norleucine	12.5	25[c]	2.0[d]	3[c]	0.24[d]
	18.8	38	2.0	5	0.27
	25.0	54	2.2	7	0.28
Angiotensin II	2.5	0.8	0.3	29	12
	12.5	1.5	0.1	154	12
Bradykinin	2.5	3.5	1.4	67	27
	12.5	5.3	0.4	262	21
Contraceptive tetrapeptide	2.5	45	18	88	35
	12.5	208	17	507	40
IgE peptide III	2.5	1.5	0.6	64	26
	12.5	3.5	0.3	178	14

Table 4 (continued)
RELATIVE FLUORESCENCE OF PEPTIDES ON REACTION WITH FLUORESCAMINE AND *o*-PHTHALALDEHYDE

Peptide	Conc. (nmol/mℓ)	Fluorescence			
		OPA[a]		FA[b]	
Liver cell growth factor	2.5	67	27	132	53
	12.5	240	19	280	22
Tuftsin	2.5	32	13	47	19
	12.5	146	12	331	26
TP 19	2.5	0.9	0.4	46	18
	12.5	3.4	0.3	114	9
TP24	2.5	0.4	0.2	49	20
	12.5	2.3	0.2	184	15
SP 20	2.5	22	8.8	23	9.2
	12.5	120	9.6	94	7.5
SP 21	2.5	13	5.2	66	27
	12.5	54	4.3	193	15
SP 22	2.5	0.1	0.04	83	33
	12.5	0.6	0.05	291	23
SP 23	2.5	0.25	0.1	22	8.8
	12.5	0.85	0.07	133	10.6
SP 24	2.5	0.6	0.2	35	14
	12.5	2.1	0.2	134	11
SP 25	2.5	40	16	63	25
	12.5	140	11	274	22
SP 26	2.5	3	1.2	21.5	8.6
	12.5	12.5	1.0	119	9.5

[a]   OPA, *o*-phthalaldehyde.
[b]   FA, fluorescamine.
[c]   Relative fluorescence.
[d]   Fluorescence/nmol.

Notes: Conditions were selected to mimic those present in a system designed to automatically monitor column effluents. The *o*-phthalaldehyde reaction was carried out at pH 10.4 and the fluorescamine reaction at pH 8.3. Fluorescence was measured relative to a blank sample containing no peptide. Flagellin peptides were obtained from a tryptic digest of the flagellar protein of *Salmonella paratyphi* B SL 877 (TP 19 and TP 24). Peptides synthesized by the solid phase method are denoted by SP. The amino acid sequences of the peptides are as follows: Angiotensin II, Asp-Arg-Val-Tyr-Ile-His-Pro-Phe; bradykinin, Arg-Pro-Pro-Gly-Phe-Ser-Pro-Phe-Arg; contraceptive tetrapeptide, Thr-Pro-Arg-Lys; IgE peptide III, Asp-Ser-Asp-Pro-Arg; liver cell growth factor, Gly-His-Lys; tuftsin, Thr-Lys-Pro-Arg; TP 19, Gly-Leu-Thr-Gln-Ala-Ser-Arg; TP 24, Leu-Ser-Ser-Gly-Leu-Arg; SP 20, Asp-Ser-Asp-Pro-Lys; SP 21, Leu-Ser-Ser-Gly-Leu-Lys; SP 22, Thr-Pro-Arg; SP 23, Thr-Asp-Pro-Arg; SP 24, Gly-Ser-Asp-Pro-Arg; SP 25, Gly-Lys-His; SP 26, Asp-Ser-Asp-Pro-Lys (Dipmoc). (Dipmoc, diisopropylmethyloxycarbonyl.)

REFERENCE

1. **Joys, T. M. and Kim, H.,** *Anal. Biochem.*, 94, 371, 1979.

Reproduced by permission of Academic Press, Inc.

Table 5
FLUORESCENCE RESPONSE
OF PEPTIDES IN AN
o-PHTHALALDEHYDE POSTCOLUMN
REACTION

Peptide	Fluorescence (MV/nmol)
Lys-Val	180.0
Tyr-Gly	3.2
Trp-Gly	3.0
Gly-Gly-Gly	12.0
MSH-INH[a,b]	1.0
Tuftsin[c]	240.0
Kentsin[d]	210.0
Met-enkephalin[e]	3.3
ERP[a,f]	42.0
Xenopsin[g]	190.0
LHRH[a,h]	2.5

[a] Amide at the carboxyl terminus.
[b-h] Sequences are as follows:
[b] Pro-Leu-Gly,
[c] Thr-Lys-Pro-Arg,
[d] Thr-Pro-Arg-Lys,
[e] Tyr-Gly-Gly-Phe-Met,
[f] Lys-Phe-Ile-Gly-Leu-Met,
[g] pGlu-Gly-Lys-Arg-Pro-Trp-Ile-Leu,
[h] pGlu-His-Trp-Ser-Tyr-Gly-Leu-Arg-Pro-Gly, (pGlu, pyrrolidone carboxylic acid.)

REFERENCE

1. **Schlabach, T. D. and Wehr, T. C.**, *Anal. Biochem.*, 127, 222, 1982.

Reproduced by permission of Academic Press, Inc.

Table 6
COLOR OF THE CADMIUM NINHYDRIN COMPLEX OF PEPTIDES

Peptide	Color
Gly-Ala	Yellow
Gly-Val	Yellow
Gly-Leu	Yellow
Gly-Ser	Yellow
Gly-Pro	Yellow
Ala-Gly	Red
Val-Gly	Red
Leu-Gly	Red
Ser-Gly	Orange
Pro-Gly	Mauve

REFERENCE

1. **Heathcote, J. F., Washington, R. J., Keogh, B. J., and Glanville, R. W.**, *J. Chromatogr.*, 65, 397, 1972.

Table 7
MINIMUM DETECTABLE AMOUNT OF ALKYLAMINONAPHTHYLENESULFONYL MIF (MELANOTROPIN INHIBITING FACTOR) AND ITS METABOLITES

Compound	Minimal amount detected (10^{-12} mol)[b]				
	Dns	Ethansyl	Propansyl	Bns	Monoisopropansyl
MIF	219	1014	1217	3388	320
Pro-Leu	30	343	365	597	99
Leu-Gly	58	762	1352	1792	446
Pro	10	49	91	80	12
Leu	19	124	182	4480	78
Gly	ND[c]	66	182	149	27
Gly-NH$_2$	562	686	4057	7261	448

[a] Abbreviations: Dns, ethansyl, propansyl, Bns, and monoisopropansyl, respectively for the *N,N*-dimethyl, diethyl, dipropyl, dibutyl, and *N*-monoisopropyl-aminonaphthylenesulfonyl compounds.
[b] Compounds were separated by TLC and detected by their fluorescence under UV light.
[c] Not determined since glycine could not be separated from Dns-hydroxide.

REFERENCE

1. **Hui, K.-S., Salschutz, M., Davis, B. A., and Lajtha, A.,** *J. Chromatogr.*, 192, 341, 1980.

Reproduced by permission of Elsevier Science Publishers B.V.

Table 8
AMINO ACID COMPOSITION OF A TRYPTIC PEPTIDE AFTER PURIFICATION IN DIFFERENT WAYS

Amino acid	Unstained[a]	Stained with ninhydrin[b]	Stained with fluorescamine[b]
Glutamic acid	2.0	2.2	2.0
Leucine	2.1	2.0	2.2
Lysine	1.0	1.2	1.0
Threonine	1.0	0.7	1.0
Valine	0.9	0.9	0.8

[a] Isolated by column chromatography and silica gel thin-layer chromatography.
[b] Isolated from silica gel thin layer chromatograms after staining.

Note: Amino acid compositions were determined using an amino acid analyzer. The values represent the number of amino acid residues per peptide. The background of a silica gel blank was deducted from the amount of each amino acid residue. Values for threonine were extrapolated to zero time.

REFERENCE

1. **Fujiki, H. and Zurek, G.,** *J. Chromatogr.*, 140, 129, 1977.

Section IV
Methods of Sample Preparation Including Derivatization

Section IV

METHODS OF SAMPLE PREPARATION INCLUDING DERIVATIZATION

This section describes procedures for the extraction, preparation, and derivatization of samples prior to chromatography. It is hoped that the methods can, in many cases, be used directly by researchers for the preparation of samples for chromatography, but they should at least serve as a guide illustrative of the kind of steps which are required for workers designing their own procedures.

1. RAPID CONCENTRATION OF URINARY PEPTIDES AND PROTEINS

A method for the rapid concentration and dialysis of urinary peptides and proteins has been described. The method, based on the use of cheap disposable dialysis coils, permits concentration of peptides and proteins in 1 to 20 ℓ of urine to less than 0.3 to 0.5 ℓ in 2 to 24 hr. Sodium bicarbonate (5 g/ℓ) and sodium azide (0.5 g/ℓ) as preservatives are added to the urine, which is filtered and stored at 4°C before dialysis. Dialysis and concentration are performed at 4°C using a disposable dialyzer intended for the treatment of kidney patients. Urine contained in a large beaker or a bucket is fed through a 5-mm internal diameter tubing to the pump, which is connected to the dialyzer. From the dialyzer the urine is led back to the bucket. The urine is dialyzed against 5 to 10 ℓ of distilled water, which is changed at 0.5- to 2-hr intervals.

Dialysis is monitored by measuring the volume of urine, absorption at 214 nm, and the sodium concentration. Recovery of proteins and large peptides is higher than 80% as long as the final volume of the dialyzed urine is greater than 300 to 400 mℓ. A higher recovery of protein is achieved if the sample is concentrated to a volume compatible with the next fractionation step. If the next step is ion-exchange chromatography, the sample can be dialyzed against the appropriate buffer.

REFERENCE

1. **Stenman, U.-H., Personen, K., and Huhtala, M.-L.,** *Anal. Biochem.,* 123, 291, 1982.

2. RECOVERY OF PROTEIN FROM LARGE VOLUMES OF SOLUTION

Mahuran et al. have described a procedure which can recover 40 to 80% of a protein from a solution containing 1 µg/mℓ.

Protein precipitation — To a 30 mℓ sample is added 120 µℓ of 2% sodium desoxycholate; the sample is mixed and placed on ice for 30 min. The concentration of the detergent may be varied from 50 to 500 µg/mℓ of sample without affecting the recovery; 10 mℓ of cold 24% trichloroacetic acid is added, and the sample is mixed and kept on ice for 1 further hr to allow the protein-detergent complex to precipitate. The suspension is centrifuged at 4°C and the supernatant is removed from the pellet by suction. The pellet is dissolved in 0.5 mℓ of 62.5 mM Tris-HCl, pH 6.8, containing 3% sodium dodecyl sulfate (SDS) made 0.5 M in NaHCO$_3$ by adding solid NaHCO$_3$ and giving a final pH of 8.8. The solution is dialyzed overnight at room temperature against 2 ℓ of 6.25 mM Tris-HCl, pH 6.8, containing 0.3% SDS and is then freeze-dried in 7-mℓ screw-top test tubes.

Detergent extraction — SDS and residual sodium desoxycholate are removed by ion pair extraction using acetone, glacial acetic acid, and triethylamine (90:5:5) as the extraction solution. The dried sample is dissolved in 1 mℓ of the extraction solution and is quickly transferred to a 3-mℓ glass conical test tube. The proteins precipitate after 1-hr incubation on ice and are collected by centrifuging at room temperature. The pellet is washed once with 1 mℓ of the extraction solution and once with 1 mℓ of acetone and is then dried under nitrogen. The initial precipitation eliminates many interfering compounds commonly found in buffers used for protein purification and the final extraction and precipitation removes the residual detergents as well as other ionic compounds.

REFERENCE

1. **Mahuran, D., Clements, P., Carrella, M., and Strasberg, P. M.,** *Anal. Biochem.,* 129, 513, 1983.

3. THE COLUMN EXTRACTION OF PEPTIDES FROM PLASMA

Four milliliters of plasma are passed through a Sep-Pak C_{18} cartridge, the bonded silica matrix is washed with 5 mℓ of 0.05 M trifluoroacetic acid (TFA), and is eluted slowly with 2 mℓ of acetonitrile-0.05 M TFA (80:20). The eluate is lyophilized, reconstituted with 150 $\mu\ell$ of 0.05 M TFA, and subjected to HPLC.

REFERENCE

1. **Schöneshöfer, M. and Fenner, A.**, *J. Chromatogr.*, 224, 472, 1981.

4. THE SELECTIVE ISOLATION OF AMINO-TERMINAL PEPTIDES FROM PROTEINS

In the characterization of a new protein it is desirable in most instances to obtain amino-terminal sequences. A method for the selective isolation and purification of amino-terminal peptides from enzymatic digests of proteins has been described. The procedure can be applied to proteins with free or blocked amino termini and is independent of the amino acid composition of the amino-terminal region.

When a protein, all of whose free amino groups have been blocked by citraconylation, is digested with a protease, all the peptides produced, except those derived from the amino terminus, have a free amino group. Reaction of such a digest with 1-fluoro-2,4-dinitrobenzene (FDNB), followed by removal of citraconyl groups by acid treatment and removal of dinitrophenyl (DNP) groups from histidine and tyrosine side chains by thiolysis, results in dinitrophenylation of all the α-amino groups derived from internal cleavages. Only peptides derived from the amino terminus lack a dinitrophenyl group. Due to the strong adsorption of the dinitrophenyl group to polystyrene, when a mixture of peptides treated in the above way is passed through a polystyrene column, only peptides from the amino terminus will elute.

Procedure — A protein sample (0.1 to 1 μmol) is dissolved in 5 mℓ of 8 M urea and citraconylated by adding 50 $\mu\ell$ of citraconic anhydride. The urea should be free from cyanate. The pH is maintained at about 8.5 by the addition of 5 M NaOH. The citraconylated protein is dialyzed against 0.5% NH_4HCO_3 and digested with a basic protease for 3 hr at 37°C. The protein-enzyme ratio should be 50:1. The digest is then freeze-dried. Chymotrypsin and aldolase may be digested with porcine elastase and cytochrome c with chymotrypsin. The enzymatically digested sample is dissolved in 1 to 5 mℓ of 10% N-methylmorpholine and a 10 to 50 $\mu\ell$ aliquot of a 50% solution (v/v) of FDNB in acetonitrile is added. The reaction mixture is shaken for 16 hr at 22°C, is acidified with HCl, extracted 3 times with 5 mℓ of ether, and is evaporated to dryness.

The DNP-peptide mixture is dissolved in 1 mℓ of 10% formic acid and allowed to stand for 1 hr to remove citraconyl groups. A polystyrene column (Porapak Q) is prepared by soaking the resin in ethanol for several hours and then packing a column, 0.6 × 10 mm. After equilibrating the column with 0.01 N HCl, the sample is added and the column washed with 10 mℓ of 0.01 N HCl. The column is then developed with 20% acetone-water and 10 mℓ of eluate collected. This fraction contains the free amino or the naturally blocked amino-terminal peptides. If an amino-terminal peptide contains histidine or tyrosine, it remains absorbed to the column on elution with 20% acetone-water. If this is the case, all the adsorbed peptides are eluted with 80% acetone-water and the solution is evaporated to dryness. Thiolysis with mercaptoethanol is performed and, after drying, the samples are again applied to a polystyrene column.

The peptides are separated on paper and are located by staining with ninhydrin or starch-iodine. They are then eluted with 6 *N* HCl, are hydrolyzed *in vacuo*, and their amino acid composition determined using an automatic analyzer.

REFERENCE

1. **Kaplan, H. and Oda, G.**, *Anal. Biochem.*, 132, 384, 1983.

5. ISOLATION OF THE COOH-TERMINAL FRAGMENT OF PROTEINS

The isolation of the COOH-terminal fragment facilitates the determination of the COOH-terminal sequence of proteins and large peptides, since the isolated fragment can be subjected to NH$_2$-terminal sequential degradation procedures. Furka et al. have described a method for the isolation of the COOH-terminal fragment in which the protein is esterified with methanolic-HCl and subsequently digested with pepsin. The peptide mixture is then submitted to paper electrophoresis in pH 2.1 buffer. The COOH-terminal peptide is identified by preparing a guide peptide map, using pH 5.5 buffer in the second dimension, when the fragment appears as an on-diagonal spot. It can be isolated by a run at pH 5.5 of the corresponding band from the first, pH 2.1, electrophoretogram. The peptide is obtained in a yield of about 40%.

Procedure — 20 mg of protein is stirred with 2 mℓ of methanolic 0.1 *N* HCl in a stoppered flask at 4°C for 48 hr. Excess reagent is evaporated in a desiccator under reduced pressure over KOH and P$_2$O$_5$ at room temperature. To the dry residue, 2 mℓ of 5% (v/v) formic acid and 0.2 mg of pepsin are added and the mixture is stirred at 37°C. After 8 hr the solution is freeze-dried. The peptic digest is then dissolved in buffer and subjected to electrophoresis.

REFERENCE

1. **Furka, A., Dibo, G., Kovacs, J., and Sebestyen, F.**, *Anal. Biochem.*, 129, 14, 1983.

6. THE SELECTIVE ISOLATION OF METHIONINE-CONTAINING PEPTIDES

Methionine-containing peptides are of particular importance in the sequence analysis of proteins. Such analysis often demands isolation from enzymatic digests of small methionine-containing peptides required to overlap fragments derived by cleavage with cyanogen bromide. The procedure of Sasagawa et al. designed to accomplish this involves recognition of methionine by ^{14}C-methylation, thiolytic regeneration of ^{14}C-labeled methionyl peptides, and rapid isolation by taking advantage of the change in mobility induced by the regeneration step.

$$R\text{-}S\text{-}CH_3 \;\rightarrow\; \underset{\overset{|}{^{14}CH_3}}{R\text{-}S\text{-}CH_3} \;\rightarrow\; R\text{-}S\text{-}CH_3$$

$$\text{or}$$

$$R\text{-}S\,{}^{14}CH_3$$

The protein is first treated with (^{14}C)methyl iodide to selectively label methionine residues as their sulfonium salts, and it is then subjected to tryptic digestion followed by preliminary reversed-phase HPLC. Fractions are pooled on the basis of the pattern of radiolabel. Thiolytic treatment of each pool regenerates an uncharged methionyl peptide, which on HPLC has a decreased mobility relative to contaminants which accompanied it in the first chromatogram.

Procedure — A protein (200 nmol) is dissolved in 400 $\mu\ell$ of 0.1 M sodium phosphate buffer, pH 4.5, containing 6 M guanidine hydrochloride. To this solution is added a 200-fold molar excess of (^{14}C)methyl iodide. The reaction mixture is placed in the dark, stirred at room temperature for 24 hr and is then desalted by dialysis against water. The S-methylated protein is treated with trypsin, 100:2, in 0.1 M NH$_4$HCO$_3$ for 6 hr at 37°C. Peptides are separated by reversed-phase HPLC on a μBondapak® C$_{18}$ column, the mobile phase being 0.1% TFA and the modifier acetonitrile containing 0.07% TFA. The concentration of acetonitrile is increased linearly at 2%/min during 30 min at a flow rate of 2 mℓ/min. Each methylated peptide elutes earlier than its nonmethylated counterpart, in accordance with the decrease in hydrophobicity accompanying the conversion of the methionyl residue to its sulfonium salt. This change in mobility serves as the basis for the separation scheme.

The sulfonium derivatives are then demethylated by treatment with 0.5 M dithiothreitol in 0.2 M borate buffer at pH 10.0 in sealed tubes under nitrogen at 37°C for 24 hr and then acidified with 0.1% TFA. Subsequent chromatography in each case then yields a radioactive methionine-containing peptide with a later retention time, a nonradioactive peak of contaminants with the original retention time, and two large peaks of dithiothreitol, one of which includes a ^{14}C-methylated derivative of dithiothreitol. Each radiolabeled product of the second chromatogram is hydrolyzed and subjected to amino acid analysis. The composition of peptides containing methionine is thus established. The method is rapid and relatively simple.

The isolated products become ^{14}C-radiolabeled without otherwise altering the chemical character of the peptides. In effect some of the methyl groups of the methionine residues have been exchanged for (^{14}C)methyl groups. As a result methionine residues can be identified by both conventional procedures and the presence of the radiolabel.

REFERENCE

1. **Sasagawa, T., Titani, K., and Walsh, K. A.**, *Anal. Biochem.*, 128, 371, 1983.

7. THE SELECTIVE ISOLATION OF TRYPTOPHAN-CONTAINING PEPTIDES

The selective isolation of tryptophan-containing peptides from complex digests has been achieved by inducing changes in their hydrophobicity and chromatographic mobility by reaction with *o*-nitrophenylsulfenyl chloride. The protein is first digested with, for example, trypsin, and subjected to preliminary reversed-phase HPLC. Tryptophan-containing peptides are located in crude fractions by their fluorescence at 348 nm, and each such fraction is reacted with *o*-nitrophenylsulfenyl chloride.

Procedure — The protein, for example, cytochrome *c*, 30 nmol, is digested with trypsin, 100:2, w/w, in 0.1 M NH$_4$HCO$_3$ for 6 hr at 37°C and the digest fractionated by HPLC. Fluorescent fractions are collected and lyophilized. Tryptophan-containing peptides are then *o*-nitrophenylsulfenylated at the 2-position of their indole rings. Each peptide of this type (30 nmol) in 200 $\mu\ell$ of glacial acetic acid is treated with 100 nmol of *o*-nitrophenylsulfenyl chloride in the dark at room temperature for 18 hr. Peptides are then separated by chromatography on a μBondapak® C18 column, the mobile phase being 0.1% TFA and the modifier acetonitrile containing 0.07% TFA. The concentration of acetonitrile is increased linearly at 2%/min during 30 min at a flow rate of 2 mℓ/min. Elution is monitored by absorbance at 210 nm and by fluorescence at 348 nm with excitation at 287 nm. The method, which is rapid and simple, separates individual tryptophan-containing peptides from each other.

REFERENCE

1. **Sasagawa, T., Titani, K., and Walsh, K. A.,** *Anal. Biochem.,* 134, 224, 1983.

8. THE CLEAVAGE OF PEPTIDES AND PROTEINS WITH BNPS-SKATOLE

BNPS-skatole [2-(2-nitrophenylsulfenyl)-3-methyl-3-bromoindole] is a reagent for the selective modification of tryptophan. It is also useful for the selective cleavage of tryptophanyl peptide bonds.[1] With a variety of tryptophan-containing peptides, a 10-fold excess of reagent in 20 hr gives 30 to 59% of cleavage. However, BNPS-skatole cleaves at sites other than tryptophanyl peptide bonds. Since cleavage is time dependent and since tryptophan reacts rapidly with the reagent, modification without any significant cleavage is to be expected when the reaction is carried out with a low excess of reagent and for shorter times. Longer reaction times and excess reagent are needed for peptide bond cleavage. The protein or peptide is allowed to react with BNPS-skatole in the presence of tyrosine. This amino acid is added to the reaction mixture as scavenger in order to prevent modification of the tyrosine residues of the protein by the excess reagent.

Procedure[2] — The procedure is described as applied to the cyanogen bromide fragment B derived from *Bacillus* amyloliquefaciens α-amylase. Twenty milligrams of the cyanogen bromide peptide B are treated with a 100-fold excess of BNPS-skatole in 80% acetic acid. To the reaction mixture, 100 equiv. of exogenous tyrosine per tyrosine residue of B are added. Incubation is allowed to proceed for 68 hr with stirring and the reaction is terminated by passing the reaction mixture through Sephadex® G-15 equilibrated with 10% acetic acid. Resolution of the peptides is achieved by passing the desalted material through a Bio-Gel® P-60 column equilibrated with 0.5 *M* formic acid.

REFERENCES

1. **Fontana, A.,** *Methods Enzymol.,* 25, 419, 1972.
2. **Detera, S. D. and Friedberg, F.,** *Int. J. Pept. Prot. Res.,* 17, 93, 1981.

9. THE CLEAVAGE OF TRYPTOPHANYL AND METHIONYL PEPTIDE BONDS BY CYANOGEN BROMIDE

Quantitative cleavage of peptide bonds adjacent to tryptophanyl and methionyl residues in proteins is achieved using cyanogen bromide in the presence of heptafluorobutyric (HFBA) and formic acids. Amino acid analysis of peptides isolated from digests obtained in this way indicates that tyrosine is modified and the derivative elutes in a position preceding lysine; however, the color constant with ninhydrin remains unaltered. Peptides containing the modified tyrosine are still susceptible to chymotryptic cleavage at this residue. Cleavage of methionyl peptide bonds can be prevented by methylene blue-sensitized photooxidation prior to treatment with cyanogen bromide and HFBA.

The procedure as applied to cytochrome b_5 is described. For cleavage of peptide bonds adjacent to tryptophan and methionine, to 100 to 500 nmol of heme-free apocytochrome, 1.0 mℓ each of 88% formic acid and anhydrous HFBA are added. After adding 700 mg of solid CNBr the sample is kept in the dark for 24 hr. The reagent and solvents are removed with a stream of nitrogen and the remaining material is suspended in 10.0 mℓ of water and lyophilized. The dried material may be dissolved in 1.0 mℓ of 88% formic acid and a sample used for NH₂-terminal analysis. For gel filtration, the lyophilized digest is dissolved in 50% acetic acid and applied to a column of Sephadex® G-75, 2.9 × 105 cm, which has been equilibrated with 50% acetic acid. Peptide fractions are identified by reaction with ninhydrin after alkaline hydrolysis.

For selective cleavage at tryptophanyl bonds the protein is photooxidized prior to CNBr/ HFBA treatment by the following procedure. A sample of 300 to 500 nmol of cytochrome b_5 is dissolved in 2.0 mℓ of 84% acetic acid and transferred to a Pyrex test tube, 1×10 cm, fitted with a glass jacket. The temperature of the protein solution is maintained at 0°C by circulating 95% ethanol through the jacket. A solution of 0.04% methylene blue in 84% acetic acid is prepared and immediately added to the protein solution in the dark to give a dye-protein ratio of 1.5:1.0 (nmol/nmol).

The reaction vessel is irradiated with a single 150-W tungsten light source at a distance of 30 cm for 8 hr while oxygen is slowly bubbled through the protein solution. After irradiation the solvent is removed using a vacuum assembly line equipped with a cold trap. The protein is dissolved in 10% acetic acid and separated from the dye by gel filtration on a column of Bio-Gel® P-2, 0.9×40 cm, developed with the sample buffer. CNBr/HFBA cleavage of the photooxidized protein is performed as described above.

REFERENCE

1. **Ozols, J., Gerard, C., and Stachelek, C.**, *J. Biol. Chem.*, 252, 5986, 1977.

10. THE PURIFICATION OF PEPTIDES

A method for the separation of peptides from amino acids and for the desalting of peptides is based on passage of the sample through a column of octadecyl-silica (ODS), which retains peptides but not amino acids and salts. In the simple chromatographic system used, a 4×0.9 cm column is filled with ODS (Porasil C 18, 50 to 80 μm). Samples are dissolved in 0.2 M acetic acid and 0.5 mℓ of the solution added to the column. The column is eluted at a flow rate of 1 mℓ/min with aqueous solvent until the effluent is free of salts and amino acids. Peptides which are retained on the column are eluted with alcoholic solvent. Pairs of aqueous and alcoholic solvents that may be used are (1) 1% (v/v) TFA in water/80% methanol and 1% TFA in water, (2) 0.2 M acetic acid/60% n-propanol in 0.2 M acetic acid, and (3) 1 M pyridine formate pH 3.0/60% n-propanol in 1 M pyridine formate pH 3.0. Peptide samples eluted with methanolic solvent are freed from methanol by means of a nitrogen stream and lyophilized. Samples eluted with n-propanol containing solvent are lyophilized directly. The method is simple and affords rapid separations while the ODS column can be reused. There is practically no limit to the volume of samples as long as the capacity of the column for retaining peptides is not exceeded.

REFERENCE

1. **Böhlen, P., Castillo, F., Ling, N., and Guillemin, R.**, *Int. J. Pept. Prot. Res.*, 16, 306, 1980.

11. PURIFICATION OF PROTECTED SYNTHETIC PEPTIDES

The importance of preparing synthetic peptides for biological or conformational studies in pure form is almost self-evident. The prospects of obtaining pure free peptides are greatly enhanced if intermediate protected peptides are carefully purified. Chromatographic purification of protected synthetic peptides by the open-column method is slow and not very efficient. The advantage of HPLC compared with open-column procedures lies in its operation at relatively high flow rates, while the use of a stationary phase of small particle size affords high plate numbers and low retention.

Until relatively recently applications of HPLC to free or synthetic peptides had dealt principally with analytical problems. Gabriel et al., however, have described a HPLC system for the rapid and efficient purification of protected synthetic peptides. Purification of these peptides on a gram scale is possible through the use of pre-packed silica gel 60 columns (Merck, heavy-wall glass size C columns, 3.8 × 43 cm, prepacked with silica gel 60 of 60 Å porosity) equipped with low dead-volume fittings. The outlet from the column is connected to monitoring instruments and then to a fraction collector. The effluent is monitored with a UV detector at 254 nm. Thin-layer chromatography may be used to control the homogeneity of fractions containing the desired product.

In most cases it is possible to purify protected peptides directly by HPLC of the crude reaction mixture without any prior purification or conventional washing procedures with base, acid, and water. Suitable solvent systems are chloroform mixed in varying degrees with isopropanol, ethanol, methanol, and acetic acid. Products up to tetrapeptide derivatives were purified and the procedure appeared to be adaptable to even larger peptides. The chromatographic system is assembled using standard instruments commonly in use in most laboratories; the original paper should be consulted for details.

REFERENCE

1. **Gabriel, T. F., Jimenez, M. H., Felix, A. M., Michalewsky, J., and Meienhofer, J.,** *Int. J. Pept. Prot. Res.,* 9, 129, 1977.

12. THE PREPARATION OF LABELED PEPTIDES

Peptides may be labeled with ^{125}I using the iodogen method. In this method a water-insoluble phase of iodogen (1,3,4,6-tetrachloro-3α,6α-diphenyl glycoluril) serves as an acceptor in the iodination reaction, making the presence of damaging reagents such as chloramine-T in the peptide-containing solution unnecessary. A polypropylene tube is coated with iodogen by evaporating 20 to 50 μℓ of a 1 mg/mℓ solution of iodogen in dichloromethane. Two micrograms of peptide in 25 μℓ 0.5 M phosphate and 0.5 mCi Na^{125}I solution are added. The tube is tapped gently at 1-min intervals, and after 10 min the reaction is terminated by aspirating the solution. Peptides labeled in this way have been used as tracers in radioimmunoassay procedures. For this purpose the labeled peptide solution is diluted with 0.02 M phosphate buffered saline containing 0.25% bovine serum albumin.

REFERENCE

1. **Loeber, J. G., Verhoef, J., Burbach, J. P. H., and Witter, A.,** *Biochem. Biophys. Res. Commun.,* 86, 1288, 1979.

13. THE PREPARATION OF DIMETHYLAMINOAZOBENZENETHIOCARBAMOYL-(DABTC)-PEPTIDES

DABTC-peptides are prepared by coupling polypeptides with dimethylaminoazobenzene isothiocyanate (DABITC) through their amino groups. These derivatives can be separated by reversed-phase HPLC and detected in the visible region at 436 nm. As little as 2 pmol of a DABTC-pentapeptide can be identified against a stable baseline with a signal-to-noise ratio of 10. DABTC-peptides can be recovered from the column and their NH$_2$-terminal amino acids and amino acid compositions and sequences analyzed at the picomole level.

Procedure — The peptide or peptide mixture (1 to 2 nmol) is dissolved in 20 μℓ of water and 40 μℓ of freshly prepared DABITC solution in pyridine (2 mg/mℓ) is added. The mixture is heated at 70°C for 1 hr and excess of DABITC and its byproducts are extracted by mixing the reaction mixture with 5 250-μℓ portions of heptane-ethyl acetate, 2:1, in a vortex mixer. After centrifuging the organic phase is removed with a 5-mℓ syringe. Hydrophilic DABTC-peptides remain soluble in the aqueous phase, while hydrophobic DABTC-peptides tend to form a precipitate at the interface. The DABTC-peptides are freeze-dried and redissolved in aqueous 50% acetonitrile for HPLC analysis. Some very hydrophobic DABTC-peptides are only soluble in aqueous 50% pyridine.

REFERENCE

1. **Chang, J.-Y.,** *Biochem. J.,* 199, 537, 1981.

14. THE REACTION OF PROTEINS WITH CITRACONIC ANHYDRIDE

Reversible masking of amino groups is a valuable procedure for protecting these groups from side reactions which might occur during modification of other functional groups in proteins and peptides. Reversible masking of amino groups is also useful for making hydrolysis with trypsin specific for cleavage at arginine residues. Several reversible masking reagents have been described. Citraconic anhydride is one of the most satisfactory of these, since on deblocking it yields homogeneous preparations identical with the native protein in biological properties and in conformational and hydrodynamic properties. Citraconic anhydride reacts with amino groups to give two reaction products.

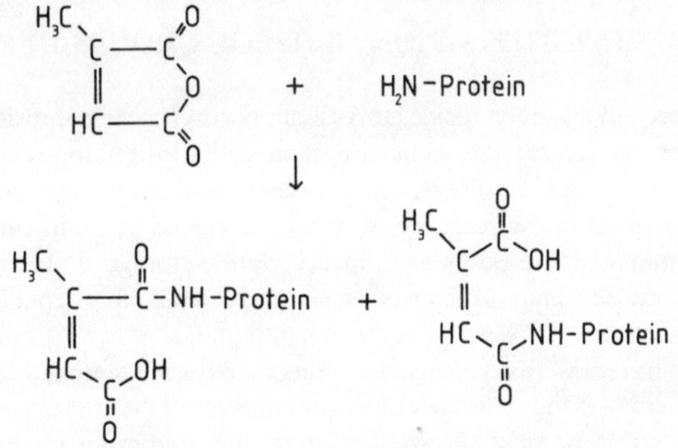

This apparent complication is not serious, since the two reaction products do not differ greatly in stability. Citraconylation is carried out at pH 8 to 9. The following reaction with lysozyme gives an example of the procedure.

Procedure — Lysozyme, 1.96 g, is dissolved in water and the pH is adjusted to 8.2. Aliquots (100 μℓ) of citraconic anhydride are added to the magnetically stirred solution at intervals of 30 min, until a total of 800 mg has been added. The reaction proceeds at room temperature and the pH is maintained at 8.2 by the addition of 5 N NaOH using a pH-stat. When reagent addition is completed the reaction mixture is stirred for 2 hr at room temperature and pH 8.2. The solution is then dialyzed at 0°C against several changes of water, preadjusted to pH 8.5 to 8.8 with NH_4OH, and is finally freeze-dried. Removal of citraconyl blocking groups is carried out at acid pH. For example, citraconyl-myoglobin (5 mg/mℓ) undergoes complete deblocking within 4 hr when kept at pH 3.5 and 30°C. Deblocking of citraconyl-lysozyme may be carried out in 0.05 M acetate buffer at pH 4.2 and 40°C.

REFERENCE

1. **Atassi, M. Z. and Habeeb, A. F. S. A.,** *Methods Enzymol.*, 25, 546, 1972.

15. THE PREPARATION OF ALKYLAMINONAPHTHYLENESULFONYL PEPTIDES AND AMINO ACIDS FOR HPLC

Akylaminonaphthylenesulfonyl chlorides are synthesized by heating 5-amino-1-naphthylenesulfonic acid, an alkyl iodide and anhydrous KF in dimethylformamide at 130 to 140°C for 18 hr with vigorous agitation, followed by chlorination of the isolated acid using phosphorus pentachloride. To prepare the N,N-dimethyl,diethyl,dipropyl,dibutyl or N-monoisopropylaminonaphthylenesulfonyl derivative, 5 $\mu\ell$ of an acetone solution of the appropriate sulfonylchloride (1 mg/mℓ) is added to 5 $\mu\ell$ of 3.6 mM amino acid or peptide solution mixed with 5 $\mu\ell$ of 0.5 M NaHCO$_3$, and the mixture is shaken vigorously and kept in the dark at 37°C for 1 hr. A 0.1 mℓ volume of ethyl acetate is added to separate the product from water and NaHCO$_3$ and the organic layer is used for chromatography. The alkylaminonaphthylenesulfonyl derivatives have strong fluorescence, which allows their detection at the level of 10^{-11} to 10^{-9} mol.

REFERENCES

1. **Davis, B. A.,** *Biomed. Mass. Spectrom.*, 6, 149, 1979.
2. **Hui, K.-S., Salschutz, M., Davis, B. A., and Lajtha, A.,** *J. Chromatogr.*, 192, 341, 1980.

16. THE AMINOETHYLATION OF THIOL GROUPS IN PROTEINS

Conversion of protein cysteinyl resides to S-(2-aminoethyl)cysteinyl residues has generally been accomplished by treating the reduced protein with ethyleneimine. One disadvantage of the use of this reagent is its hazardous character; another is slow polymerization that ethyleneimine undergoes in the presence of water. Traces of water in commercial samples of ethyleneimine may catalyze polymer formation during storage. Polyethyleneimine binds strongly to many proteins and such bound material can mask thiol-containing regions and prevent aminoethylation.

Schwartz et al.[1] have described an alternative method for the aminoethylation of cysteinyl residues, employing N-(β-iodoethyl)trifluoroacetamide (IE-TFA). Synthesis of IE-TFA was accomplished in two steps. First, β-bromoethylamine hydrobromide salt was reacted with trifluoroacetic anhydride to yield N-(β-bromoethyl)trifluoroacetamide. An iodo/bromo exchange reaction was then conducted to yield IE-TFA. Protein modification is carried out as follows.

Protein is dissolved (1 mg/mℓ) in 6 M guanidine hydrochloride buffered at pH 8.1 with 0.2 N ethylmorpholine acetate. Distilled water used in the preparation of buffer is deoxygenated by boiling, followed by purging with nitrogen. Dithiothreitol to give a 20-fold molar excess with respect to protein disulfide is added, and the solution is blanketed with nitrogen and allowed to stand at room temperature for 4 hr. Alkylation is conducted at pH 8.1 or 8.6, the pH being adjusted or maintained by adding 2 N NaOH using a pH-stat assembly. The reaction mixture is thermostatted at 50°C, maintained under nitrogen, and stirred magnetically. IE-TFA in methanol is added so as to constitute a 25-fold molar excess of reagent with respect to total thiol in the reduction mixture. The volume of methanol is used to give a final concentration of 10% (v/v) in the reaction mixture. The reagent solution is added in two portions, the second being added after the thiol concentration determined by Ellman's[2] method has been reduced to 25% or less of the initial value. This occurs after approximately 1 hr at pH 8.6 or 2 hr at pH 8.1.

The reaction is allowed to proceed until the thiol concentration is reduced to approximately 1% of the original value. A total reaction period of 2 to 2.5 hr at pH 8.6 or 4 to 4.5 hr at pH 8.1 is usually required.

The solution is then brought to pH 5.5 with 2 N acetic acid and is dialyzed against 0.005 M acetic acid. The protein solution is then lyophilized and a portion of the product hydrolyzed with 6 N HCl at 110°C for 24 hr for amino acid analysis. Essentially complete conversion of cysteine to aminoethylcysteine occurs and residues other than cysteine are unaffected.

REFERENCES

1. **Schwartz, W. E., Smith, P. K., and Royer, G. P.,** *Anal. Biochem.,* 106, 43, 1980.
2. **Ellman, G. L.,** *Arch. Biochem. Biophys.,* 82, 70, 1959.

17. THE RADIOIODINATION OF PROTEINS USING A SOLID-STATE REAGENT

Radioiodination provides a very sensitive method for the detection of proteins and polypeptides during purification and biochemical analysis, and is useful for quantification in radioimmunoassays. The availability of solid-state reagents facilitates the use of radioiodine by simplifying the separation of iodinated products from iodinating agent. The reagent Iodo-beads (Pierce Chemical Co.) consist of the oxidant N-chloro-benzene-sulfonamide immobilized on 2.8-mm-diameter nonporous polystyrene spheres. The Iodo-beads have an oxidative capacity of 0.45 μmol per bead for tyrosine-containing peptides. Their use in radiolabeling antiporcine insulin antiserum is described.

Procedure — On the day of use Iodo-beads are washed twice in PBS using approximately 1 mℓ per bead in a small beaker and are then blotted dry with filter paper. PBS is Dulbecco's phosphate buffered saline without divalent cations, which consists of 0.8% NaCl, 0.02% KCl, 0.215% $Na_2HPO_4\cdot7H_2O$ and 0.02% KH_2PO_4 at pH 7.2. 100 μg of antiserum in 0.5 mℓ of PBS is added to a 5-mℓ reaction tube together with 1 mCi of $Na^{125}I$. The iodination reaction is initiated by adding 1 Iodo-bead and is allowed to proceed at room temperature (22°C) in an iodinating hood. The iodinated antiserum is then transferred with a glass or plastic Pasteur pipette to a second tube, leaving the Iodo-bead in the reaction tube. For maximal recovery the bead is washed twice with 1.0 mℓ of PBS and the washes are added to the second tube.

Carrier iodide (0.5 mol of $Na^{127}I$) is added to the second tube to facilitate removal of unreacted radioiodide by gel filtration or by acid precipitation of the protein, and also as a safety precaution to reduce adsorption of radioiodide to glassware. The iodination reaction mediated by Iodo-beads is also effective in small volumes. Iodination of proteins is mediated over a broad pH and temperature range and in buffers containing urea and high salt concentrations.

In an alternative procedure, Iodo-beads are loaded with ^{125}I before the initiation of an iodination. The Iodo-beads are washed twice with 50 mM Na_2HPO_4 buffer, pH 7.4, using about 1 mℓ buffer per 2 beads, and are then dried on Whatman 52 paper. Two beads are added to a 10 μℓ ^{125}I solution, about 1 mCi, diluted with 200 μℓ of the phosphate buffer, and left at room temperature for 5 min. Protein, 5 μg/5 μℓ in 500 mM sodium acetate, pH 5.6, diluted to 20 μℓ with the phosphate buffer, is added to the preloaded Iodo-beads. After incubation for 2 to 15 min, depending on the protein or level of incorporation desired, the reaction contents are transferred directly to an equilibrated Sephadex®-G 100 superfine column and the radiolabeled protein is recovered. The reaction time for maximal incorporation is determined by taking aliquots at intervals, diluting, precipitating with trichloroacetic acid and centrifuging. Both the pellet and supernatant are then counted.

REFERENCES

1. **Markwell, M. A. K.**, *Anal. Biochem.*, 125, 427, 1982.
2. **Hearn, M. T. W.**, Pierce Chemical Company (Box 117, Rockford, Ill.), Preview November 1982.

18. DERIVATIZATION OF COMPOUNDS WITH PRIMARY AMINO GROUPS ON THIN-LAYER CHROMATOGRAPHY (TLC) PLATES WITH FLUORESCAMINE

The derivatization reagent is prepared by dissolving 10 mg of fluorescamine in 20 mℓ acetone and then adding 80 mℓ of hexane. The solution is stable for at least 1 week. Using the predevelopment method a 1-$\mu\ell$ sample in borate buffer, pH 9.0, is applied with a volumetric micropipet 1.5 cm from the lower edge of a TLC plate. After air drying the plate is placed in a tank containing a 1-cm depth of derivatization reagent. After the solvent front has moved at least 10 cm, the plate is removed and the solvent is evaporated with a hairdryer. The plate is then developed with an appropriate solvent system and fluorescence is observed under a long wave (366 nm) UV lamp after air drying. The prelabeling of the plates is carried out in tanks wrapped in aluminum foil to protect the samples from light.

Using the predipping method, the lower 2 cm of the spotted plate is dipped in the derivatizing reagent for 30 min, after which the plate is dried and developed. Of numerous amines, amino acids, and peptides tested, only β-naphthylamine was displaced from the origin during labeling. The method has been applied to the TLC analysis of peptides and is recommended by the authors for its sensitivity and simplicity.

REFERENCE

1. **Nakamura, H. and Pisano, J. J.**, *J. Chromatogr.*, 121, 33, 1976.

19. THE PREPARATION OF DIPEPTIDE DERIVATIVES FOR GAS CHROMATOGRAPHY/MASS SPECTROMETRY (GC/MS)

This technique has been applied to mixtures of dipeptides obtained on hydrolysis of polypeptides by dipeptidylaminopeptidase I/IV, the digest then being fractionated by HPLC. The HPLC eluent containing the dipeptides is transferred to a Reacti-vial equipped with a PTFE-lined screw cap and lyophilized to dryness. After adding approximately 200 $\mu\ell$ of dry methanol, the mixture is cooled in a dry ice-ethanol box and 40 $\mu\ell$ of thionyl chloride is slowly added. The solution is heated in a constant temperature block at room temperature for 20 min, and after removal of the reagents *in vacuo,* 100 $\mu\ell$ of perfluoropropionic anhydride (PFPA) is added. After reaction at room temperature for 15 min the PFPA is removed *in vacuo* and the residue is dissolved in 50 $\mu\ell$ of dry acetone. 5 $\mu\ell$ of the solution is used for GC/MS analysis. Dipeptides containing arginyl residues, require additional derivatization of the guanidine moiety, the arginyl residue being converted to the (4,6-dimethylpyrimid-2-yl) ornithine derivative for GC/MS analysis.

REFERENCE

1. **Lin, S.-N., Smith, L. A., and Caprioli, R. M.,** *J. Chromatogr.*, 197, 31, 1980.

20. THE PREPARATION OF TRIMETHYLSILYL (TMS) DERIVATIVES OF DIPEPTIDES FOR GC

A sample of about 0.5 mg of each dipeptide is placed in a Teflon-capped Hypo vial (Pierce Chemical Co.) and trimethylsilylated with 0.4 mℓ of a mixture of BSTFA (bis(trimethylsilyl) trifluoroacetamide) and acetonitrile (1:1) by heating for 30 min at 120°C in a sand bath.

REFERENCE

1. **Dizdaroglu, M. and Simic, M. G.,** *Anal. Biochem.*, 108, 269, 1980.

21. THE DETECTION OF RACEMIZATION DURING PEPTIDE SYNTHESIS

Racemization can be a serious problem in the chemical synthesis of peptides. Unequivocal and sensitive methods for detecting racemization are necessary for the synthesis of peptides to be used in biological studies. A highly sensitive enzymic method, based on the generally low side-chain selectivity and absolute L-specificity which is found under suitable conditions for leucine amino peptidase, has been described. Peptides with the structure L-Ala-D-Ala-L-Ala-L-Ala- . . . in which the next to the last residue from the amino end is of the D-configuration, resist enzymic attack completely.

A dipeptide with the L-D configuration is coupled with an all-L-alanine peptide, the blocking groups are removed, and the peptide digested with leucine amino peptidase. Any racemization during the coupling produces a peptide with all-L configuration which in consequence is split to free alanine residues on enzymic digestion. The peptide with unchanged configuration resists hydrolysis by leucine amino peptidase. The amount of alanine liberated by the enzyme corresponds to a multiple of the degree of racemization occurring during the coupling. The alanine is determined by amino acid analysis. The sensitivity of the method depends on the amplification factor introduced by the amino fragment, the resolving power of the amino acid analysis, and the optical purity of the D-amino acid used in the synthesis of the carboxyl fragment. Under some conditions as little as 0.2% racemization can be detected. No manipulations which may lead to additional racemization are necessary before the enzymic hydrolysis. Due to the broad specificity of leucine amino peptidase, the procedure may be extended to an array of different peptides. An example of the technique is described.

Procedure — Preparation of Z-L-Ala-D-Ala (Z = benzyloxycarbonyl): D-alanine (2.67 g, 30 mmol) and NaHCO$_3$ (5.05 g, 60 mmol) are dissolved in 75 mℓ of water. Z-L-Ala-ONSu (–ONSu = succinimide-oxy-), (8.9 g, 30 mmol) are dissolved in 75 mℓ of dioxane and are added to the solution, which is allowed to stand at room temperature overnight. The mixture is concentrated to about 50 mℓ, is brought to pH 1 with 6 N HCl and extracted 4 times with a total of 500 mℓ of ethyl acetate. The extract is washed twice with 100 mℓ of water and saturated NaCl solution, and is dried over MgSO$_4$. Evaporation leaves an oily residue which solidifies in the presence of a little diisopropyl ether. Recrystallization from ethyl acetate/diisopropyl ether gives 5.9 g (67%) of colorless dipeptide, m.p. 115 to 116°C.

Fragment condensation — Preparation of Z-L-Ala-D-Ala-L-Ala-L-Ala-ONb (–ONb = *p*-nitrobenzyloxy): Z-L-Ala-D-Ala (295 mg, 1 mmol), HBr.H-L-Ala-L-Ala-ONb (395 mg, 1 mmol) are dissolved in 6 mℓ of dimethylformamide. The hydrobromide is neutralized with 1 or 2 equiv. of a tertiary amine (triethylamine or *N*-ethylmorphine) and the mixture cooled in an ice-salt bath to −10°C. *N,N'*-dicyclohexylcarbodiimide (206 mg, 1 mmol) is added in solid form to the well-stirred solution, stirring being continued at 4°C overnight. The reaction mixture is filtered and the protected tetrapeptide precipitated from the filtrate by the addition of 20 vol of 0.05 *N* HCl. The crude product is filtered off, washed with cold water, and dried *in vacuo* over P_2O_5.

Catalytic hydrogenation — The blocked peptide is dissolved in acetic acid (1 to 5 g/ 100 mℓ, according to solubility). Slight heating may be required to obtain a clear solution. Pd/C (10%, 200 mg/g peptide) suspended in water is added and hydrogen is bubbled through the stirred mixture for 3 to 4 hr. The catalyst is filtered off, the filtrate evaporated, taken up in 20 to 50 mℓ of water, and again evaporated to complete dryness. The residue is suspended or partially dissolved in water (about 5 mℓ/g peptide) and 10 to 20 vol of acetone are added. After 2 hr at 4°C the free peptide is collected by filtration, washed with cold acetone, and dried *in vacuo* over P_2O_5. Enzymic digestion with leucine aminopeptidase. The stock solution of substrate is 30 mmol in water. The commercial preparation of leucine aminopeptidase, delivered as a suspension, is diluted to 0.05 mg/mℓ with 50 mmol veronal-HCl of pH 8.6 containing $MnSO_4$ (c = 10 m*M*) and is preincubated at 37°C for 1 hr immediately before use. Equal volumes (usually 0.2 mℓ) of substrate and enzyme solutions are mixed in a small test tube covered by a piece of parafilm and the mixture is incubated at 37 ± 1°C for 3 hr. An aliquot of 50 µℓ is pipetted into 0.2 mℓ of citrate buffer (0.2 *M*, pH 2.2) to give the sample for amino acid analysis.

Amino acid analysis is performed on the long column of an automatic analyzer, with elution buffers of pH 3.25 and 4.25 and a temperature of 55°C. The hydrolysis-resistant peptides are eluted after the alanine peak. The separation of free amino acids from intact peptides is sufficient to permit heavy overloading of the column and determination of at least 1 ‰. The percentage of racemization is calculated as 200 × A/P × amplification factor, where A is the amount of free amino acid determined in the analysis and P the original amount of peptide contained in the same sample. The latter value is based on the nitrogen analysis of the free peptide.

REFERENCE

1. **Bosshard, H. R., Schechter, I., and Berger, A.,** *Helv. Chim. Acta,* 56, 717, 1973.

22. THE PREPARATION OF A BONDED TRIPEPTIDE STATIONARY PHASE

Bonded optically active tripeptides have been applied as stationary phases for liquid chromatography.[1] Initial chloride groups for peptide attachment are first placed on the silica gel. Ten grams of Bio-Sil® A and 12 g of reagent Y-5918 [1-trimethoxysilyl-2-(4-chloro-methylphenyl)-ethane] are placed in 50 mℓ of benzene and shaken overnight. The material is then rinsed with benzene and chloroform and dried. Analysis shows 10% bonded C and 2.69% Cl. The silica gel CT (3.5 g) is placed with 7 g of Y-5918 in 50 mℓ of dry benzene overnight and is then refluxed for 3 hr. The material is washed with benzene and dried. The first amino acid is then attached to the modified surface. The silica gel and Boc-amino acid (Boc = *tert*-butoxycarbonyl) are placed in 150 mℓ of absolute ethanol containing 5 mℓ of triethylamine and the mixture is refluxed with constant stirring by heating in an oil bath at 90°C. Refluxing is continued for 48 hr. The material, after washing and drying, is transferred to a 5 g Merrifield-type reaction vessel where the *tert*-Boc protecting group is removed by 4 *N* HCl-dioxane. The second and third Boc-amino acids are attached via a standard diimide coupling schedule with appropriate washing and cleaning procedures.[2]

REFERENCES

1. **Kitka, E. J., Jr. and Grushka, E.,** *J. Chromatogr.*, 135, 367, 1977.
2. **Stewart, J. M. and Young, J. D.,** *Solid State Peptide Synthesis*, W. H. Freeman, San Francisco, 1969.

23. PURIFICATION OF ACETONITRILE FOR HPLC

Acetonitrile is one of the most widely used solvents for HPLC. To be used for this purpose, the acetonitrile must be of high purity. A simple method of purifying the commercially available solvent by column chromatography has been described. The absorption at 220 nm or 230 nm of the purified acetonitrile is decreased to values which provide a constant baseline in HPLC, making the analysis of peptides and proteins easier.

Procedure — Acetonitrile is purified chromatographically over aluminum oxide 90 (active, neutral, particle size 0.063 to 0.200 mm, for column chromatography, Merck). For analysis columns, 1 × 20 cm, filled with about 20 g of aluminum oxide, are used. For preparative purposes columns, 2.5 × 50 cm, containing about 200 g of aluminum oxide are employed. A flow rate of 200 to 400 mℓ/hr is used.

REFERENCE

1. **Braunitzer, G., Rücknagel, P., and Oberthür, W.,** *Hoppe-Seyler's Z. Physiol. Chem.*, 363, 485, 1982.

24. THE CHOICE OF A SOLVENT SYSTEM FOR PREPARATIVE HPLC

Gabriel et al. have described the factors to be taken into consideration when choosing a solvent system for the chromatographic purification of protected synthetic peptides on silica gel 60. Crude samples applied to the column, 3.8 × 43 cm, varied in amount from 100 mg to 7.5 g. The guiding principle for the choice of a suitable system was the solvent polarity. The most useful solvents, in order of increasing polarity, were 1-chlorobutane, chloroform, methylene chloride, isopropanol, ethanol, methanol, acetic acid, and water. In a satisfactory system the desired product must be separated from contaminants as well as being eluted in as short a time as practical, i.e., generally in less than 10 column volumes of eluent. Column volumes on silica gel columns are related to R_F values of thin layer chromatograms on silica gel G plates by the equation $1/R_F - 1 = k'$, in which the capacity factor k' equals column volumes beyond the void volume, V_o. TLC of the material to be purified can thus serve as a guide in the choice of an appropriate solvent system. The relationship between R_F and k' is given below:

$$R_F \quad 0 \quad 0.1 \quad 0.2 \quad 0.3 \quad 0.4 \quad 0.5 \quad 0.6 \quad 0.7 \quad 0.8 \quad 0.9 \quad 1.0$$

$$k' \qquad 9.0 \quad 4.0 \quad 2.3 \quad 1.5 \quad 1.0 \quad 0.7 \quad 0.4 \quad 0.2 \quad 0.1 \quad 0$$

A substance with an R_F of 0.1 in a given solvent system would require 10 column volumes ($k' = 9 + V_o$) of that system for solution from the column, and a substance with an R_F of 0.7 would require 1.4 column volumes ($k' = 0.4 + V_o$) for elution. The best results were obtained when compounds eluted between 2 and 5 column volumes, i.e., k' between 1 and 4 and R_F between 0.2 and 0.5. The peak width of compounds eluted from the preparative silica gel 60 column was generally between 0.4 and 1.0 column volume, which meant that adjacent peaks should be separated by at least 0.5 k' to minimize overlap. The values above show a k′ difference of 0.9 between R_F 0.5 and 0.9, but between R_F 0.1 and 0.5 there is a k′ difference of 8 and components in this range will be better resolved. It was found that substances generally eluted somewhat faster from the silica gel columns than expected from their R_F values. It was therefore good practice to commence with a slightly less polar solvent and add increasing amounts of alcohols or acetic acid as required.

REFERENCE

1. Gabriel, T. F., Jimenez, M. H., Felix, A. M., Michalewsky, J., and Meienhofer, J., *Int. J. Pept. Prot. Res.*, 9, 129, 1977.

25. TRACE ANALYSIS OF THE MIF ANALOGUE PARAPEPTIDE IN BLOOD PLASMA

The analysis of a pharmacologically active peptide in body tissues presents problems due to the relatively low concentration involved and the presence of other peptides, proteins, and amino acids. Krol et al. have described a procedure for the determination of a synthetic analogue of MIF (melanocyte stimulating hormone release-inhibiting factor). This analogue is parapeptide, L-prolyl-N-methyl-D-leucyl-glycinamide. Parapeptide has no natural fluorescence or strong UV absorption and the only readily derivatized group is the secondary amine of the proline moiety.

Krol et al., therefore, investigated HPLC of a fluorescent parapeptide derivative, the reagent used for reaction with the prolyl secondary amine group being 7-chloro-4-nitrobenzyl-2-oxa-1,3-oxadiazole (NBD-Cl). The advantages of NBD-Cl over dansyl chloride are a low fluorescent excess reagent background and higher solubility and stability of the derivative in aqueous solution.

The following method is used for the derivatization of parapeptide in plasma. Saturated dibasic potassium phosphate solution, the pH of which is adjusted to 11.5 with 50% KOH, is added to deproteinized plasma until pH 9.5 is reached. One volume of the pH 9.5 plasma is then mixed with 2 vol of an acetonitrile solution of NBD-Cl. Since the water-acetonitrile mixture yields a two-phase system, derivatization is facilitated by continuous tumbling of the capped vials in a 50°C water bath for 40 min. A rotating wheel device is used for this purpose. The vials are the removed from the bath and acetonitrile is evaporated in a nitrogen stream. To reduce overloading of the reversed-phase column and prolong its life, the aqueous phase of the reaction mixture is then extracted with 2 mℓ of pentane-diethyl ether (1:1) solution. The partially purified aqueous phase is then extracted with 3 2-mℓ portions of ethyl acetate and the organic layers are combined.

This procedure is carried out in a centrifuge tube so that the samples can be spun, thus affording better separation of the aqueous and organic phases. The ethyl acetate layer is then blown to dryness with a stream of nitrogen and the residue is dissolved in the chromatographic solvent (1 or 2 mℓ depending on the amount of parapeptide NBD derivative). The parapeptide derivative is separated on an octadecylsilane bonded column and quantitated using a short wavelength excitation fluorometric detector. The detection limit is 5 ng/mℓ of plasma.

REFERENCE

1. Krol, G. J., Banovsky, J. M., Mannan, C. A., Pickering, R. E., and Kho, B. T., *J. Chromatogr.*, 163, 383, 1979.

Section V
Products and Sources of Chromatographic Materials

Section V

PRODUCTS AND SOURCES OF CHROMATOGRAPHIC MATERIALS

This section contains descriptions and sources of widely used chromatographic materials. Materials not mentioned may serve equally well in many cases as those listed. Other sources may be available for the materials in addition to those given. The sources described are not necessarily intended as the author's endorsement. Interested readers will find extensive lists of chromatography instruments, accessories, supplies, manufacturers, and dealers in publications such as *The Journal of Chromatographic Science, Analytical Chemistry Lab Guide*, and the *American Association for the Advancement of Science Guide to Scientific Instruments*. Technical information was obtained from the catalogs and literature of the various companies, whose help is gratefully acknowledged.

MATERIALS FOR GAS CHROMATOGRAPHY COLUMNS

A variety of materials has been used for gas chromatography (GC) columns; the following describes some of the relevant factors to be considered in making a choice.

Glass: Glass is generally recommended for the trace analysis of drugs, steroids, and pesticides, and hydrogen bonding compounds such as amines, phenols, and fatty acids. Glass columns are more efficient than those made of metal. To make the inner column walls more inert they should be silanized with 5% dimethyldichlorosilane in toluene and then rinsed with anhydrous toluene and methanol before use.

Nickel: Nickel is more inert than stainless steel and should be used to replace it. Nickel may be used in the low-level analysis of biomedical compounds and amines and for the analysis of highly corrosive halogen compounds. Nickel tubing should be rinsed with organic solvents and dilute nitric acid before use.

Stainless steel: Stainless steel is reasonably inert and is widely used, but is not suitable for the trace analysis of polar compounds, drugs, biomedical compounds, and water. The performance of these columns is improved by rinsing with organic solvents and dilute nitric acid before use.

Teflon®: Teflon® is used in the trace analysis of sulfur dioxide, hydrogen sulfide, mercaptans, and water, and in the analysis of corrosive and reactive inorganic halides, COS, HF, H_2S, F_2, etc. However, diffusion of air through the column walls can adversely affect the response of an electron capture detector or shorten the life of thermal conductivity detector filaments.

Aluminum: The inner walls of aluminum columns are coated with aluminum oxide which can react with certain compounds. Aluminum columns, however, have been used successfully for analyzing certain chlorinated pesticides.

Copper: The inner walls of copper columns contain copper oxide which can be reactive. Oxygen must therefore be excluded. Copper columns are good for trace water analysis but must not be used in the analysis of unsaturated compounds.

Technical information from Foxboro/Analabs catalog.

LIQUID PHASES FOR GC

A number of attempts have been made to define the selectivity of GC phases. Kovats in 1959 introduced the concept of retention index values (I), a method that derives a representative number for an organic compound from a plot of the adjusted retention time of the compound vs. the carbon number (X100) of a series of normal alkanes.[1] I values provide the basis for the development of McReynold's constants. McReynold's used test probes to classify the selectivity of liquid phases for different classes of compounds.[2] To produce the table of constants (ΔI), the Kovats indexes (I) of the test compounds are related to the indexes of a standard nonpolar liquid phase, squalane, thus

$$\Delta I = I_{phase \ x} - I_{squalane}$$

ΔI values are usually tabulated in order of increasing liquid phase polarity. This is achieved by adding the values of five probes and listing the phases in order of increasing Σ values. In using McReynold's constants it must be remembered that the higher the ΔI value for a given probe the longer will that class of compounds be retained on a GC column prepared with that liquid phase, or the greater the value of Σ the more selective is the phase. The different classes of compounds represented by each probe are listed in Table 1.

If it is desired to separate a complex mixture of alcohols and ketones, the I values for 1-butanol and 2-pentanone should be examined. McReynold's table shows that only one group of phases, the trifluoropropylsilicones, elute alcohols before ketones. This specificity for ketones could be an advantage if the sample to be examined only contains trace levels of alcohols. McReynold's constants can be used to determine whether a liquid phase will separate one class of chemical compounds from another, but give no information about the performance of the phase for a homologous series or isomers.

Although the probes used differ in polarity, all McReynold's constants are based on the ultimate nonpolar series of *n*-alkanes used to determine retention indexes. The McReynold's system, or variations on it, is designed to characterize the liquid stationary phase, but not a packed column, unless it has been freshly prepared for this experimental purpose. After being in use for some time, there is no simple way of determining the loading of stationary phase remaining in the column. Moreover, nearly 9% of the medium-molecular-weight polyethylene glycol 6000 may evaporate from a support after only 2 hr at 100°C.

McReynold's values also imply the use of the same concentrations of test and standard stationary phases in similar columns. In practice, polysiloxanes are usually prepared with a 2 to 5% loading, while polyglycols and polyesters are often used at 3 to 15% of the weight of the column packing. Betts et al.[3] therefore considered that a practical system of values was needed to express the relative polarities of the columns kept by each laboratory. This automatically took account of the history and possible misuse of individual columns. These authors made a study of 5 of their packed columns, 3 of which contained polysiloxanes. Three C_{10} volatile oil substances which had short retention times and which represented a range of chemical structures were selected as probes. They are (−)-linalool, estragole, and (+)-carvone.

Each of the 5 columns, prepared at different times and used for a variety of purposes, was allowed to stabilize under standard operational conditions and then received injections of (1) mixtures of the 3 probes in alcohol, (2) mixtures of appropriate *n*-alkanes, (3) mixtures of *n*-aldehydes in alcohol, and (4) mixtures of *n*-alcohols in alcohol. Very consistent results for the retention indexes were obtained against alkanes on all the polysiloxanes, against alcohols on SP-2250 and PEG 20 M, and against aldehydes on OV-1, SP-2250, and DEGS. In other cases, results showed a surprising variation for what should have been a standard procedure. The consistent results obtained with the SP-2250 column led to its selection by Betts et al.[3] as the reference base column for their laboratory.

The two fully methyl polysiloxane columns, OV-1 and SP-2100, which would be expected to behave identically, showed different retention indexes. This was least apparent when they were rated against the aldehyde series, but when compared using alkanes and alcohols they were dissimilar, one being almost twice as nonpolar as the other. A DEGS column was the most polar, but the apparent polarity of the PEG column changed with the reference series used. Betts et al. suggested that another series of polar compounds such as alcohols should be used together with alkanes as standard series. They further suggested that their method would enable an individual chromatographer to characterize his own set of columns.

There is considerable evidence that the number of liquid phases for GC in use exceeds a desirable level, and suggestions have been made that the use of impure (mixtures) or undefined compounds (polymers and polyesters) should be discouraged.[4] The methyl silicones represented by SE-30, e-301, OV-1, L-46, DC 200, OV-101, DC 410, DC 401, and silicone oil (May and Baker) have been considered to be essentially identical chromatographically.

Leary et al.[5] applied a nearest-neighbor classification to determine groups and similarities in the data of McReynolds, and proposed a system for reducing the number of gas chromatographic liquid phases. A set of 12 preferred liquid phases was presented, and which of these phases could best be substituted for most of the phases currently in use was tabulated. Other lists of standard or preferred phases have been prepared.[6] Table 2 lists the McReynold's constants for a number of commonly used phases.

Table 1
PROBES USED IN THE CLASSIFICATION OF THE SELECTIVITY OF LIQUID PHASES

Probe	Represents chemical class
Benzene	Aromatics, olefins
1-Butanol	Alcohols, phenols
2-Pentanone	Ketones, aldehydes, esters
1-Nitropropane	Nitro, nitrile compounds
Pyridine	Bases, part aromatic *N*-heterocycles
2-Methyl-2-pentanol	Branched chain compounds
1-Iodobutane	Halogenated compounds
2-Octyne	Olefins, acetylenic compounds
1,4-Dioxane	Ethers, bases
cis-Hydrindan	Terpenes, naphthalenic compounds

Table 2
MCREYNOLD'S CONSTANTS

Phase	Probe										
	1	2	3	4	5	6	7	8	9	10	Σ_1^5
Squalane	0	0	0	0	0	0	0	0	0	0	0
Apolane-87	21	10	3	12	25						71
Apiezon M	31	22	15	30	40	12	32	10	28	29	138
Apiezon L	32	22	15	32	42	13	35	11	31	33	143
SF-96	12	53	42	61	37	31	0	21	41	6	205
Apiezon N	38	40	28	52	58	25	41	15	43	35	216
SE-30	15	53	44	64	41	31	3	22	44	−2	217
OV-1	16	55	44	65	42	32	4	23	45	−1	222
UCW-982	16	55	45	66	42	33	4	23	46	−1	224
DC-200 (12,500)	16	57	45	66	43	33	3	23	46	−3	227
OV-101	17	57	45	67	43	33	4	23	46	−2	229

Table 2 (continued)
MCREYNOLD'S CONSTANTS

Phase	Probe

Phase	1	2	3	4	5	6	7	8	9	10	Σ_i^5
Versilube F-50	19	57	48	69	47	36	7	23	50	−1	240
SE-52	32	72	65	98	67	44	23	36	67	9	334
SE-54	33	72	66	99	67	46	24	36	68	10	337
OV-73	40	86	76	114	85	57		39			401
OV-3	44	86	81	124	88	55	39	46	84	17	423
OV-105	36	108	93	139	86	74		29			462
Dexsil 3PP	41	83	117	154	126						521
OV-7	69	113	111	171	128	77	68	66	120	35	592
Dexsil 400	60	115	140	188	174						677
Diisodecyl phthalate	84	173	137	218	155	133	83	59	130	24	767
OV-61	101	143	142	213	174	99		86			773
OV-11	102	142	145	219	178	100	103	92	164	59	786
Dinonyl phthalate	83	183	147	231	159	141	82	65	138	18	803
OV-1701	81	172	156	245	165	127	80	63	130	40	819
Poly 1-110	115	194	122	204	202	152	116	55	165	76	837
Dexsil 410	85	165	170	240	180						840
OV-17	119	158	162	243	202	112	119	105	184	69	884
Poly A-103	115	331	149	263	214	221	137	62	149	78	1072
OV-22	160	188	191	283	253	133	152	132	228	99	1075
Poly A-101A	115	357	151	262	214	233	136	64	152	76	1099
Didecyl phthalate	136	255	213	320	235	201	126	101	202	38	1159
OV-25	178	204	208	305	280	144	169	147	251	113	1175
PPE (5-ring)	176	227	224	306	283	177	169	135	266	103	1216
Poly A-135	163	389	168	340	269	282					1329
QF-1	144	233	355	463	305	203	136	53	280	59	1500
OV-202	146	238	358	468	310	202	139	56	283	60	1520
OV-210	146	238	358	468	310	208	139	56	283	60	1520
OV-215	149	240	363	478	315	208		56			1545
Triton® X-100	203	399	268	402	362	290	181	145	304	83	1634
OV-330	222	391	273	417	368	284		158			1671
XE-60	204	381	340	493	367	289	203	120	327	94	1785
OV-225	228	369	338	492	386	282	226	150	342	117	1813
Igepal Co-880	259	461	311	482	426	334	227	180	362	112	1939
Triton® X-305	262	467	314	488	430	336	229	183	366	113	1961
Hi-EFF 8BP	271	444	330	498	463	346	252	175	396	127	2006
Carbowax® 20M	322	536	368	572	510	387	282	221	434	148	2308
Carbowax® 20M-TPA	321	537	367	573	520	387	281	220	435	148	2318
Carbowax® 8000	322	540	369	577	512	390	282	222	437	147	2320
Carbowax® 3350	325	551	375	582	520	399	285	224	443	148	2353
OV-351	335	552	382	583	540						2392
Silar 5CP	319	495	446	637	531	379	320	216	470	175	2428
XF-1150	308	520	470	669	528	401	302	174	471	156	2495
Carbowax® 1000	347	607	418	626	589	449	306	240	493	161	2587
Hi-EFF 2AP	372	576	453	655	617	462	325	250	546	177	2673
Reoplex 400	364	619	449	647	671	482	317	245	540	171	2750
Hi-EFF 1AP	378	603	460	665	658	479	329	254	554	176	2764
Carbowax® 1450	371	639	453	666	641	479	325	255	534	172	2770
Silar 7CP	440	638	605	844	673	492	401	268	603	225	3200
EGSS-X	484	710	585	831	778	566	412	316	713	237	3388
Silar 9CP	489	725	631	913	778	566	459	292	696	256	3536
Hi-EFF 1BP	499	751	593	840	860	595	422	323	725	240	3543
Silar 10C	523	757	659	942	801	584	480	298	722	267	3682
THEED	463	942	626	801	893	746	427	269	721	254	3725
TCEPE	526	782	677	920	837	621	444	333	766	237	3742

Table 2 (continued)
MCREYNOLD'S CONSTANTS

Phase	Probe										
	1	2	3	4	5	6	7	8	9	10	Σ_i^5
Hi-EFF 2BP	537	787	643	903	889	633	452	348	795	259	3759
TCEP	593	857	752	1028	915	672	503	375	853	267	4145
OV-275	781	1006	885	1177	1089						4938
Squalane Absolute I values	653	590	627	652	699	690	818	841	654	1006	

Reproduced by permission of Applied Science (1983 catalog).

Table 3
STATIONARY PHASES FOR GC

Silar Phases

Silar phases are a combination of phenyl and cyanoalkyl functional groups linked to a polysiloxane chain. They combine the high polarity required for the GC separation of saturated and unsaturated compounds with the high temperature stability of silicones. They are stable at temperatures of up to 275°C and show selectivity at temperatures above those useful for polyester phases. The available Silar phases encompass the whole range of polarities provided by polyesters:

Silar-5CP
Silar-7CP
Silar-9CP
Silar-10C

Available from Silar Labs, Inc., 10 Alplaus Road, Scotia, N.Y.

Apolane-87

Apolane-87 has the chemical composition 24,24-diethyl-19,29-dioctadecylheptatetracontane. This well-characterized phase of known molecular weight is an ideal nonpolar reference phase for McReynold's constants. The working temperature range of 30 to 260°C permits quantitative polarity comparisons of high temperature phases at their practical temperatures of use. The phase shows no batch to batch variation.

Available from Applied Science, P. O. Box 440, State College, Pa.

Durawax

Durawax can replace Carbowax® for a large number of applications. It has a higher temperature limit (250°C isothermal) and a higher efficiency than Carbowax®. Durawax has a polarity range between methyl silicones and Carbowax®. It is useful for essential oils, alcohols, and volatile fatty acids. Durawax phases available are DX-1, DX-2, DX-3, and DX-4.

Supelco SP-Phases

SP-Silicones
SP-400

SP-400 is a chlorophenylmethyl silicone. It is a gas chromatographic version of DC-560 and has better thermal stability.

SP-2100

SP-2100 is a methyl silicone fluid which can be best used at temperatures from 0 to 350°C. It has the advantages of low bleed at high temperatures and of low viscosity over its usable range. These properties of SP-2100 make it possible to produce columns of high efficiency, which can be used for a wide range of application, from hydrocarbons to steroids.

SP-2250

This phase is a methyl phenyl silicone (50% phenyl). It is similar to OV-17, but has rather lower viscosity and better thermal stability. When used for analysis of polar compounds it must be used with glass columns and the support used should be AW-DMCS treated.

SP-2300 series (SP-2300, 2310, 2330, and 2340)

The SP-2300 series includes four cyanosilicone stationary phases which may be used up to 275°C. The four vary from moderately polar for SP-2340 with two intermediate stages of polarity obtained with SP-2310 and SP-2330. These four cyanosilicones are similar to the polyester stationary phases which have been used to separate fatty acid methyl esters.

SP-2401

SP-2401 is a fluoropropyl silicone similar to QF-1 and OV-210 but with lower viscosity and better thermal stability. The low viscosity allows the preparation of columns with higher efficiency.

Other SP Phases

SP-216-PS

SP-216-PS is a high polarity polyester stationary phase stabilized with phosphoric acid. It was developed for the separation of free fatty acids and their methyl esters in the C12-C20 range.

SP-222-PS

This is a highly polar polyester stationary phase, stabilized with phosphoric acid. This phase was developed to separate fatty acid methyl esters rather than the free fatty acids. The separation obtained differs slightly from that obtained with SP-216-PS.

SP-300

SP-300, *n*-lauroyl-L-valine-*t*-butylamide was developed to separate D- and L-isomers of amino acids.

SP-1000

SP-1000 is a moderately polar stationary phase prepared from Carbowax® 20M and a derivative of terephthalic acid. It is similar to FFAP but has greater thermal stability. It will separate fatty alcohols in order of the degree of unsaturation and also functions as a general purpose stationary phase.

SP-1200

SP-1200 is a low polarity ester type stationary phase, originally developed to separate the C2-C5 free acids in aqueous solution. The phase along with 1% H_3PO_4 is used for C2-C5 free acids. It can also be used with Bentone 34 to separate *o*-, *m*-, and *p*-xylene, styrene, etc.

SP-1220

This is a low polarity stationary phase developed for the separation of C2-C5 fatty acids and is used to identify anaerobes. It is similar to but not identical to SP-1200.

Available from Supelco, Inc., Supelco Park, Bellefonte, Pa.

Ultra-Bond

Ultra-bond describes bonded GC phases with unique characteristics. Three types are available in two mesh ranges for each type. It is important to use oxygen-free carrier gas with Ultra-bond packings.

Ultra-Bond 20M

Ultra-bond 20M is prepared by extensive deactivation of Chromosorb® W followed by coating with Carbowax® 20M and heat treatment. Exhaustive solvent extraction then yields a 0.2% loaded, physically bonded GC phase. It may also be used by itself. Ultra-bond 20M is an ideal support for lightly loaded (1 to 3%) packings. The upper temperature limit for Ultra-bond 20M is 270°C.

Ultra-Bond PEGS

Ultra-bond PEGS is the first bonded polar phase that is stable up to 300°C. The phase retains the polar nature of PEGS (polyethylene glycol succinate). A stable 0.2% load is present. As with Ultra-bond 20M the "zero-bleed" makes Ultra-bond PEGS suitable for GC/MS applications.

Ultra-Bond 11

Ultra-bond 11 is an economy grade of Ultra-bond 20M. It does not have the superior qualities of Ultra-bond 20M and is useful for general purpose analysis.

Available from RFR Corporation, 1 Main Street, Hope, R.I., 02831.

OV Silicones

The OV liquid phases are highly refined materials and extreme care is required in handling and in preparing coated column packings. Glassware and equipment must be acid rinsed to eliminate all possible caustic contamination which may cause decomposition of the silicone liquid phase at increased temperatures. The data on the maximum operating temperature should be taken as a guideline only, since this depends on factors such as the liquid loading level, the length and type of column, and the care taken in preparing the packing and packing the column. A carrier flow through the column should be maintained at all times when it is being heated. Oxygen at increased temperatures causes decomposition of silicones. The types of OV silicones are as follows:

Type	Solvent	Temp. limits (°C)
OV-1		
Dimethylsilicone gum	Toluene	325—375
OV-3		
Phenylmethyldimethyl-silicone 10% phenyl	Acetone	325—375
OV-7		
Phenylmethyldimethyl-silicone 20% phenyl	Acetone	325—375
OV-11		
Phenylmethyldimethyl-silicone 35% phenyl	Acetone	325—375
OV-17		
Phenylmethyl-silicone 50% phenyl	Acetone	350—375
OV-22		
Phenylmethyldiphenyl-silicone 65% phenyl	Acetone	350—375
OV-25		

Type	Solvent	Temp. limits (°C)
Phenylmethyldiphenyl-silicone 75% phenyl OV-61	Acetone	350—375
Diphenyldimethyl-silicone 33% phenyl OV-73	Acetone	325—375
Diphenyldimethyl-silicone gum 5.5% phenyl OV-101	Toluene	325—350
Dimethylsilicone fluid OV-105	Toluene	325—375
Cyanopropylmethyl-dimethylsilicone OV-202	Acetone	275—300
Trifluoropropylmethyl-silicone OV-210	Chloroform	250—275
Trifluoropropylmethyl-silicone OV-215	Acetone	275—350
Trifluoropropylmethyl-silicone gum OV-225	Chloroform	250—275
Cyanopropylmethylphenylmethyl-silicone OV-275	Acetone	250—300
Dicyanoallylsilicone OV-330	Acetone	250—275
Phenylsilicone-Carbowax®-copolymer OV-351	Acetone	250—275
High quality FFAP	Chloroform	250—270

Available from Ohio Valley Specialty Chemical Co.

HIGH TEMPERATURE CHROMATOGRAPHIC PHASES

Dexsil

Dexsil polymers are linear molecules incorporating both carborane and siloxane units in the chain. The meta-carborane consists of 10 boron atoms and 2 carbon atoms in a 3-dimensional structure. Meta-carborane acts as an "energy sink" stabilizing its surroundings against the disruptive forces of heat at very high temperatures.

Dexsil 300 GC

Dexsil 300 GC is used for separations where great stability is necessary at the highest temperature levels. It has low bleed properties and is used as a general column for the analysis of organic compounds and also to prepare inert surfaces in GC such as in the analysis of trace amounts of metal chelates.

Dexsil 400 GC

Dexsil 400 GC shows little tailing in the separations of compounds containing secondary amine groups. It is excellent for the analysis of polynuclear aromatic compounds and is very stable in the routine analysis of drugs and vitamins.

Dexsil 410 GC

Dexsil 410 GC has a 2-cyanoethyl group incorporated in the polymer. This group gives unique selectivities in high temperature separations of compounds containing Pi-electrons such as esters, ketones, and aromatic compounds. Dexsil 410 GC efficiently separates compounds containing secondary amine groups.

Alternatives to Dexsil Polymers

Alltech has developed some phenyl silicone polymers that can be used in place of Dexsil for most applications. The upper temperature limit, however, is lower for these polymers than for the Dexsils. In place of Dexsil 300, Alltech PS-300 is recommended and for Dexsil 400 and Dexsil 410, Alltech PS 400 and Alltech PS 410, respectively, are suggested.

Available from Dexsil Chemical Corporation.

Pennwalt Amine Packings

28% Pennwalt 223 Amine Packing

This packing was designed for the analysis of amines. It offers excellent selectivity and gives symmetrical peaks for amines, diamines, and alcohols. The packing consists of 28% Pennwalt 223 and 4% KOH on 80/100 Gas Chrom R. The upper limit of temperature is 200°C.

Pennwalt 231 GC Packing

This packing was developed for the analysis of ammonia, monomethylamine, methanol, dimethylamine, and trimethylamine in process streams. Aqueous solutions are suggested for the amine analysis for ready solution of monomethylamine. After the run water may be driven off by heating the column at 120°C for 15 min. Pennwalt 231 GC Packing should not be employed above a temperature of 120°C. Overnight conditioning at 120°C with 40 mℓ/min carrier flow is recommended.

Available from Pennwalt Corporation, Pennwalt Building, Three Parkway, Philadelphia, Pa.

Table 4
SUPPORT MATERIALS FOR GC

In GC the solid support should provide a large inert surface area. In practice a completely inert support is not possible. Most gas chromatographic support materials are of natural origin and without suitable treatment would contribute undesirable chemical activity for many separations.

Diatomaceous Earth Supports

The most popular GC supports are prepared from diatomaceous earth, also known as diatomaceous silica or kieselguhr. Diatomaceous earth consists of about 90% silica and is composed of the skeletons of fossilized diatoms, which are microscopic algae containing only one cell. The skeletons are extremely small, very porous, and have a relatively high surface area. Diatomaceous earth is used to produce two types of support for GC, white colored and pink colored. The white colored supports are prepared from diatomaceous earth which has been calcined above 900°C with sodium carbonate as flux. In this process the diatomaceous earth fuses and is held together by sodium silicate glass, the silica being partly converted to crystalline cristobalite. The product is white in color due to the conversion of iron oxide to a colorless complex of sodium iron silicate. The most inert supports are prepared from these products. Excessive handling of these fairly fragile supports during sieving, coating, and column packing abrades the particles, producing "fines" and tending to decrease column efficiency.

The pink colored supports are prepared from crushed firebrick in which the diatomaceous earth has been calcined with a clay binder above 1000°C. The metal impurities remaining form complex oxides which contribute to the pink color of the supports. The pink supports are denser than white supports because of the greater destruction of the diatom structure during the calcining treatment. They are also harder and less friable than white supports, and capable of holding larger amounts of liquid phase (up to 30%). The surface of the pink supports is generally more adsorptive and the support is not recommended for use in the analysis of polar compounds, but is efficient in the analysis of hydrocarbons and organic compounds of low polarity.

Neither type of support gives generally acceptable analysis without further treatment. With more polar compounds severe peak tailing (due to the presence of adsorptive and catalytic centers) may be observed, particularly with the dense pink supports. These adsorptive sites are attributed to metal oxides (Fe, Al) on the surface, which catalyze degradation, and to surface silanol groups, which can form hydrogen bonds with polar compounds. Washing with HCl removes mineral impurities and provides the acid washed supports, which are less adsorptive. Even with acid washing the pink supports are still more adsorptive toward polar compounds than the white type supports. Subsequent base washing does little toward removing metal impurities, but is a good pretreatment for supports to be used in analyzing basic compounds.

Neither acid nor base washing reduces peak tailing due to hydrogen bonding with surface silanol groups. These groups are masked by reaction with chlorosilanes to produce a partially silanized surface. Dimethyldichlorosilane is often used; this reagent is more effective in producing inert surfaces than hexamethyldisilazane or trimethylchlorsilane which were used earlier.

The silanized supports have hydrophobic rather than hydrophilic surfaces, which are found on the nonsilanized diatomite supports. In consequence, polar GC phases such as polyesters, DEGS and EGS, and silicones with a high content of cyano groups such as OV-275 do not wet the surface of the silanized supports, but do wet the surface of the nonsilanized support to provide a more evenly coated surface. Nonsilanized acid-washed supports perform efficiently with the most polar phases.

Technical information from Foxboro/Analabs catalog.

Chromosorb® Supports

Chromosorb® A

Chromosorb® A was developed for use in preparative scale GC. It has a good capacity for the liquid phase, a porous structure that does not readily break down on handling, and a surface that is not highly adsorptive. Chromosorb® A is a flux-calcined diatomite support, whose physical appearance and structure are similar to those of Chromosorb® P, but whose surface and chromatographic properties are closer to those of Chromosorb® W. Chromosorb® A is inherently more adsorptive than Chromosorb® W or G, but since it is designed to be used at 25% maximum loading, the adsorptive effects are minimized. The alkaline pH of its surface reduces catalytic effects which sometimes occur due to the acid surface of Chromosorb® P.

Chromosorb® G

Chromosorb® G was developed for the GC analysis of polar compounds. It has a low surface area and compared to other diatomite packings is nonfriable. Chromosorb® G will not produce fines during coating or packing procedures. This support, which is 2.5 times as heavy as Chromosorb® W, is limited to a maximum liquid loading of 5%. It has the good handling characteristics of Chromosorb® P with the comparatively nonadsorptive surface of Chromosorb® W.

Chromosorb® P

Chromosorb® P is a calcined diatomite. It is orange-pink in color and relatively hard. It gives high column efficiencies for hydrocarbons, while its surface is more adsorptive than other Chromosorb® diatomite grades due to its high surface area.

Chromosorb® W

Chromosorb® W is the most popular grade of support used for GC. It is flux-calcined diatomite, white in color, and friable compared to other diatomite grades. Care must be taken when handling this material during coating and packing procedures in order to achieve maximum performance. Chromosorb® W is relatively nonadsorptive and is recommended for the separation of polar compounds.

Chromosorb® 750

Chromosorb® 750 is flux-calcined and has better handling characteristics than Chromosorb® W. It is prepared from high purity diatomite crude with exhaustive acid washing and silane treatment. Its chemical inertness makes Chromosorb® 750 ideal for biomedical and pesticide analysis.

Chromosorb® is available in various surface treatments.

Nonacid Washed (NAW)

Chromosorb® grades A., G., P., and W. are available in an untreated form described as nonacid washed (NAW).

Acid Washed (AW)

Chromosorb® grades G., P., and W. are available with the acid washed treatment, which removes mineral impurities from the support surface and reduces surface catalytic activity. Hydrochloric acid is used to remove soluble iron and the support is then washed to nearly neutral with deionized water.

Acid Washed Dimethyldichlorosilane Treated (AW-DMCS)

Chromosorb® grades G., P., and W. are available with the combined acid washed and DMCS treatment. The DMCS treatment converts the surface silanol (Si-OH) groups to silyl ethers which reduces surface activity of the support and decreases peak tailing. The AW-DMCS treatment gives the most inert surface for these 3 types of Chromosorb®.

Hexamethyldisilazane Treated (HMDS)

Chromosorb® grades P. and W. are available with HMDS treatment on nonacid washed supports. This treatment deactivates the support by converting the surface silanol groups to silyl ethers. Tailing is thereby considerably reduced. The HMDS-treated Chromosorb® P and W. are not as inert as the AW-DMCS-treated forms but are useful where a high degree of deactivation is not necessary.

Chromosorb® H.P.

Chromosorb® H.P. is a high quality acid-washed and silanized flux-calcined diatomite with superior inertness and column efficiency, no catalytic surface activity, and short column conditioning time. It was developed for use with steroids, bile acids, alkaloids, and the analysis of pharmaceuticals, medical, and toxicological compounds. Chromosorb® W. and Chromosorb® G. are available in the H.P. grade.

Chromosorb® R-6470-1 for SCOTS Columns

Chromosorb® R-6470-1 is a diatomaceous silica developed for use in GC to coat the inside walls of capillary columns. It is produced from flux-calcined Celite diatomaceous silica and has a surface area of 5 to 6 m²/g. Chromosorb® R-6470-1 has a predominant particle size of 1 to 4 µm and retains the Celite diatom structure and purity of Chromosorb® W.

"Prep-Scale" Chromosorb® Grades

Chromosorb® A., P., W., and G. preparative grades are more modestly priced supports, intended for use in large amounts where accurately sized and graded materials are not necessary. These supports are roughly 30/60 mesh, but they contain smaller and larger particles as well as some dust. They should not be used for analytical work.

Acid Washed Celite 545 (Reagent Grade)

This is a flux-calcined diatomaceous silica filter aid specially prepared for laboratory use in chromatography and for other specialized applications. It is a dry white powdered filter aid of high purity, inertness, and uniformity and is processed by a suitable combination of acid treatment and leaching. Acid washed Celite 545 (reagent grade) is recommended for use as a support in column chromatography.

Chromosorb® Century Series

Porous Polymer Supports

Chromosorb® porous polymers are of polyaromatic-type, cross-linked resins. These resins have a uniform rigid structure of a distinct pore size and are infusible and insoluble in most solvents. Porous polymers provide excellent separations of a variety of gases and compounds with boiling points up to about 250°C. They are also recommended for analyzing aqueous solutions containing organic compounds capable of hydrogen bonding such as amines, alcohols, glycols, and free acids. These polymer beads, when packed in long columns and operated at subambient temperatures, provide separations similar to those provided by molecular sieve columns. Chromosorb® polymer beads are rigid in structure and can be packed into columns using normal techniques. The polymer beads cannot be crushed or fractured with normal handling. The column should be conditioned with inert gas, such as helium or nitrogen, with flow for at least 2 hr at 25°C below the maximum recommended isothermal temperature. This conditioning process removes light organic materials such as monomers, solvents left in the polymer matrix from the manufacturing process, and any materials adsorbed from the atmosphere.

Chromosorb® 101

Chromosorb® 101 has a relatively low affinity for hydroxyl-containing compounds. As the hydroxyl content of the compound increases, or the hydroxyl groups are moved closer together, the retention time decreases. This support shows no interaction, that is no tailing with oxygenated compounds, especially hydroxyl compounds. Chromosorb® 101 effectively separates hydrocarbons, alcohols, fatty acids, esters, aldehydes, ketones, ethers, glycols, vinyl chloride, bischlormethyl ether, styrene, acrylonitrile, and chlorinated solvents. It can be used to analyze alcohols in blood, cosmetics, and beverages, metabolites in urine, quaternary ammonium compounds in biological fluids, and to determine the carboxylic acids content of nonvolatile esters.

Chromosorb® 102

Chromosorb® 102 is a porous styrene-divinylbenzene copolymer with a surface area of about 300 to 400 m²/g. Samples can readily penetrate its pores. Columns prepared from it can be operated isothermally up to 250°C; somewhat higher temperatures are possible during a temperature-programed run. Because of its organic composition, the chromatographic characteristics of Chromosorb® 102 are similar to those of a liquid phase in gas-liquid chromatography. Retention times on the column are relatively high. Chromosorb® 102 will separate light and permanent gases, as well as lower molecular weight compounds such as acids, alcohols, glycols, ketones, esters, and hydrocarbons. It efficiently separates vinyl chloride, nitrogen, oxygen, and argon. Chromosorb® 102, 200/325 mesh, can be used as column packing material for liquid chromatography. Polymer packings have the advantage of using mobile phases having a pH range 0 to 14.

Chromosorb® 103

Chromosorb® 103 is a polyaromatic porous packing material developed for the separation of amines and basic compounds. It will not handle acidic materials, glycols, or other compounds as acidic as glycols; these are absorbed totally. Methylamine is readily separated from light gases such as ammonia. The retention time of diamines decreases as the amino groups move closer together, 1,2-propane diamine preceding 1,3-propanediamine. Although Chromosorb® 103 will handle both alcohols and amines it shows greater selectivity for basic compounds. Thus, *t*-butylamine elutes before *t*-butyl alcohol. At relatively low temperatures the water peak tails somewhat, but this is less noticeable above 150°C.

Chromosorb® 104

Chromosorb® 104 is an acrylonitrile-divinylbenzene type resin with a highly polar surface. It is efficient for gas analysis at subambient, ambient, and higher temperatures and effectively separates isomeric xylenols, alcohols, ketones, nitriles, aldehydes, and hydrocarbons. Important characteristics of Chromosorb® 104 are its ability to separate sulfur-containing compounds at low levels, aqueous ammonia and hydrogen sulfide at low levels, and gases of various types. Retention times are longer on Chromosorb® 104 than on other Chromosorb® century series porous polymers, and Chromosorb® 104 also has the highest polarity in the series. An interesting characteristic of Chromosorb® 104 is that a saturated analogue comes out first, followed by the unsaturates.

Chromosorb® 105

Chromosorb® 105 is a white, nonfriable, insoluble cross-linked polyaromatic resin with an effective surface area of 600 to 700 m²/g. It will effectively separate formaldehyde from water and methanol, acetylene from lower hydrocarbons and most other classes of organic compounds of different polarity having a boiling point up to 200°C. Chromosorb® 105 can be used to analyze quaternary ammonium compounds in biological fluids, and for determining the carboxylic acid content of esters. This support is inert and has high retentive capacity to collect volatiles from headspace and aqueous solutions.

Chromosorb® 106

Chromosorb® 106 is a cross-linked styrene-divinylbenzene type resin with a surface area between 700 and 800 m²/g. It is a hard granular solid. On a GC column its nonpolar surface enables it to retain nonpolar organic compounds longer than polar compounds. Chromosorb® 106 is stable up to 250°C. It gives excellent separations of gases and low boiling compounds and resolves C_2-C_5 fatty acids from C_2-C_5 alcohols. Chromosorb® 106 may also be used as an adsorbent to trap organics from air or water.

Chromosorb® 107

Chromosorb® 107 is a moderately polar cross-linked acrylic ester type resin with a surface area between 400 and 500 m²/g. It is a hard granular solid with different retention characteristics than other century series polymers. It efficiently separates various classes of compounds. Chromosorb® 107 elutes water, even at high concentrations, after sulfur dioxide. It is the preferred adsorbent for trapping and analyzing vinyl acetate.

Chromosorb® 108

Chromosorb® 108 is a moderately polar cross-linked acrylic ester type resin having a surface area between 100 and 200 mg^2/g. It resembles Chromosorb® 107 in surface characteristics but differs in surface area. When properly handled it is stable to 250°C. Chromosorb® 108 gives good separations of gases and aqueous compounds. Water is retained much longer on Chromosorb® 108 than on other Chromosorb® polymer grades except Chromosorb® 104. The retention characteristics of Chromosorb® 108 differ from those of other century series polymers.

Chromosorb® T

Chromosorb® T is a support for GC made from Teflon® 6, which is a high molecular weight TFE-fluorocarbon resin. Teflon® 6 is used as a support in applications in which a highly inert surface is required to avoid tailing problems encountered with compounds such as water, hydrazine, and sulfur dioxide. The TFE resin particle consists of an aggregate of very small (0.2 to 0.5 μm) particles, which cause it to have a fairly high surface area. The material is relatively soft and fragile. It develops a static charge quite readily, causing it to lump and also adhere to its container; this is particularly marked when the humidity is low. Temperature affects the physical properties of TFE resin, which melts at 327°C. At prolonged operation above 250°C, individual particles begin to fuse. TFE resin undergoes structural changes at 19 and at 30°C the change is only a tenth as large as the 19°C change. TFE resin is insoluble in common solvents and is highly resistant to chemical attack. Chromosorb® T should be handled gently. If force is applied it tends to compress and becomes a compact solid rather than remaining porous. Chilling Chromosorb® T can improve its handling properties.

Available from Johns-Manville Corporation, Ken-Caryl Ranch, Denver, Colo.

Anakrom Support

Anakrom A — Acid Washed

This support is prepared by treatment with HCl to remove trace metals, followed by exhaustive washing with deionized water. Residual water is removed by drying at elevated temperatures for 6 hr. Anakrom A is a general purpose support which can accept liquid phase loading to 25%. It sometimes provides better separations when coated with more polar liquid phases than the acid-washed, silanized support. Anakrom A is also suitable for high temperature applications.

Anakrom AS — Acid Washed and Silanized

The dried support is treated with dimethyldichlorosilane which provides an inert surface. Residual HCl, as well as unreacted-SiCl groups, are removed by washing with methanol. Anakrom AS is prepared by silanizing Anakrom A and is a general purpose support with a maximum phase loading capacity of 25%.

Anakrom Q — Special Acid Washed and Silanized

This is an acid-washed, silanized support. The acid washing procedure is similar to that used in the preparation of Anakrom A, but a special acid is used which produces a more inert surface. The inert surface decreases sample tailing and decomposition of sensitive compounds. Anakrom Q is recommended for the analysis of bile acids, steroids, drugs, and pesticides, especially those containing phosphorus.

Anakrom C22 Series

This series is prepared from diatomaceous earth calcined with clay binder at a temperature above 1000°C, producing pink supports. The particles are less friable than white supports and can hold up to 30% of liquid phases. Anakrom C22 (untreated) efficiently separates low-polarity hydrocarbons, but is not recommended for the separation of compounds containing oxygen, amine, or hydroxyl groups. Anakrom C22A (acid washed) shows rather less peak tailing for compounds of medium polarity, but is not recommended for analysis of compounds of medium to higher polarity containing oxygen and nitrogen. Anakrom C22AS (acid-washed, silanized) has a low surface activity produced by the additional silanizing treatment. Medium polar compounds can be analyzed and phase loading can be as high as 30%.

Available from Anakrom Supports, the Foxboro Company, 80 Republic Drive, North Haven, Conn.

Gas Chrom S.A.P.Z. Series

Fresh white diatomaceous earth forms the starting material for the series of supports.

Gas Chrom S
This support is carefully screened but has no further treatment.

Gas Chrom A
Gas Chrom A is made by acid washing Gas Chrom S to remove the bulk of the inorganic contaminants and then water washing to neutrality.

Gas Chrom P
Gas Chrom P is made by base washing Gas Chrom A to remove organic contaminants and then water washing to neutrality.

Gas Chrom Z
Gas Chrom Z is made by treating Gas Chrom P with dimethyldichlorosilane to eliminate active sites on the surface of the support.

Gas Chrom R Series

Gas Chrom R is red, being prepared from Johns-Manville's C-22 firebrick material. Made from diatomaceous earth it differs from the white diatomaceous earth supports in being calcined without a flux, possessing a larger surface area and generally being more adsorptive. Gas Chrom R can achieve the highest efficiency of all diatomaceous supports and is very suitable for the separation of hydrocarbons and other nonpolar compounds.

Gas Chrom Q
Gas Chrom Q is a white diatomaceous earth that has been carefully screened, acid and base washed and silanized. It shows maximum inertness, low adsorption effect, and high permeability and is used for the trace analysis of sensitive compounds. Gas Chrom Q must be handled carefully during coating and column packing. It is recommended for the analysis of steroids, pesticides, alkaloids, pharmaceuticals, and toxic pollutants.

Gas Chrom Q II
Gas Chrom Q II is a support based on saltwater diatomaceous earth. For most chromatographic purposes Gas Chrom Q and Gas Chrom Q II have very similar properties and can be used interchangeably, though in some specialized applications one support will "out perform" the other. Gas Chrom Q II has the highest efficiency over the complete range of

phase polarities, while Gas Chrom Q has optimum efficiency in the low to medium polarity range. Gas Chrom Q II has better coating characteristics for polar liquid phases.

Available from Applied Science, P. O. Box 440, State College, Pa.

Supelcoport

Supelcoport is an acid washed DMCS-(dimethyldichlorosilane) treated diatomite support. Fines are eliminated, cutting down dust to a minimum. The support requires little or no conditioning once it is coated. When using high temperature phases like PPE-20, OV-1, etc. the column can be used within an hour after placing in the chromatograph.

Available from Supelco, Inc., Supelco Park, Bellefonte, Pa.

Phase Sep W

Phase Sep W is derived from flux calcined diatomaceous earth. It is more fragile than other "W" type supports and has a slightly lower packing density, but is less adsorptive for polar compounds, particularly when treated with an anti-tailing compound. In the latter instance exceptional efficiency is obtainable.

Phase Sep P

Phase Sep P is prepared from Johns Manville Sil-o-cel C 22 insulation brick. It is similar to Chromosorb® P.

Available from Phase Separation Ltd., Deeside Industrial Estate, Queensferry, Clwyd, CH2LR, U.K.

Spherocarb

Spherocarb is a spherically shaped carbon molecular sieve. It has been used for the gas chromatographic analysis of permanent gases, C_1 to C_4 hydrocarbons, stack gas mixtures, such as NO_2, SO_2, and H_2S and other pollution gas mixtures. Spherocarb has been recently employed as an adsorbent trap in separating xenon and krypton from air. Its spherical shape, uniform size, and nonfriable character make it easy to pack GC columns, which are reproducible and consistent.

Tenax-GC

Tenax-GC is a porous polymer column packing material that is based on 2,6-diphenyl-*p*-phenylene oxide. It is suitable for separating high boiling polar compounds such as alcohols, polyethylene, glycol compounds, diols, phenols, monoamines, diamines, ethanolamines, amides, aldehydes, and ketones. The choice of column material is often determined by the polarity of compounds to be separated. Glass columns are generally preferable to metal. The properties of Tenax-GC columns are influenced by the particle size of the polymer. Best results are obtained with 60/80 and 35/60 mesh sizes. Tenax-GC packing must adjust itself to the mixture to be separated. The mixture should therefore be injected several times before an analysis is made. Separations may be affected at relatively high temperatures. However, since column efficiency increases with decreasing maximum column temperature, the lowest possible conditioning and operating temperature should be chosen for analysis. To obtain maximum efficiency, the carrier flow should be between 5 and 20 mℓ/min.

Available from Akyo Research Laboratories.

Graphpac

Graphpac is a graphitized carbon adsorbent used as an inert support for GC. It has a nonpolar surface and is usually modified with small amounts of liquid phase to reduce tailing

and enhance separation. Careful handling and column packing is required when using graphitized carbon packings. A slurry technique is used in preparing packings. The column should
be heated at a temperature 25°C below the maximum temperature of the liquid phase for 8
to 16 hr. Carrier gas flow should not be interrupted while the column is hot. Recommended
applications are C_2-C_5 acids, alcohols, cresols, ethylene oxide, C_1-C_5 hydrocarbons, ketones,
phenols, and solvents.

Available from AllTech Associates, Inc., 2051 Waukegan Road, Deerfield, Ill., 60015.

Chirasil-Val III Capillary Columns

Chirasil-Val is a family of polysiloxane stationary phases used for the separation of optical
enantiomers. Chirasil-Val is a chiral phase that may be used for determining the optical
purity of amino acids, particularly for monitoring the extent of racemization in peptide
synthesis. Both borosilicate and fused silica tubing coated with Chirasil-Val III are available.
The columns can be used repeatedly without deterioration. Almost all protein amino acid
enantiomers can be separated within 40 min. Temperatures of up to 240°C with temperature
programming can be employed. The phase has low volatility and high thermal stability
compared with other available chiral phases.

Available from Applied Science.

Durabond Capillary Columns

In Durabond columns the stationary phase is chemically bonded to the walls of the fused
silica capillary, minimizing bleed, and extending the temperature stability of the column.
The columns will tolerate large sample to solvent volumes. Durabond columns are available
in 3 film thicknesses, 0.1 μm for high boiling solutes, 0.25 μm for general purpose work,
and 1.0 μm for volatiles. Sample amounts as large as 500 ng per component can be
chromatographed using the 1.0-μm film without significant loss of resolution. Chemically
bonded phases available are DB-1, which is similar to SE-30 and DB-5, which is similar
to SE-52, SE-54, and DB-1701.

Available from J. & W. Scientific, Inc., 3871 Security Park Drive, Rancho Cordova,
Calif., 95670.

Durapak

Durapak materials consist of a liquid phase coating permanently bonded to a core material.
Durapaks eliminate column bleed liquid puddling, limited packing life, and very long column
preconditioning times. They provide sharp, symmetrical peaks, high relative retentions,
greater loading, better reproducibility, and high efficiency separations. Durapaks are de-
graded by trace amounts of moisture.

OPN/Porasil C

This Durapak has oxypropionitrile chemically bonded to Porasil C. It is a medium polarity
packing that can be used to separate compounds of widely differing polarities.

n-Octane/Porasil C

This packing is the most polar of the Durapak series, as a result of residual Si-OH groups
on the surface and adsorption effects. It is used to separate C_1 to C_5 hydrocarbons in a
relatively short time and by reducing the flow rate, complete resolution of all C_4 isomers is
possible.

Available from Waters Associates, 34 Maple Street, Milford, Mass.

Super Pak 20M

Super Pak 20M is a general purpose GC packing possessing high efficiency, increased thermal stability, and an inert surface. This packing is a diatomaceous earth, which after special treatment, is coated with a low loading of Carbowax® 20M and treatment at high temperatures.

Available from Foxboro/Analabs.

Gamma Bond Fused Capillary Columns

Fused silica capillary columns with gamma radiation-induced cross-linked silicone phases (gamma bond) have been developed. *In situ* cross-linking via gamma radiation can produce stable and reproducible films without imparting brittleness to the fused silica tubing. The use of gamma radiation avoids chemical additives which may cause residual activity in the column. Gamma bond capillaries produce nonextractable films with excellent thermal stability and reproducible thickness. The columns have high efficiency and long lives. Gamma bond capillaries are available in 0.25 and 0.32 mm internal diameter, with the following phases: GB-1 equivalent to OV-1 and SE-30, GB-5 equivalent to SE-52 and SE-54, GB-17 equivalent to OV-17, and GB-1701 equivalent to OV-1701.

Available from Foxboro/Analabs.

Table 5
PACKINGS FOR HPLC

This table includes ion-exchange materials and some specialized packings.

E. Merck

LiChrosorb®

LiChrosorb® column packing materials are high-efficiency HPLC sorbents (irregular-shaped particles). Most are available in particle sizes of 5, 7, and 10 μm. There is a complete range of polar, medium-polarity, and reversed-phase sorbents. Complete surface coverage in all derivatives ensures purely partition mechanisms. They provide optimum reproducibility.

Description

LiChrosorb® Si 40[a]	Neutral silica gel
LiChrosorb® Si 60	Neutral silica gel
LiChrosorb® Si 100	Neutral silica gel
LiChrosorb® Alox T	Basic aluminum oxide 150 (type T)
LiChrosorb® RP-2	Aliphatic hydrocarbon phase (C_2), stable against hydrolysis, chemically bonded to silica gel (reversed-phase)
LiChrosorb® RP-8	Aliphatic hydrocarbon phase (C_8), stable against hydrolysis, chemically bonded to silica gel (reversed-phase)
LiChrosorb® RP-18	Aliphatic hydrocarbon phase (C_{18}), stable against hydrolysis, chemically bonded to silica gel (reversed-phase)
LiChrosorb® DIOL	Vicinal alcoholic hydroxyl function at the end of an aliphatic hydrocarbon chain; stable against hydrolysis (hydrophilic)
LiChrosorb® NH$_2$	Amino phase, chemically bonded to silica gel; stable against hydrolysis at eluent water contents of less than 30% by volume (hydrophilic, weakly basic)

LiChrosorb® CN — Silica gel with terminal cyano group on aliphatic spacer group

LiChrosorb® AX — Silica gel with quaternary ammonium groups and an aliphatic spacer group (anion exchanger)

LiChrosorb® CX — Silica gel with sulfonic acid groups and an aliphatic spacer group (cation exchanger)

The number used in the product name, e.g., Si 100 indicates the approximate mean diameter of the silica gel matrix in Angstrom units.

Perisorb

The Perisorb series comprises superficially porous column packing materials of low specific surface area and particle size 30 to 40 μm. They are ideal for very fast separations. In addition to materials based in silica gel, the range includes a cation exchanger and an anion exchanger. The capacity of these ion exchangers is les than that of exchangers based on totally porous silica gels.

	Structure of the chromatographically active layer	Notes
Perisorb A	Silica gel 60	Maximum liquid phase load, 3% (w/w)
Perisorb RP-2	Silica gel silanized	For reversed-phase chromatography
Perisorb RP-8	Hydrocarbon phase (C_8) chemically bonded to silica gel: stable to hydrolysis	For reversed-phase chromatography
Perisorb RP-18	Hydrocarbon phase (C_{18}) chemically bonded to silica gel: stable to hydrolysis	For reversed-phase chromatography
Perisorb KAT (0.05 meq/g)	Silica gel with sulfonic acid groups	Strongly acid ion exchanger: stable between pH 1 and 9
Perisorb AN (0.03 meq/g)	Silica gel with quaternary ammonium groups	Strongly basic ion exchanger: stable between pH 1 and 9
Perisorb PA6	Polyamide 6	Polycaprolactam layer, about 2 μm

Florisil

Florisil is a white, hard granular or powdered magnesia-silica gel often referred to as a magnesium silicate. It is a highly selective adsorbent used in preparative and analytical chromatography. Some of its principal applications are sample preparation, the isolation of steroids, sex hormones, and related compounds and alkaloids. It is used in the separation of chlorinated pesticides, lipids, and nitrogen compounds from hydrocarbons and for the clean-up of pesticide residues, the purification of pharmaceuticals, the assay of vitamins, and the decolorization of oils, fats, and waxes by percolation or contact treatment. Florisil is produced in standard mesh sizes.

Florisil PR is specially activated for separation of chlorinated pesticides.

Florisil TLC is a special grade available for TLC. The normal activation temperature is 650°C, but where less active material is desired, Florisil activated at 260°C should be specified.

Available from Floridin Co., 3 Penn Center, Pittsburgh, Pa.

Du PONT

Zorbax® Packings

Zorbax® is a pure siliceous particle, produced from very small, extremely uniform colloidal silica sol beads (e.g., 80 Å) which are agglutinated in a polymerization process to form spherical particles. The pore size of the final particle is determined by the size of the silica sol beads. After the polymerization step, controlled heating removes the organic polymer and sinters the particles, giving a mechanically stable pure silica material.

Zorbax® ODS

Zorbax® ODS has a high retentivity due to its C-18 monolayer bonded phase that forms 20% of the packing. It is suitable for nonpolar to moderately polar compounds. Typical applications include polynuclear aromatic hydrocarbons, esters (phthalates), fat-soluble vitamins, steroids, hydroquinones, and alcohol-soluble natural products. Triglycerides and high-molecular-weight hydrocarbons and lipids generally may be analyzed by nonaqueous reversed-phase chromatography.

Zorbax® C-8

The shorter chain length of this bonded-phase packing makes it suitable for separating compounds of moderate to high polarity which are water soluble. The product is excellent in reversed-phase ion-pairing systems, using additives such as hexane sulfonic acid for basic solutes and tetrabutyl ammonium hydroxide for acid solutes. Typical applications include analgesics, anticonvulsants, antihistamines, and other groups of highly polar pharmaceuticals, nucleosides, PTH-amino acids, and purine bases.

Zorbax® TMS

Zorbax® TMS is produced by chemically bonding trimethylsilane groups to the Zorbax® support. It is suited to reversed-phase separations of highly polar compounds. Zorbax® TMS, like Zorbax® ODS and C-8, equilibrates rapidly after mobile phase changes.

Zorbax® SIL

This silica column used in the adsorption mode gives selectivity for many kinds of compounds with polar functional groups. Zorbax® SIL particles have approximately 350 m^2/g of surface area with closely controlled particle and pore size. Typical applications are fat-soluble vitamins, steroids, aromatic compounds, and esters. Acetonitrile, ethanol and isopropanol are generally used at low levels, e.g., 1% in less polar organic solvents to achieve desired retention of compounds. Water saturation or 50% water saturation of the mobile phase is generally necessary to stabilize column activity and obtain repeatable retention times.

Zorbax® CN

Zorbax® CN is formed by the monolayer bonding of a cyanopropyl moiety to Zorbax® SIL. Zorbax® CN columns may be used in both normal and reversed-phase chromatography. In the normal-phase mode, with relatively nonpolar solvents, the column will separate many of the same polar compounds as Zorbax® SIL. An advantage of Zorbax® CN is its rapid equilibration with mobile phases making it more suitable than Zorbax® SIL for gradient elution separation. Zorbax® CN has less tendency to foul with noneluted compounds than an adsorption column. Typical applications are fat-soluble vitamins, organic esters, water-soluble vitamins, amine compounds, aromatic alcohols, and pesticides.

Zorbax® NH₂

This amino bonded-phase packing is a multipurpose product which functions in normal phase, weak anion exchange, and reversed-phase modes. In the normal phase mode Zorbax® NH₂ separates polar compounds such as substituted anilines, esters, chlorinated pesticides. Carbohydrates can be separated in the reversed-phase mode with good column stability. Organic acids of several types, for example, dicarboxylic acids, may be analyzed in the ion-exchange mode. For ion exchange, buffers such as acetates and phosphates, can be used in conjunction with organic compounds, for example, acetonitrile. In this mode the pH should be maintained between 2 and 7. Zorbax® NH₂ should be preferably used with only one mode of operation, since long equilibration times make it difficult to rapidly change from one mode to another.

Zorbax® SAX

Zorbax® SAX is a strong anion-exchange packing with a permanently bonded phase that terminates in a fully quaternized function. Anion exchange is possible over the pH range 2 to 7. The use of a guard column is recommended for Zorbax® SAX. Organic modifiers such as methanol or acetonitrile may be used together with an aqueous phase to enhance solubility in some systems. Zorbax® SAX is effective in both the isocratic and gradient modes. The use of ionic strength modifiers, for example, phosphate salts, to control elution is recommended in gradient work.

Zorbax® 300 SCX

Zorbax® 300 SCX is a polar bonded-phase packing used for cation-exchange HPLC. The packing consists of an aromatic sulfonic acid moiety chemically bonded to 7 to 8 μm Zorbax® PSM 300 (300 Å, porous silica microspheres) through Si-O-Si bonds. A trifunctional organosilane reagent is used in the production of Zorbax® 300 SCX in order to maximize bonded-phase stability with aqueous mobile phases. Typical applications include those normally used in the separation of basic, water-soluble compounds. The retention of basic substances is influenced by pH and ionic strength, a decrease in either or both of them tending toward greater sample retention. Common buffered elements such as citrate or phosphate may be used. The pH of the mobile phase should be maintained in the range 2 to 6.5. Organic modifiers such as methanol can be used together with aqueous solutions to improve solubility and achieve better separations.

Chromosorb®ᵃ

Chromosorb® LC-1

This is an inert diatomite-based support compatible with all liquid phases. It provides fast and efficient separation of a wide variety of polar compounds.

Chromosorb® LC-2

This is a pellicular packing with active porous silica layers permanently bonded to a spherical solid glass core. Various classes of compounds may be separated by using the appropriate solvent system.

Chromosorb® LC-3

This is a pellicular packing with alumina bonded to the surface of diatomite, which can be used for the separation of polycyclic hydrocarbons, alkaloids, amines, aldehydes, ketones, surfactants, and plasticizers.

Chromosorb® LC-4

Chromosorb® LC-4 is a pellicular reversed-phase packing with a monomolecular layer of octadecyltrichlorosilane chemically bonded to Chromosorb® LC-2. It provides excellent separations of fused ring aromatic hydrocarbons, quinones, herbicides, barbiturates, steroids, and food additives.

Chromosorb® LC-5

Chromosorb® LC-5 is a pellicular reversed-phase packing with a monomolecular layer of diphenyldichlorosilane chemically bonded to Chromosorb® LC-2. It will separate a wide variety of pharmaceutical and biomedical compounds.

Chromosorb® LC-6

Chromosorb® LC-6 is a fully porous microparticulate silica of 10 μm particle size, which affords high efficiency, capacity, resolution, and speed of analysis. Close polarity compounds can be separated with an isocratic solvent system and wider polarity range mixtures with gradient elution.

Chromosorb® LC-7

This is a reversed-phase microparticulate HPLC column packing material in which an octadecyl group is chemically bonded to Chromosorb® LC-6 via a Si-O-Si-C type bond. The resulting C_{18} nonpolar stationary phase performs well in partition or ion-pair chromatography modes. It will separate a wide variety of compounds used in pharmaceutical and food products.

Chromosorb® LC-8

Chromosorb® LC-8 is a micro-particulate polar-bonded phase made by chemically bonding a cyano-type group to Chromosorb® LC-6. It complements Chromosorb® LC-7 for reversed-phase HPLC, is highly polar, and efficiently separates polar compounds.

Chromosorb® LC-9

This is a polar-bonded HPLC packing prepared by chemically bonding an amino-type functional group to a 10-μm silica particle. It can be used as a weak anion exchanger or a polar reversed-phase packing. Chromosorb® LC-9 will separate carbohydrates, polyhydroxy compounds, amines, aromatics, and nitro compounds.

Available from Johns-Manville Corporation.

WHATMAN

Partisil® Packings

Partisil® Silica Base

Silica gel is the most popular base material found in modern HPLC because of its high structural strength and controllable purity. The high degree of rigidity of silica gel allows columns to be operated at high flow rates and pressures with little likelihood of collapse. In addition, silica gel is inert to most common organic HPLC solvents. Partisil® silica gel is a highly uniform material of exceptional purity. Its surface area of more than 350 m²/g is greater than that of most silica gels and provides enhanced selectivity and loading capacity. As a support for bonded phases, the increased surface area of Partisil® permits higher effective carbon loads and allows sufficient space between alkyl chains to minimize interactions.

Partisil® for Adsorption Chromatography

Silica gel is the most widely used adsorbent in liquid chromatography and is most effective when separating materials with dissimilar polarities. The surface hydroxyl groups (silanols) attract molecules of different polarities to different degrees, thus effecting separation. Retention is greatly affected by the degree of hydration of the silanols. Absolutely dry or highly activated silica is a strong adsorbent and a certain amount of moisture, usually present in the solvent, is used to moderate the adsorbent quality. This affords relatively rapid elution with minimal peak tailing.

Partisil® has been used in both isocratic and gradient elution modes. Partisil® 5 columns are recommended for separations requiring maximum resolution with moderate back pressure. Partisil® 10 columns are most useful for routine analysis where the greater column life resulting from lower pressure drops is advantageous.

Partisil® PAC

Partisil® PAC is a polar-bonded phase used in the normal phase mode and is effective in separating highly polar compounds. Partisil® PAC consists of alkyl groups containing aminocyano groups (2:1) bonded to the surface of 5- and 10-μm particles of Partisil® silica gel. Its Si-O-Si bonds are stable both chemically and thermally. The chemical nature of the bonded group depends on the chemical composition of the mobile phase. In an acidic aqueous mobile phase, NH groups are quaternized, creating a weak anion-exchange column. In an acidic nonaqueous mobile phase the NH groups are again quaternized but behave as an ion pair with a unique selectivity. This selectivity control can be obtained because of the fast re-equilibration of the phase. In addition to the normal phase partition involving aminocyano groups, Partisil® PAC shows some adsorption involving the residual silanol sites.

Partisil® 10 Carbohydrate

Partisil® 10 Carbohydrate is a system for separating a complete spectrum of mono-, di-, and oligosaccharides. The system contains an analytical column packed with aminocyano media and a solvent preconditioning column. The medium consists of 10-μm Partisil® silica bonded with alkyl chains containing 2 secondary amines and a terminal cyano. Its mode of separation is normal phase partition involving the aminocyano groups, and adsorption involving the residual silanol sites. Monosaccharides are eluted first, followed by disaccharides, and finally oligosaccharides. Separations can be carried out at room temperature and the media re-equilibrates rapidly after a solvent change. An acetonitrile-water mobile phase containing ammonium hydroxide (CH_3CN-H_2O-NH_4OH, 80:20:1 v/v) is most effective for monosaccharide separation. The ammonium hydroxide deprotonates any of the amine functionality which might be present in the NH_2^+ form. Without this action some stereoisomeric hexoses would not be resolved.

Partisil® ODS-3

In Partisil® ODS-3 a trifunctional octadecyl (C_{18}) silane is bonded to the Partisil® base to give a polymeric bonded phase. The bonded medium is then capped with a trimethylsilane reagent. This produces a surface almost totally free of chromatographically accessible silanol sites for true single-mode (reversed-phase) separations. The density of alkyl chains is 10.5% carbon by weight. Partisil® ODS-3 is available in 5-, 10-, and 40-μm particle sizes. The moderate back pressures and high efficiencies associated with Partisil® 10 ODS-3 make it suitable for a wide range of reversed-phase separations. Partisil® 5 ODS-3 provides the same selectivity as Partisil® 10 ODS-3 but with higher performance.

Partisil® CCS/C₈

Partisil® CCS/C$_8$ differs from other reversed-phase Partisil® media. Its octyl (C$_8$) groups give it a selectivity which differs from that of octadecyl (C$_{18}$) bonded phases and make it particularly suited for more polar solutes often encountered in biochemical analysis. Its bonding chemistry is bonded on a monofunctional octyldimethylsilane, which gives a monomeric (brush) phase when bonded to Partisil® silica. Partisil® CCS/C$_8$ is then capped with a trimethylsilane reagent, producing a reversed-phase medium with a 9% carbon load. Unlike polymeric phases, monomeric phases have considerable rotational freedom. This presents no apparent problem when using solvent systems containing organic modifiers, since the nonpolar modifier associates with alkyl chains maintaining full extension. In highly aqueous solvent systems, however, there is minimal interaction between the alkyl chains and solvent, and after some time interaction between the chains occurs producing "collapse" of the bonded phase. The chromatographic inaccessibility of the collapsed chain leads to a slow downward drift of k' to a level below that predicted for the carbon load. The process of controlled chain spacing (CCS) used in manufacturing Partisil® CCS/C$_8$ uniformly spaces the alkyl chains across the surface of the silica base. The spacing between chains is controlled to prevent chain collapse. Intervening silanols are then capped with trimethylsilyl groups. The resulting surface has excellent stability toward solvents.

Partisil® 10 ODS-2

Partisil® 10 ODS-2 is a polymeric octadecyl (C$_{18}$) bonded phase produced by reacting Partisil® 10 μm silica gel with a trifunctional octadecyl silane. The resulting 15% carbon load, as well as increasing retention, reduces the available silanols to approximately 25%. These residual silanols allow some contribution of adsorption to the final selectivity. Partisil® 10 ODS-2 is particularly useful where solutes are only slightly retained on less heavily loaded materials or where a high level of organic modifier is required. It is very effective for preparative chromatography.

Partisil 10® ODS

Partisil® 10 ODS is a reversed-phase packing with a 5% carbon load produced by bonding a trifunctional octadecyl (C$_{18}$) silane to 10-μm particle diameter Partisil® silica gel. This reagent creates a bonded phase of polymeric nature. Almost 50% of the residual silanols remain uncapped, a level which allows Partisil® 10 ODS to be used as a reversed-phase partition medium as well as a normal-phase adsorption medium.

Available from Whatman Chemical Separation, Inc., 9 Bridewell Place, Clifton, N.J.

Protesil (Protein Separation Media)

Protesil describes chromatographic media designed specifically for protein separations, based on a partition rather than a size exclusion mechanism. The bonded-phase media, Protesil 300 Diphenyl and Protesil 300 Octyl, are based on a 300-Å pore size silica gel. Most available HPLC separation media have an average pore size of 85 Å, which limits accessibility to proteins of up to 50,000 mol wt. In contrast, the Protesil 300 Å media has pores which allow full interaction with components up to 150,000 mol wt. Protesil 300 Diphenyl is manufactured by bonding diphenyl groups via an Si-O-Si bond to 10-μm particles of silica gel, to give the maximum achievable density of diphenyl groups on the surface of the silica. The material is then capped with a trimethylsilyl reagent, producing a greater than 95% reduction in silanol sites. The final carbon load is 8%. The predominant mode of separation is reversed-phase partition. Column selectivity can be influenced by solvent changes. To prepare Protesil 300 Octyl, octyl (C$_8$) chains are bonded to a 10-μm silica base via an Si-O-Si bond and capped with a trimethylsilyl reagent. The carbon load is approximately 7.5%. Protesil 300 Octyl has the same versatility as its diphenyl counterpart, but a different selectivity. Both Protesil media can be operated at temperatures up to 50°C and within a pH range of 1.5 to 7.0. They show little "solvent memory" and re-equilibrate quickly for rapid gradient analysis.

Whatman Ion Exchangers

Most HPLC ion exchangers consist of a silica gel base to which an organic group is covalently bonded. This group is then chemically modified to give the immobilized ion. The eluent most commonly used in ion exchange is an aqueous buffer containing an organic or inorganic salt. Retention is controlled by the type of buffer and the concentration and pH of the buffer. Precautions must be taken to protect the silica-based packing when operating in an ion-exchange mode. When possible, conditions which promote dissolution are to be avoided. The lowest possible buffer concentration should be used and the pH maintained between 2.5 and 7.0.

Partisil® 10 SAX

Partisil® 10 SAX is a widely used strong anion exchanger for HPLC. Quaternary ammonium groups ($-NR_3^+$) are covalently bonded to the surface of 10-μm Partisil® silica. It is used to separate a complete range of anionic species, including analgesics, steroid conjugates, antibiotics, and preservatives. It is best known, however, for the separation of nucleotides. Partisil® 10 SAX is most commonly used with aqueous buffers within a pH range of 1.5 to 7.5.

Partisil® 10 SCX

Partisil® 10 SCX is a strong cation exchanger manufactured by bonding aromatic benzene sulfonic acid functional groups to 10-μm particles of Partisil® silica gel. The Si-O-Si-C bond is exceptionally stable both chemically and thermally. It is used extensively for the separation of nucleic acids, amino acids, polyamines, drugs, and other cationic species. It may also be loaded with specific metallic cations such as Cu^{2+} or Ag^{2+} for use in ligand-exchange chromatography. Partisil® 10 SCX is most commonly used with aqueous buffers within a pH range of 1.5 to 7.0.

Sephadex® Ion Exchangers

Sephadex® ion-exchange gels are derivatives of Sephadex® G-25 and G-50; anion and cation exchangers are denoted by A and C, respectively. Strong exchangers are suitable for compounds which are only charged at extremes of pH, while weak exchangers are suitable for most applications. Sephadex® ion exchangers have very high capacities and can be used at relatively high ionic strengths. A-25 and C-25 types are most suitable for small molecules such as amino acids, peptides, nucleotides, and very large molecules; A-50 and C-50 types are particularly useful for molecules in the mol wt range 30,000 to 200,000. These ion exchangers are stable in the pH range 2 to 10.

DEAE-Sephadex®   — Weakly basic anion exchanger.
QAE-Sephadex®   — Strongly basic anion exchanger.
CM-Sephadex®   — Weakly acidic cation exchanger.
SP-Sephadex®   — Strongly acidic cation exchanger.

Available from Pharmacia Fine Chemicals AB, Uppsala, Sweden.

Sepharose Ion Exchangers

Sepharose ion-exchange gels are derived from cross-linked agarose gel Sepharose CL-6B. They afford excellent resolution where a weakly basic or acidic exchanger is needed and are particularly useful for purifying large macromolecules. Their physical and chemical stability allows these gels to be employed at high flow rates in solvents such as 6 M urea and in eluents containing nonionic detergents. The gels are stable in aqueous solution over the pH range 3 to 10, can be used at temperatures up to 70°C, and may be employed for large-scale applications.

DEAE-Sepharose CL-6B   — Weakly basic anion exchanger.
CM-Sepharose CL-6B   — Weakly acidic cation exchanger.

DEAE-Sephacel®
DEAE-Sephacel® is a cellulose ion exchanger in a rigid bead form, which allows it to be easily packed in columns and prevents the generation of fines. The absence of fines gives improved flow rates and better resolution and recovery. DEAE-Sephacel® has applications in all fields of biochemistry and is recommended for fractionation of serum proteins. It has an exclusion limit for globular proteins of 1,000,000 and is stable in aqueous solution over the pH range 2 to 12.
Available from Pharmacia Fine Chemicals AB, Uppsala, Sweden.

BECKMAN

Ultrasphere

The base silica of Ultrasphere is given a maximum surface covering of C-18 or C-8 and is subsequently end capped to remove unreacted silanols. The silica matrix has an extremely uniform pore structure, from which micropores have been virtually eliminated. Ultrasphere series columns are packed with 5-μm silicas which give higher column permeabilities than irregular particles. Smaller particles also lead to faster analysis times.

Ultrasphere-ODS (C-18)

Ultrasphere-Octyl (C-8)
Ultrasphere-ODS and octyl are designed for reversed-phase liquid chromatography, i.e., separations of solutes based on the size and structure of alkyl groups. Reversed-phase chromatography is also useful for separations based on functional group differences through alterations of mobile phase solvent composition and type. With aqueous mobile phases, solute retention is greater on the C-18 phase than on the C-8. Changes in elution order, i.e., selectivity between C-8 and C-18, can occur for several solutes.

Ultrasphere-IP
Ultrasphere-IP is designed for ion-pairing reversed-phase HPLC. In this method counter ions are added to the mobile phase to neutralize the charge of the solutes.

Ultrasphere-Si
Ultrasphere-Si is used in normal phase (adsorption) chromatography. It is useful for separation of nonpolar to moderately polar solutes. Mixtures of organic solvents are typically used.

Ultrasphere-Cyano
Ultrasphere-cyano is designed for use where a moderately polar bonded-phase material is desired. It can be used in both normal and reversed-phase modes.

3-μm Ultrasphere
The 3-μm Ultrasphere column gives a greatly improved speed of analysis.

Ultrapore RPSC Column

The Ultrapore RPSC (reversed-phase short chain) column is packed with large pore (30 nm), 5-μm, spherical silica which is maximally bonded with a C_3 alkyl chain and end capped to increase column life. The combination of particle and pore size and surface chemistry endows this column with excellent performance for separation of large proteins. The column's short length reduces analysis times and solvent consumption. The Ultrapore semipreparative column provides the loading capacity for purifying large quantities of protein.
Available from Beckman Instruments, Inc., 2500 Harbor Blvd., Fullerton, Calif.

PHASE SEPARATIONS LTD.

Spherisorb

Spherisorb is a porous spherical silica with very high specifications on pore shape, volume, size, and size distribution, particle shape, size, and size distribution, and specific surface area. Spherisorb is a family of three particle sizes, S 3 W, S 5 W, and S 10 W, with mean particle sizes of 3, 5, and 10 μm, respectively. Bonded phases are based on the same basic silica material but are synthesized with organic modifiers in monomolecular layer over the silica surface.

Spherisorb ODS 1
Spherisorb ODS 1 (octadecyl 1) has a total carbon content of 7%; the major silanols are capped.

Spherisorb ODS 2
Spherisorb ODS 2 (octadecyl 2) is a high capacity ODS with a total carbon content of 12%; the silanols are fully capped.

Spherisorb CN (Nitrile)
Spherisorb nitrile, carbon 0.6 mmol/g of cyanopropyl, will operate with aqueous or nonaqueous phases, finding application in the separation of polar compounds. It has little tendency to retain materials irreversibly and equilibration with the mobile phase is more rapid. This is of advantage with gradient elution.

Spherisorb NH$_2$ (Amino)
Spherisorb amino, carbon 0.6 mmol/g aminopropyl is useful in carbohydrate separations as it is much less critical with respect to the water content of the mobile phase. It can even be used as a weak anion exchanger, but should not be employed with aldehydes or ketones.

Spherisorb C8 (Octyl)
Spherisorb octyl is a fully capped material with carbon, 0.6 mmol/g (total carbon 6%). It gives high efficiency and good peak symmetry. It is particularly useful for high polarity molecules or for samples with a wide range of polarities.

Spherisorb C6 (Hexyl)

Spherisorb hexyl is a fully capped material, carbon 0.6 mmol/g. It is complementary to C8.

Spherisorb P (Phenyl)

Spherisorb phenyl has a coverage of 0.3 mmol/g with only the major silanols capped.

Spherisorb General Purpose (GP)

Spherisorb GP is available as plain silica or bonded with ODS or nitrile. It is suitable for low cost, semipreparative or precolumn applications.

Spherisorb S5 SAX

Spherisorb S5 SAX is a strong anion exchanger. The anion exchanger phase is *N*-propyl *N,N,N*-trimethyl ammonium ligand chemically bonded to the surface of 5-μm Spherisorb. The monolayer coverage of the surface ensures mass rapid transfer, giving much higher efficiencies than can be obtained from materials based on organic-based ion exchangers or silica-based materials with polymeric phases. The bonding is stable to aqueous, polar, and very low pH conditions.

Spherisorb S5 Cl

Spherisorb S5 Cl is a methyl bonded phase based on 5-μm Spherisorb silica. Trimethylsilane is fully reacted with silica to ensure maximum coverage of the surface with the minimum possible free silanol groups remaining. The resulting Cl reversed-phase material allows the use of more highly polar mobile phases than is possible with other reversed-phase materials. The lower retention associated with the short alkyl chain permits the analysis of samples which cannot be analyzed on other reversed-phase materials with strong retentions. The short chain of the bonded phase allows very rapid mass transfer, producing very efficient separations. Spherisorb S5 Cl can often be substituted for other reversed-phase materials to achieve phase separations.

Hypersil® Packing Materials

Hypersil®, a porous, spherical silica developed specially for HPLC forms the basis of a family of high performance packing materials. The spherical shape of Hypersil® offers several advantages over irregular materials. These are greater permeability and the consequent lower back pressure, no risk of column blockage due to abrasion-generated fine particles, and better and easier column packing due to improved particle uniformity. Hypersil® has a pore size of 120 Å, which is optimum for the efficient separation of small molecules. The material has a narrow pore size distribution and an absence of very small pores which can degrade performance. The pore volume of 0.7 cm²/g of Hypersil® gives a very robust particle, and the surface area of 170 m²/g gives excellent retentivity characteristics.

Bonded Hypersils®

Bonded Hypersils® are manufactured from Hypersil® under controlled conditions to produce a monolayer coverage of organic material. A minimum of free accessible silanol group remains which gives the materials true reversed-phase characteristics. All groups are attached via siloxane bonds to give a very stable product. The bonded Hypersils® have excellent mass transfer properties which lead to high performance and shallow plate height curves.

Hypersil®, Silica

This is a high performance, fully porous, spherical silica adsorbent.

ODS-Hypersil®, C-18

This is a fully capped, C-18 reversed-phase material with monolayer coverage and an exceptionally low level of residual silanol groups. The material exhibits true reversed-phase characteristics.

MOS-Hypersil®, C8

This is a reversed-phase material with a monolayer coverage of dimethyloctylsilyl groups. It is suitable for the separation of molecules of moderate polarity or where sample components have a range of polarities.

APS-Hypersil®, NH_2

This is a versatile amino phase which can be used as a weak anion exchanger with aqueous buffers, a carbohydrate phase with acetonitrile-water mixtures, and as a normal-phase material where its selectivity is complementary to Hypersil®.

SAS-Hypersil®, Cl

SAS-Hypersil®, Cl is a trimethyl reversed-phase material with ideal selectivity for polar and multifunctional compounds. It is used as a stationary phase for ion-pair chromatography and also in the normal phase mode.

APS-Hypersil® 2, NH_2

APS-Hypersil® 2 is manufactured by a slightly different process than APS-Hypersil®, and although its range of potential uses is the same, its selectivity may differ in some applications.

CPS-Hypersil® CN

This is a cyanopropyl phase which can be operated in both normal and reversed-phase modes. In the normal phase it is complementary to Hypersil® and APS-Hypersil®. It has the advantage over silica of more rapid phase equilibration and is not deactivated by traces of water in the mobile phase. In reversed-phase chromatography it complements the alkyl chain phases.

Phenyl-Hypersil®

This material is recommended for the reversed-phase separation of moderately polar compounds. It is complementary to the alkyl reversed phases and the cyano phase when the latter is operated under reversed-phase conditions.

Wide Pore Hypersil® Materials

Hypersil® materials are available in a wide pore size for separation of biopolymers. The larger pore size is important in accommodating the large size of many biopolymers, especially when in the solvated state. The materials are robust and can withstand pressures normally obtained in analytical HPLC for small molecules.

Hypersil®-WP 300

This is a wide pore silica material for normal-phase adsorption chromatography that is especially useful for synthetic polymers with mol wt of up to 1 million.

Hypersil®-WP 300-Octyl

This is a fully end capped C8 material for the separation of peptides, proteins, and polynucleotides.

Hypersil®-WP 300-Butyl

This is a fully end capped C4 material for reversed-phase operation and is more suited to higher molecular weight materials. Its selectivity complements that of the C8 material.

Available from Shandon Southern Instruments, Inc., 515 Broad Street, Sewickley, Pa., 15143.

WATERS

HPLC Stainless Steel Columns

µBondapak® C₁₈ Column — Reversed-Phase

This column has a wide range of selectivity, moderate capacity, and provides very fast analysis. In addition to a 10% C₁₈ carbon load, a layer of monochlorsilane is permanently bonded to the µPorasil packing material. The column will separate acidic, neutral, and basic compounds.

µBondapak® Phenyl Column — Reversed-Phase

This column has an aromatic bonded phase and is fully end capped. The level of hydrocarbons bonded to the µPorasil particles is lower than in the µBondapak® C₁₈ column, which results in lower retention times and less solvent consumption for neutral compounds. The µBondapak® phenyl column is also used when sample components differ in "polar functional" rather than aliphatic groups. It provides excellent selectivity for acids, neutrals, and especially bases.

µBondapak® NH₂ Column — Normal and Reversed-Phase

This versatile column is normally used for adsorption with nonpolar solvents to analyze samples with different polarity groups. Aldehyde and ketone groups should be avoided; acid amines will react with the packing material to form Schiff bases. In order to avoid oxidation of the NH₂ group, peroxides should not be used. The column can also be used in the reversed-phase mode. When an acidic mobile phase is used the NH₂ groups are protonated, giving the surface a positive charge. The column then functions as a weak ion exchanger. In this mode it is stable over the pH range 2 to 8.

µPorasil Column — Normal Phase

This silica column, with a particle size of 10 µm, has high efficiency, moderate capacity, and provides rapid analyses. It is used in conjunction with nonpolar organic solvents.

µBondapak® CN Column — Reversed-Phase

This column has 9% CN group bonded to µPorasil. It is used to separate both polar and nonpolar compounds through normal phase (adsorption) and reversed-phase (partitioning), being recommended especially for reversed-phase uses.

µBondapak® CN Column — Normal Phase

This column is intended for normal phase use with low to intermediate polarity solvents.

Resolve C₁₈ Column — Reversed-Phase

This column combines the high efficiency of a 5-µm spherical particle with the fast throughput of a 15-cm steel column. The column is used to separate low to moderately polar compounds. The high percentage of bonded C₁₈ produces a column which is highly retentive for nonpolar compounds.

Resolve Silica Column — Normal Phase

This 5-μm silica column is used for adsorption phase separations requiring the highest efficiency and rapid analysis; it is used to separate very polar compounds containing polar functional groups.

Protein Analysis Columns

These columns achieve rapid separation and purification of proteins, typically in minutes. Microgram to milligram amounts of protein can be isolated, with high recoveries of total protein and bioactivity. The rigid hydrophilic porous silica can be used with organic solvents and aqueous buffers of pH 2 to 8. Proteins are separated on protein analysis columns on the basis of their effective size in solution, hydrophobic nature, and ionic characteristics. The separation can be enhanced by other factors such as the ionic strength of elution buffer, interactions with buffer salts, the pH, and hydrophobic interactions with the column packing surface. By varying these parameters, separations can be effected that are not achievable by conventional gel filtration techniques.

Carbohydrate Analysis Column

This column is used for the separation of sugars and polyhydroxy compounds. Typical applications are carbohydrates, substituted sugars, polyglycerols, glucoalkaloids, high fructose syrup, and licorice extract. The 10-μm particle size affords high efficiency and high capacity for analytical and milligram scale preparative separations. Analysis is rapid. Simple saccharides elute before polysaccharides. Water and acetonitrile solvent combinations are employed.

Sugar-Pak 1 Column

This microparticulate cation-exchange resin column achieves rapid separations of sugar products and processing liquors. It separates common sugars obtained from cane, corn, and beef sources, and common alcohols produced in fermentation processes. Using water as the mobile phase, polysaccharides are separated in a peak, while common mono- and disaccharides in each type of sample are separated. Polysaccharides elute before simple sugars.

Fatty Acid Analysis Column

This column separates free fatty acids directly, thus avoiding derivative formation. Its mean 10-μm particle size provides high efficiency and high capacity for both analytical and milligram-scale preparative separations. Typical applications are free fatty acids, fatty acid derivatives, and some fatty amides, i.e., saponified coconut oil, tall oils, fatty alcohols, and fatty acid isomers. Straight chain saturated acids are eluted before their saturated parent compounds, and for a given carbon number and chain configuration, the greater the unsaturation, the earlier the elution.

Triglyceride Analysis Column

This column provides rapid triglyceride separations at room temperature. Typical applications include triglycerides and some mono- and diglycerides, e.g., soybean and peanut oil, oleo margarine, and butter. It has high efficiency and capacity for analytical and milligram-scale preparative separations. In the preparative mode, concentration of components allows convenient collection for mass spectrometer and/or infrared analysis. High molecular weight components elute later than low-molecular-weight components, and unsaturated compounds elute earlier than their saturated counterparts.

Energy Analysis (NH₂) Column

This column is designed specifically for hydrocarbon group separations. Typical applications include hydrocarbon separations (saturates, aromatics, and polars) in crude oil, synthetic fuels, and coal liquefaction products. It incorporates a specialized packing with an aminopropylsilane chemically bonded to 10-μm fully porous silica particles. Analysis is rapid and gives high recoveries. A single solvent, usually hexane or heptane, is employed. Both analytical and milligram-scale separations are possible.

Available from Waters Associates, Inc.

Vydac Packings

Vydac packings are spherical in shape, hard and incompressible, and superficially porous. They consist of a solid core of glass beads of 30 to 44 μm diameter having active shells. They allow columns to be packed dry without difficulty.

Vydac-101 Si — Adsorption

This packing consists of inert glass beads with a 1-μm silica gel layer type 60. The specific surface is 12 m²/g.

Vydac-201 RR — Reversed-Phase

This packing is similar to Vydac-101 Si but with a hydrocarbon C_{18} phase chemically bonded to the silica gel.

Vydac-301 SB — Anion Exchanger

Vydac-301 SB is similar to Vydac-101 Si but with quaternary ammonium groups R-$N(R_1)_3^+$ chemically bonded to silica gel. It is supplied in the chloride form and is strongly basic with a capacity of 0.1 meq/g.

Vydac-401 SA — Cation Exchanger

Vydac-401 SA is similar to Vydac-101 Si but with sulfonic acid groups R-SO_3H chemically bonded to silica gel. It is supplied in the hydrogen form and is strongly acid with a capacity of 0.1 meq/g.

Vydac-501 PP — CN-Phase

This packing is similar to Vydac-101 Si but with oxynitrile groups Si-R-O-CH_2-CH_2-CN chemically bonded to silica gel.

Available from The Separation Group, 16695 Spruce Street, Box 867, Hesperia, Calif., 92345.

Spherogel TSK IEX Ion-Exchange Columns

Spherogel TSK IEX high performance weak ion-exchange columns (DEAE and CM) provide high efficiency in a short time. The principles of ion-exchange chromatography, including the use of gradients of pH and ionic strength, are directly transferable to these columns. The optimal surface chemistry provides maximal ion-exchange selectivity. Spherical bonded silica affords minimum unwanted hydrophobic interaction and long column life. Spherogel TSK IEX columns are available in pore sizes of 13 nm for small proteins and 25 nm for proteins larger than 20,000 mol wt. The columns have exchange capacities greater than 0.3 meq/g.

μ-Spherogel Carbohydrate Column

This column is a 7.5% cross-linked column for carbohydrate analysis. The packing material is calcium loaded sulfonated polystyrene divinyl benzene resin 7.5% cross-linked. The mean particle size is 8 μm, and the operational temperature range is 80 to 90°C. As solvent high purity HPLC grade water with a maximum 35% acetonitrile is suitable. No other mobile phase modifier is recommended.

Available from Beckman Instrument Co.

PIERCE CHEMICAL CO.

Controlled Pore Glass (CPG)

Controlled pore glass is a support material for use in liquid chromatography and related processes. It consists of rigid glass particles honeycombed with precisely controlled pores. The glass is produced from a borosilicate base material which has been heated to separate the borates and leave the silicates. The borates are etched from the material, leaving the porous structure. Before etching the material is ball milled and screened to its finished mesh size. The size of the pores is determined by the heating process. Controlled pore glass is an improvement over traditional gel supports and is compatible with water and nearly all solvents. It has high mechanical strength, gives high flow rates, and is suitable for high pressure operation. CPG is available with particle sizes of 37 to 74, 74 to 125, and 125 to 177 μm, and in four nominal pore sizes, 40, 100, 240, and 500 Å. Because CPG is nearly pure silica it is extremely durable, although the high surface area of the glass contributes to increased solubility.

Affinity and Immobilization Supports

Aminoaryl

Aminoaryl supports contain aromatic amine groups attached to the supports through amide linkages. They may be activated by diazotization or conversion to other reactive groups, and have been used for immobilizing enzymes, antigens, antibodies, drugs, and hormones.

Aminopropyl (AP)/Long Chain Alkylamine (LCAA)

AP and LCAA support materials contain covalently bonded extension arms which furnish primary amine groups at their terminal ends. The longer arm of the LCAA facilitates attachment of bulky ligands where steric effects and surface repulsion cause problems. Both can be derivatized to give an isothiocyanate support or an aldehyde support.

Carbonyl Diimidazole (CDI) Glycophase

This support is an activated matrix that is ideal for immobilizing enzymes and affinity ligands through amino functions. The glass is initially treated to incorporate the hydrophilic, nonionic glycophase layer which removes the ionic and denaturing properties of glass. The glycophase layer is then activated with carbonyl diimidazole, giving a support with high capacity and low nonspecific adsorption.

Carboxyl

This material has a covalently attached aliphatic extension arm ending in an active carboxyl group. The 10 Å length of the arm allows immobilization of most compounds containing free primary amine groups such as proteins, polypeptides, amino acids, substrates or inhibitors, hormones, and drugs under mild conditions.

N-Hydroxysuccinimide (NHS) Glycophase

On this support the activated ester intermediate is a covalently attached aliphatic extension arm, ending in an active *N*-hydroxysuccinimide glycophase extension arm. The activated support immobilizes compounds containing free primary amine groups such as proteins, polypeptides, amino acids, substrates or inhibitors, hormones, and drugs under mild conditions. The glycophase extension arm eliminates nonspecific adsorption.

Stable Diazonium Salt

This support enables coupling by the diazo linkage to be performed. The material provides an aromatic amine spacer arm terminating in an active diazonium borofluoride group. It remains stable in a dry form, while conventional materials diazotized with HCl yield a diazonium salt solution which must be used for coupling shortly after its preparation.

Thiol

CPG/Thiol contains a covalently attached hydrocarbon extension arm about 10 Å long ending in a reactive thiol group. Coupling with dicyclohexylcarbodiimide or 1-ethyl-(3-dimethylaminopropyl)carbodiimide in organic or aqueous systems is possible. Ligands containing a free carboxyl group can be covalently attached using either of the carbodiimides. The thiol ester bond is stable under physiological conditions, but can be split by exposure to neutral hydroxylamine or thiols. Ligands containing an alkyl halide group react with the thiol to give a thiol ether bond. The free thiol groups can be used to remove heavy metals from solution or act as an insoluble reducing agent.

Chelating and Reducing Supports

8-Hydroxyquinoline (8-HQ) Support

8-HQ chelating support is prepared on diazotized arylamine controlled pore glass. Its principal application is the concentration of metal ions from dilute solutions. Concentrations into 1% of the original volume are possible with two passes through the chelating column. Conventional methods cannot match the sensitivity and accuracy of trace metal determinations using this system of concentration.

Lipoamide

Dihydrolipoamide is covalently coupled to controlled pore glass to yield an immobilized reducing agent. The material has significant advantages over dithiothreitol and reduces biological disulfides of all types. Filtration removes both excess reactant and oxidized byproducts. The immobilized lipoamide material is reusable.

Exclusion and Permeation Supports

Glycophase

These supports incorporate a hydrophilic nonionic carbohydrate layer which is covalently bonded. This "glycerol" coating covers the active sites on the glass, minimizing their influence on the materials to be separated. This material is independent of the buffer and SDS gradients from 0 to 2% or urea gradients from 0 to 8 M have no effect on it. Although the supports are resistant to strong acids, solvents, and high salt concentrations, they should be used cautiously above pH 9. Strong oxidizing agents are to be avoided unless oxidation is required for an immobilization procedure.

Uncoated CPG

Uncoated controlled pore glass has been described above. Its surface contains many hydroxyl groups which have a negative charge in aqueous solution. Although neutral and acidic proteins polynucleic acids and polysaccharides elute easily from CPG, some neutral proteins, viruses, and strong basic proteins show a tendency to adsorb.

Ion-Exchange Supports

Carboxymethyl (CM) Glycophase

CM glycophase is a weak acid cation exchanger prepared on glycophase. It is used in reactions involving amines, amino acids, metals, purines, nucleosides, and other compounds which form cations in aqueous solutions.

Diethylaminoethyl (DEAE) Glycophase

DEAE glycophase is a weak basic anion exchanger prepared on glycophase. It is used for the chromatography of carboxylic acids, sulfonic acids and phosphoric acid and is useful in reactions involving nucleotides, nucleic acids, peptides, proteins, and other compounds forming anions in aqueous solutions.

Quaternary Triethylammonium Chloride (QAE) Glycophase

QAE glycophase is a strong basic anion exchanger prepared on glycophase. It is useful in reactions involving amines, amino acids, and other compounds forming anions in aqueous solution, especially at high pH values.

Sulfonic (SP) Glycophase

SP glycophase is a strong acid cation exchanger prepared on glycophase. It is useful in the chromatography of amines, amino acids, metals, purines, nucleosides, and compounds forming cations in aqueous solution, particularly at lower pH values.

Available from Pierce Chemical Company, Box 117, Rockford, Ill.

Preparative Chiral Stationary Phase

''Baker'' chiral stationary phase (CSP) separates enantiomers of a wide range of racemic mixtures. CSP is comprised of (*R*)-*N*-3,5-dinitrobenzoyl-phenylglycine of high enantiomeric purity ionically bonded to specially treated 40-μm aminopropyl silica. The resulting chiral support can achieve separations using low pressure liquid chromatography systems. Owing to the use of 40-μm silica particles, the efficiency of preparative columns packed with this CSP is lower than that of Bakerbond analytical chiral HPLC columns packed with 5-μm particles. The separation factor, α, however, is essentially the same for both packings. The analytical column allows the conditions and feasibility of a preparative resolution to be determined. The mobile phase used for separations is hexane containing sufficient 2-propanol (0.5 to 20%) to optimize retention times. To ensure a long life for the column, the presence of water should be avoided by using dry, HPLC grade solvents. Mobile phase polarity should be kept to a minimum. The CSP can be dry- or slurry-packed in a wide range of column sizes. Some of the compound classes which can be resolved by this chiral column include secondary benzyl alcohols, mandelic acid analogues, aryl hydroxy phosphonates, α-indonal and α-tetralol analogues, propranolols, aryl sulfoxide, aryl phthalides, aryl lactams, aryl succinimides, aryl hydantoins, bis-β-naphthol analogues, aryl acetamides, cyclic alcohols, and nitrogen heterocycles.

Available from J. T. Baker Chemical Co., 222 Red School Lane, Phillipsburg, N.J.

Hamilton Resin Columns

PRP-1 Reversed-Phase Columns

PRP-1 resin is a rigid macroreticular styrene-divinylbenzene copolymer with adsorptive properties and pore characteristics that produce an excellent stationary phase for HPLC. It is suitable for normal- as well as reversed-phase separations and functions well in gel permeation chromatography for mol wt of 2000 to 100,000. PRP-1 is a spherical macroporous adsorbent available in nominal 10-μm particle size, which results in low back pressures. The resin can be used over the pH range of 1 to 13, shows little or no band broadening, and has a high surface area. It works efficiently with salt concentrations up to 0.5 N. PRP-1 reversed-phase columns have been used to separate nucleosides and bases, antiepileptic and sulfa drugs, chlorophenols, carboxylic acids, catecholamines, analgesics, water-soluble vitamins, phenoxyacetic acid, and in ion pairing of organic and inorganic anions.

Available from Hamilton Company, P. O. Box 10030, Reno, Nev.

Donor-Acceptor Columns

Donor-acceptor complex chromatography (DACC) has become popular for the separation of aromatic molecules. A variety of stationary phases can attract solutes by donor-acceptor or charge-transfer mechanisms in normal or reversed-phase solvent systems. Their retention of donor molecules such as aromatic hydrocarbons, aromatic amino acids, peptides, vitamins, or nucleotides is probably due to their strong acceptor properties. Three DACC columns in high-coverage bonded phases, TCI (tetrachlorophthalimidopropyl), DNAP (dinitroanilinopropyl), and TENF (tetranitrofluoreniminopropyl) are available. TCI is useful for group-type separation of aromatic hydrocarbons in coal liquids and crude oil. DNAP and TENP silica show strong affinity for aromatics and have different selectivity than TCI.

Available from Applied Science.

PEI Coated Silica Gels

Two cross-linked polyethyleneimine (PEI) coated silica gel supports have recently been produced. PEI-Xama and PEI-glutaraldehyde silica gels readily achieve the low pressure quantitative separation of nucleotides. Baseline resolution is achieved for a mixture of adenosine, AMP, ADP, and ATP. Nucleotides resolved on these supports are recovered quantitatively. cAMP is not retarded on these PEI coated silica gels when AMP, ADP, and ATP bind strongly, an observation which forms the basis for an improved adenylate cyclase assay. The new supports are also excellent media for the separation of proteins.

PEI-Xama silica gel has high anion-exchange capacity, accessible exchange sites, and is predominantly hydrophilic in character. It is particularly useful for both nucleotide and protein separations.

PEI-glutaraldehyde silica gel can effect the same separations as PEI-Xama silica gel but larger column volumes are required due to a reduced ion-exchange capacity and slightly increased hydrophobicity. These features may be of advantage in difficult separations.

Available from Pierce Chemical Co.

Pirkle Chiral HPLC Column

The Pirkle Type 1-A column is packed with a spherical 5-μm-aminopropyl packing which has been modified by treatment with the N-3,5-dinitrobenzoyl derivative of D-phenylglycine. This column will separate optical isomers of alcohols, sulfoxides, bi-β-naphthols, and β-hydroxysulfides as well as the enantiomers of hydantoins, succinimides, and β-adrenergic agents related to propranolol. Using this column the enantiomeric purity and the absolute configuration of samples can be determined down to the subnanogram range. To prepare pure enantiomers, up to 10 mg of sample may be applied to the column at one time.

Pirkle Covalent Phenylglycine Columns

Methanol or even water may be used as solvent for the sample or as mobile phase with phenylglycine chiral separations. The efficiency of the covalent column equals or exceeds that of the ionic column. The covalent column separates not only the carbinol racemates with which ionic column level success was expected, but also other racemates with which such success was not expected. The covalent column may yield chiral separations that may not be as effective in some cases as those using the ionic column, but is available when solute solubilities demand a polar mobile phase. The change to a mobile phase may in some cases improve the relative retention.

Available from Phase Separation.

Table 6
PACKINGS FOR GEL PERMEATION CHROMATOGRAPHY

HIGH PERFORMANCE SIZE EXCLUSION COLUMNS

Zorbax® PSM Packings

DuPont produces microparticulate silica-based packings for size-exclusion separations. The Zorbax® PSM packings yield outstanding precision and accuracy in determining molecular weight parameters, i.e., number average molecular weights. They provide a linear calibration of log mol wt 10^2 to 10^6. This performance is achieved by connecting two columns filled with packings with properly matched pore sizes: Zorbax® PSM 60 and Zorbax® PSM 1000. The columns are available in unsilanized and silanized forms, the silanized forms being wettable with organic solvents. Zorbax® PSM 60 (silanized) is excellent for the separation of oligomers and other low molecular weight species. Other size exclusion columns are available. The PSM 300 and PSM 300S may be used for polymers with intermediate molecular weights up to 300,000. The SE 4000 is used for the analysis of polymers with very high molecular weights (10^5 to 10^7).

Typical applications are the determination of molecular weight parameters for natural and synthetic polymers: polyolefins, polyesters, amides, epoxies, rubbers, and acrylics. The separation of small molecules from polymeric matrixes and sample clean-up are other uses for these packings. All common organic solvents are compatible with these columns; aqueous mobile phases and alcohols may also be used. The silanized Zorbax® PSM packings are not wettable by aqueous mobile phases and the unsilanized packings should be employed with aqueous phases.

Available from E. I. DuPont de Nemours.

Gel Permeation Chromatography

Over a considerable period separation of macromolecules of biological and synthetic origin has been conducted based on their differences in molecular size. This technique is described as gel permeation chromatography or gel filtration chromatography. Soft hydrophilic gels have been used for the separation of biopolymers. Unfortunately these materials have relatively weak mechanical strength; they have a tendency to swell or shrink when the mobile phase is changed and the particles are too big to allow fast separations. Rigid small hydrophilic particles for aqueous gel permeation chromatography would therefore seem desirable. Chemically modified silica gels and other materials are now available for this purpose; those manufactured by Toyo Soda of Japan are TSK-Gel SW and TSK-Gel PW.

TSK-Gel SW

SW-type gel is a rigid, hydrophilic, spherical, and porous gel. Columns using this gel provide measurements at high flow rates; they have high theoretical plate numbers and large gel capacities. The gel is suitable for use in the analytical and preparative separation of proteins, enzymes, and saccharides. SW-type gel hardly shrinks or swells in solvents, and may thus be used with polar solvents other than water. The surface of the gel is covered with hydroxyl groups, which makes use as a base material, for example, as a carrier for affinity chromatography possible. The gel is used in the pH range 2.5 to 7.5, and the temperature at which it is employed should be kept below 45°C. Analytical type columns available are G 2000 SW, G 3000 SW, and G 4000 SW; preparative columns available are G 2000 SWG, G 3000 SWG, and G 4000 SWG.

TSK-Gel PW

In comparison with conventional soft gels, TSK-gel PW shows high mechanical strength, is resistant to pressure, and has a small particle size of 10 to 20 μm and a narrow size distribution. It has excellent physical and chemical stability. In comparison with TSK-gel SW it covers a wider molecular range, shows lower adsorption of water-soluble synthetic polymers, and is more resistant to high pH. TSK-gel PW is a fully porous, spherical semirigid gel of high porosity whose surface is covered with hydroxyl groups. It can be used over the pH range 2 to 12. PW-type columns are more strongly adsorptive to proteins than SW-type columns, but are more weakly adsorptive to water-soluble synthetic polymers. Aqueous solvents commonly used in conventional gel permeation are used with PW type columns. Analytical type columns available are G 1000 PW, G 2000 PW, G 3000 PW, G 4000 PW, G 5000 PW, and G 6000 PW, while preparative column types are G 1000 PWG, G 2000 PWG, G 3000 PWG, G 4000 PWG, G 5000 PWG, and G 6000 PWG.

BIO-RAD

Bio-Sil TSK Columns

Classical gel filtration materials are inert and capable of very high efficiency, but their major drawback is their low mechanical strength. This has prevented their use in the small particle sizes needed for high-speed high performance work. the alternatives to gels have been silica- or glass-based materials of controlled pore size distribution. These materials are not inert and the presence of adsorptive silanol groups means that they are unsuitable for most biochemical gel filtration work. The introduction of Bio-Sil TSK gel filtration columns for HPLC has resolved this conflict between the inert but soft and rigid but adsorptive materials. The principal advantage of Bio-Sil TSK is the increased speed of separation, which in most cases is reduced to less than 1 hr with no loss of resolution. Bio-Sil TSK packings use a hydrophilic bonded phase which eliminates residual silanol activity. Recoveries of protein from these packings is commonly greater than 95%. Column types TSK-125, TSK-250, and TSK-400 are available.

Bio-Gel® TSK Columns

Gel filtration of polycations is best carried out on Bio-Gel® TSK columns. These are packed with a hydroxylated polyether-based material comparable in mechanical strength and efficiency to Bio-Sil TSK, but with a slight positive rather than a slight negative charge. Column types available are TSK-10, TSK-20, TSK-30, TSK-40, TSK-50, and TSK-60.

Available from Bio-Rad Laboratories, 2200 Wright Avenue, Richmond, Calif.

MERCK

LiChrospher Si

LiChrospher Si types are completely synthetic silica gels with spherical particles. The series comprises types of graded pore diameters, which means that these materials may be used in high performance gel permeation chromatography. They have the advantage over organic polymers of being nonswellable. The LiChrospher Si types may also be used in adsorption of partition chromatography.

Available from E. Merck, Darmstadt, Federal Republic of Germany.

	Exclusion limit[a]
LiChrospher Si 100[b]	$5\text{---}8 \times 10^4$
LiChrospher Si 300	$1.5\text{---}3 \times 10^5$
LiChrospher Si 500	$3\text{---}6 \times 10^5$
LiChrospher Si 1000	$0.5\text{---}1.5 \times 10^6$
LiChrospher Si 4000	$2.5\text{---}8 \times 10^6$

[a] The exclusion limits, given in terms of molecular weights, were determined using linear polystyrenes and chloroform as the eluent and are intended as reference values.

[b] The number used in the type description, e.g., Si 100, indicates the approximate mean pore diameter of the silica gel lattice in Angstrom units.

Fractogel TSK

Fractogel TSK describes hydrophilic gels for gel permeation chromatography in aqueous media (gel filtration). The matrix consists of hydrophilic vinyl polymers, while the pore walls are formed from interwoven polymeric agglomerates. The surface of the gel particles, including the pore wall, has highly hydrophilic properties due to ether bonds and hydroxyl groups. Substances such as proteins dissolved in aqueous buffer and salt solutions are not generally adsorbed. The high mechanical stability of Fractogel TSK allows higher pressures to be applied in order to increase the flow rate, even in large pore gels. This is important in the gel filtration of labile substances as well as in longer columns. The small particle size and narrow particle size distribution affords high resolving power and sharp chromatographic peaks with only slight sample dilution. Fractogel TSK is chemically stable for pH 1 to 14 as well as in solutions containing urea, guanidinium chloride, or dodecyl hydrogen sulfate. It can also be used with hydrophilic organic solvents as eluents.

The Fractogel TSK series comprises gels of various pore diameters intended for the separation of biological substances ranging from small molecules up to very large biopolymers. Individual types are generally available in two particle size ranges which are characterized by narrow particle size classification, a minimum of 80% being within the declared limits.

Type F — Particle Size (Moist with Water) 0.032 to 0.063 mm and Type S — Particle Size (Moist with Water) 25 to 40 μm

Type S provides greater resolving power than type F at identical elution rates (necessitating a slightly higher pressure). Conversely, type S permits separations to be performed at a much higher elution rate with roughly comparable resolving power, providing that a correspondingly higher pressure is applied.

Fractogel TSK types available are as follows.

Fractogel TSK HW-40 (F) and (S)

Exclusion limit PEG (polyethylene glycol), 3000 ± 30%. Uses: gel chromatography of peptides and oligosaccharides. Fractionation range for proteins M = about 10^2 to 10^4 g/mol.

Fractogel TSK HW-50 (F) and (S)

Exclusion limit PEG, 18,000 ± 30%. Uses: gel chromatography of proteins (enzymes). Fractionation range M = about 5×10^2 to 10^5 g/mol.

Fractogel TSK HW-55 (F) and (S)

Exclusion limit PEG, 150,000 ± 30%. Uses: gel chromatography of proteins (enzymes). Fractionation range M = about 10^3 to 10^6 g/mol.

Fractogel TSK HW-65 (F)

Exclusion limit PEG, 1,0000,000 ± 30%. Uses: gel chromatography of nucleic acid and large proteins. Fractionation range for proteins M = about 5×10^4 to 5×10^6 g/mol.

Fractogel TSK HW-75 (F)

Exclusion limit PEG, more than 5,000,000. Uses: gel chromatography of large biopolymer molecules, particles, viruses, etc. Fractionation range for proteins M = about 5×10^5 to 5×10^7 g/mol.

BECKMAN

Spherogel TSK SW Size Exclusion Columns

Spherogel TSK SW columns provide minimal chemical interaction between the column and the compound of interest with a corresponding minimal loss of biological activity. The packing of these columns consists of small rigid spherical particles of bonded silica with a very narrow size distribution, which gives fast and high efficiency separations.

Column	Exclusion limits (mol wt)
Spherogel TSK 2000 SW	7.0×10^4
Spherogel TSK 3000 SW	3.0×10^5
Spherogel TSK 4000 SW	1.0×10^6

Spherogel TSK SWG columns are similar to TSK SW columns, but are preparative columns with increased loading capacity.

Available from Beckman Instrument Co.

WATERS

Ultrastyragel, μStyragel, and Styragel

These packing materials are made by cross-linking styrene and divinylbenzene to form small, rigid, gel particles with carefully controlled pore sizes. The pore size ratings from 100 to 10^6 Å cover a wide range of molecular weights. Separations achieved with Ultrastyragel and µStyragel columns include organic compounds (mol wt 100 to 2000) for sample clean-up, natural products, and synthetic organic reaction mixtures. The columns are also used for separating polymers such as acrylics, alkyds, amides/imides, butyl rubber, cellulosics, natural rubber, neoprene, nonionic surfactants, nylon, phenolics, polyacrylonitrile, polybutadiene, polycarbonates, polyethylene, polyglycols, polyisobutylene, polyisoprene, polypropylene, polystyrene, polysulfones, polyurethane, and silicones. Typical solvents for these columns include tetrahydrofuran, methylene chloride, toluene, and chloroform.

Ultrastyragel columns provide fast analysis and reproducibility. They are available in a variety of pore sizes to suit specific applications. The pore sizes of Ultrastyragel columns and µStyragel columns correspond exactly and calibration curves are equivalent. Ultrastyragel columns are compatible with solvents used with µStyragel columns. The increased resolving power of Ultrastyragel columns opens up new areas of application such as the baseline separation of closely related hydrocarbons. Other applications are oligomers, polymer additives, natural products, and polymers.

µStyragel columns are used for both analytical and milligram-scale preparative separations. Sample loading limit is 10 mg per column. These columns are recommended for use in high temperature analyses of polyolefins (polyethylene, polypropylene). In all other cases Ultrastyragel columns are recommended due to their higher efficiencies and more rapid analyses.

Styragel columns are available in a variety of prepacked column sizes.

µBondagel

µBondagel columns are fully porous silica gel with a bonded organic ether stationary phase. They offer a variety of pore sizes to separate a range of molecular weights and can be used with aqueous solvents. Typical solvents range from isooctane to water. The small column volume combined with the high efficiency of the packings permits high resolution in a few minutes. Typical applications include molecular weight characterizations of polymers such as dextrans, carboxymethylcellulose, film-forming polyesters, keratin proteins, lignin and ligno sulfonates, polyvinylalcohols, polyacrylic acid and sodium salts, polyamides polysaccharides (e.g., amylopectins), polysulfones, and sulfonated polystyrenes.

µPorasil 60 Å

This column is used to extend the range of size separations to very small molecules and is used to complement µBondagel columns. Although it forms part of the µPorasil silica family, once deactivated with water the µPorasil 60 Å column is used in the gel permeation chromatography mode.

PHARMACIA

Sephadex® G

Sephadex® is a bead-form cross-linked dextran gel which swells in aqueous salt solutions and in water. Gels of different porosities and with different fractionation ranges are available. Sephadex® is stable between pH 2 and 12 and in buffers containing urea and detergents. It will fractionate mixtures of proteins, peptides, and smaller nucleic acids and polysaccharides and can be used for the estimation of molecular weights, the estimation of equilibrium constants, and for desalting and buffer exchange. Sephadex® is produced in different bead sizes for different applications.

Sepharose

Sepharose is a bead-form agarose gel obtainable in three fractionation ranges, the fractionation range being determined by the agarose concentration in the beads. Sepharose is stable over the pH range 4 to 9 and is compatible with eluents containing high concentrations of salt, urea, and guanidine hydrochloride. The open structure of the agarose matrix makes Sepharose a good medium for the fractionation of protein complexes and polysaccharides. It is employed for immobilization of enzymes, antibodies, and hormones for affinity chromatography.

Sepharose CL

Sepharose CL is a cross-linked agarose gel for the separation of large molecules under strongly dissociating conditions. It is derived from Sepharose by cross-linking the individual chains in the matrix, resulting in greater physical and mechanical stability. Sepharose CL has a practically unlimited life in eluents like 6 M guanidine hydrochloride and can be used in organic solvents and at temperatures up to 70°C. It is stable over the range pH 3 to 14.

Sephacryl®

Sephacryl® is prepared by covalently cross-linking allyl dextran with N,N'-methylenebis acrylamide to give a very stable matrix. Gels are produced with exclusion limits from approximately 250,000 (Sephacryl® S-200) to several hundred million (Sephacryl® S-1000). In an additional step, residual active centers are removed, ensuring good recoveries of proteins and other biopolymers. Sephacryl® can be used in aqueous buffers, in concentrated urea or guanidine hydrochloride, and in some organic solvents. Sephacryl® S-200 and S-300 are useful for most proteins; Sephacryl® S-300 may be used for fractionating serum proteins. Sephacryl® S-400 and S-500 are excellent for polysaccharides, while Sephacryl® S-1000 fractionates restriction fragments of DNA, very large polysaccharides, proteoglycans, and membrane-bounded vesicles up to 300 to 400 nm in diameter.

Sephadex® LH-20 and LH-60

Sephadex® LH-20 and LH-60 are prepared from Sephadex® G-25 and G-50, respectively, by hydroxypropylation. They are stable in the range pH 2 to 10 in all aqueous and organic solvents in the absence of strong oxidizing agents. Sephadex® LH types have wide applicability in the fractionation of lipids, steroids, fatty acids, hormones, vitamins, and other small molecules. Apart from gel filtration in organic solvents, Sephadex® LH types are useful for adsorption and partition chromatography.

Sephasorb HP Ultrafine

Sephasorb is an adsorbent prepared by hydroxypropylation of cross-linked dextran. It has rigidity and a narrow range of bead sizes (10 to 23 μm). Sephasorb is ideal for preparative separations of a wide variety of small organic compounds, including aldehydes, phenols, ketones, nucleotides, and aromatic compounds. It can be used for either straight phase or reversed-phase chromatography.

Available from Pharmacia Laboratories, Inc.

Table 7
PREFORMED LAYERS FOR TLC

WHATMAN

Plate types are layer formulations designated by the letter K plus a numerical suffix, e.g., K5, K6, KC$_{18}$, etc. The type establishes product specifications. Most TLC plates are available with or without a zinc silicate phosphor for visualization under UV irradiation at 254 nm. Compounds absorbing UV radiation at this wavelength produce a dark spot in the brilliant green fluorescent background. Plates incorporating the fluorescent indicator are suffixed F, e.g., K6F. Analytical plates have a layer thickness of 250 μm, while preparative plates, which have the prefix letter P have a layer thickness of 500 or 1000 μm. L preadsorbent TLC plates incorporate a preadsorbent spotting area, introduced originally as the linear-K concept. These TLC plates are precoated glass with both an analytical layer (standard and high performance silica gel) and a preadsorbent area, or strip, a few centimeters wide. The preadsorbent strip accepts the sample. Sample components then undergo ''clean-up'' by extraction from insoluble substances in the specimen and the solvent front carries the concentrated samples to the silica gel area, presenting them as ''ideally'' applied samples.

Plate type	Adsorbent	Description
LK6	K6 silica gel	Analytical, with preadsorbent strip
LK6F	K6 silica gel + F	Same as LK6 plus fluorescent indicator
LK6D	K6 silica gel	Same as LK6 except scored into 8-cm wide channels
LK6DF	K6 silica gel + F	Same as LK6D plus fluorescent indicator
LK5	K5 silica gel	Analytical with preadsorbent strip
LK5F	K5 silica gel + F	Same as LK5 plus fluorescent indicator
LK5D	K5 silica gel	Same as LK5 except scored into 8-cm wide channels
LK5DF	K5 silica gel + F	Same as LK5D plus fluorescent indicator
PLK5	K5 silica gel	Preparative
PLK5F	K5 silica gel	Same as PLK5 plus fluorescent indicator
KC$_{18}$	ODS on silica gel	Analytical C$_{18}$ groups bonded (Si-O-Si-C) to special silica gel
KC$_{18}$F	ODS on silica gel + F	Same as KC$_{18}$ plus fluorescent indicator
MKC$_{18}$F	ODS on silica gel + F	Same as KC$_{18}$F except 1 × 3″ size
LKC$_{18}$	ODS on silica gel	Analytical, same as KC$_{18}$ plus preadsorbent strip
LKC$_{18}$F	ODS on silica gel + F	Same as LKC$_{18}$ plus fluorescent indicator
PLKC$_{18}$F	ODS on silica gel + F	Preparative
HP-KF	Special silica gel + F	HP analytical; high performance TLC plates; silica gel is ultra high purity, with an average particle size of 4.5 μm; it has an extremely narrow particle size distribution
LHP-K	Special silica gel	HP analytical with preadsorbent strip
LHP-KF	Special silica gel + F	Same as LHP-K plus fluorescent indicator
CS5	ODS/silica gel	Research analytical; surface has strip of C$_{18}$ groups bonded to silica gel, remaining surface is K5F silica gel; both adsorption and reversed-phase TLC are possible on a single plate; development can be 1- or 2-dimensional
K6	K6 silica gel	Analytical; inert organic binder
K6F	K6 silica gel + F	Same as K6 plus fluorescent indicator
MK6F	K6 silica gel + F	Same as K6F except 1 × 3″ size
PK6F	K6 silica gel + F	Preparative
K5	K5 silica gel	Analytical; wide polarity range separations
K5F	K5 silica gel + F	Same as K5 plus fluorescent indicator
PK5	K5 silica gel	Preparative
PK5F	K5 silica gel + F	Preparative

Plate type	Adsorbent	Description
K2	Cellulose	Analytical; exceptional speed, linearity, uniformity; no binder
K2F	Cellulose + F	Same as K2 plus fluorescent indicator
PK2F	Cellulose + F	Preparative

Available from Whatman Chemical Separation, Inc.

Microamide TLC Sheets

Microamide TLC sheets are used to achieve extremely sharp separations of dansyl-amino acids within extremely small distances. The microamide adsorbent (membrane) is available supported by aluminum (A 1700) or plastic (F 1700). The adsorbent, which contains no binder, covers both sides of the support, enabling separations and analyses to be done on both sides of the sheets. Sensitivity of 10^{-12} mol per amino acid is achieved on the F 1700. Excellent clarity of spots is achieved using the A 1700 because of the fluorescent-poor adsorbent and the fluorescent-free aluminum. The sensitivity of detection for A 1700 is 0.1 pmol (10^{-13} mol) for dansyl amino acids. Sample volumes can be below 0.1 μm and two-dimensional chromatograms can be done on a 1×1 cm foil.

Available from Schleicher and Schuell, Inc., 543 Washington Street, Keene, N.H.

MERCK

TLC Plates Pre-Coated

Silica gel 40 F_{254}
Silica gel 60 F_{254}
Silica gel 60 (without fluorescent indicator)
Silica gel 60/Kieselguhr F_{254}
Silica gel 100 F_{254}
Aluminum oxide 60 F_{254} (Type E)
Aluminum oxide 150 F_{254} (Type T)
Cellulose F
Cellulose (without fluorescent indicator)
PEI-cellulose F
PEI-cellulose (without fluorescent indicator)
Kieselguhr F_{254}
Silica gel 60 silanized (without fluorescent indicator)
Silica gel 60 F_{254} silanized
Polyamide 11 F_{254}

Available from E. Merck, Darmstadt, Federal Republic of Germany.

Table 8
ADSORBENTS FOR TLC

The principal adsorbents used in TLC are silica gels and aluminum oxides, followed by cellulose, magnesium silicate, and kieselguhr. In the Merck range the letter G denotes an adsorbent containing 10 to 15% of gypsum as binder. Grades designated H contain a fine-particulate silicon dioxide, but neither gypsum nor organic binder, those designated P are intended for preparative layer chromatography, while those designated R are particularly pure adsorbent. The letter F indicates the addition of a fluorescent indicator, the excitation wavelength of which is given as a subscript. The inorganic fluorescent indicator is a manganese-activated zinc silicate with an excitation wavelength of 254 nm. For normal use the indicator may be added with intensive mixing to the indicator-free adsorbent in 2% concentration. The fluorescent indicator F_{366} is an organic indicator with an excitation wavelength of 366 nm. This indicator may be partly eluted by some solvents. The particle size range for adsorbents for TLC is 5 to 40 μm. Adherence to a narrow particle size distribution during manufacture ensures optimum separations. In the range of Merck graded silica gels, differentiation is made between 3 standard types; silica gel 40, 60, and 100. The figure quoted directly after the adsorbent type shows the mean pore diameter in Angstroms. The special silica gel types 200, 500, and 1000 macroporous represent a progression of the series silica gel 40, 60, and 100; the figures again refer to the mean pore diameter.

Aluminum oxide 60 G neutral (type E)
Aluminum oxide 60 G F_{254} neutral (type E)
Aluminum oxide 60 H basic (type E)
Aluminum oxide 60 HF_{254} basic (type E)
Aluminum oxide 150 basic (type T)
Aluminum oxide 150 neutral (type T)
Aluminum oxide 150 acidic (type T)
Cellulose microcrystalline Avicel
Cellulose native
Silica gel 60 G
Silica gel 60 G F_{254}
Silica gel 60 H
Silica gel 60 HF_{254}
Silica gel 60 HF_{254} + 366
Silic gel 60 HR extra pure
Silica gel 200 macroporous
Silica gel 500 macroporous
Silica gel 500 macroporous
Silica gel 1000 macroporous
TLC Silica gel 60 G, mean particle size 15 μm
TLC Silica gel 60 GF_{254}, mean particle size 15 μm
TLC Silica gel 60 H, mean particle size 15 μm
TLC Silica gel 60 HF_{254}, mean particle size 15 μm
Silica gel 60 H silanized
Silica gel 60 HF_{254} silanized
Kieselguhr G

SCHLEICHER AND SCHUELL

Powders with a suffix designation /LS254 contain a fluorescent indicator which has a pale green fluorescence at 254 nm. Strongly acidic solvents such as mineral acids render the fluorescent indicator ineffective.

Cellulose double acid washed, 142 dg, particle size 10 to 50 μm
Cellulose micro-crystalline, 144 and 144/LS254, particle size about 20 μm
Cellulose natural fiber, 180 and 180/LS254, particle size 2 to 25 μm

Cellulose natural fiber, acid washed, 180a, particle size 2 to 25 μm
Cellulose acetylated 180/21 ac and 180/45 ac, particle size about 50 μm
Silica gel, 150 and 150/LS254, particle size less than 40 μm
Silica gel with gypsum, 150 G and 150G/LS254, particle size less than 40 μm; 12% gypsum is included as binder
Silica gel with starch, 150S and 150S/LS254, particle size less than 40 μm; 3% starch is included as binder

Ion-Exchange Powders
   The Schleicher and Schuell ion-exchange celluloses are finely powdered material which can be spread to a uniformly thin coating. The particle size is less than 30 μm and primarily less than 20 μm.

Anion Exchangers
DEAE cellulose, 66, capacity 0.9 ± 0.1 meq/g
ECTEOLA cellulose, 67, capacity 0.3 ± 0.1 meq/g
TEAE cellulose, 83, capacity 0.9 ± 0.1 meq/g
PEI cellulose, 84, capacity 0.3 ± 0.1 meq/g
QAE cellulose, 85, capacity 0.9 ± 0.1 meq/g

Cation Exchangers
CM cellulose, 68, capacity 0.7 ± 0.1 meq/g
P cellulose, 69, capacity 0.9 ± 0.1 meq/g

Table 9
CHROMATOGRAPHY PAPERS

SCHLEICHER AND SCHUELL

Analytical Grade
204a. This is a smooth-surfaced paper with a weight of 85 g/m² and a capillary rise of 110 mm.[a] It is used for general chromatographic work, especially descending chromatography.

2043a. This is a paper with an especially smooth, uniform surface, with a weight of 85 g/m² and a capillary rise of 110 mm. It is suitable for most chromatographic work.

2043b mgl. This is a smooth-surfaced paper with a weight of 120 g/m² and a capillary rise of 110 mm. It is suitable for most chromatography work, including circular chromatography and evaluation by the elution method.

591-C. This is a thin, smooth-textured paper with a weight of 82 g/m² and a capillary rise of 90 mm. It is used for electrophoresis and one and two-dimensional chromatography.

589 White Ribbon-C. This is a rough-surfaced paper with a weight of 89 g/m², a capillary rise of 88 mm, and a very low ash content. It is suitable for most chromatographic and electrophoretic work as well as for circular chromatography.

589 Black Ribbon-C. This is a rough-surfaced paper with a weight of 87 g/m², a capillary rise of 80 mm, and a very low ash content. It is suitable for chromatography when the R_F values are well separated.

[a]   Average rise in millimeters in distilled water in 30 min.

589 Orange Ribbon-C. This is a smooth-surfaced paper with a weight of 126 g/m^2, a capillary rise of 80 mm and a very low ash content. It is used for general chromatography and electrophoresis.

2045a. This is a smooth-surfaced paper with a weight of 90 g/m^2 and a capillary rise of 75 mm. It is used for circular and ascending chromatography with low absorption heights.

Preparative Grade
2668. This is a smooth-surfaced paper with a weight of 250 g/m^2 and a capillary rise of 320 mm. It is used for the separation of large molecules in electrophoresis.

470-C. This is a thick, smooth-surfaced paper with a weight of 335 g/m^2 and a capillary rise of 240 mm. It is used for electrophoresis and ascending chromatography of amino acids.

903-C. This is a soft, medium-thick, smooth-surfaced paper with a weight of 182 g/m^2 and a capillary rise of 200 mm. It is used for electrophoresis and chromatography.

2727. This is a thick, smooth-surfaced paper with a weight of 700 g/m^2 and a capillary rise of 200 mm. It is used for heavy loadings for preparative chromatography.

598-C. This is a rough-surfaced, moderately thick, open-textured paper with a weight of 157 g/m^2 and a capillary rise of 145 mm. It is used for preparative work.

589. Green Ribbon-C. This is a thick, rough-textured paper with a weight of 162 g/m^2, a capillary rise of 128 mm, and a very low ash content.

593-C. This is a smooth-surfaced, moderately thick paper with a weight of 171 g/m^2 and a capillary rise of 105 mm. It is used for the chromatographic and electrophoretic separation of amino acids and peptides.

2316. This is a smooth-surfaced paper with a weight of 165 g/m^2 and a capillary rise of 4 mm. It is used for the preparative separation of large sample quantities.

Impregnated Papers
2043 b hy and 2043 a hy. These are smooth-surfaced papers with weights of 120 g/m^2 and 85 g/m^2 and capillary rises of 30 and 27 mm, respectively. They are impregnated with silicone 1107 (Dowex®) and are used for reversed-phase paper chromatography of lipophilic substances such as amines, amino acids, fats, fatty acids, glycerides, peroxides, phenazines, and vitamins.

WHATMAN

Whatman chromatography papers are made from high purity alpha cellulose. Its low ash content is an indication of high purity and the absence of metallic contaminants, especially iron and copper. It contains no fluorescent particles under UV light since these impurities are excluded from the cellulose and the manufacturing process water. Chromatographic testing establishes standards of water flow linearity, uniformity of capillary action, and spot formation.

1 Chr. This is a thin paper, 0.16 mm thickness. Flow rate, 130 mm/30 min for water. White, smooth, normally hard surface. It is recommended for general purpose chromatography.

2 Chr. This is a thicker (0.18 mm) slightly slower paper than No. 1 Chr. Flow rate 115 mm/30 min for water. White, smooth, normally hard surface. It is recommended for optical and radiometric scanning.

3 Chr. This is a thick (0.38 mm) paper. Flow rate 130 mm/30 min for water. It has a white, rougher surface than No. 3 MM Chr. This paper is frequently used to separate inorganics and for electrophoresis. Its wet strength is higher than that of other chromatography papers, except No. 3 MM Chr, which is similar.

3MM Chr. This is a relatively thick (0.33 mm) paper with a flow rate of 130 mm/30 min for water. It has a white surface and high wet strength. This paper is widely used for electrophoresis.

4 Chr. This is the fastest thin paper (0.22 mm) with a flow rate of 180 mm/30 min for water. It has a white, smooth, normally hard surface and is suitable for routine and/or repetitive chromatography where loadings are relatively low.

17 Chr. This is a very thick (0.8 mm) paper with a flow rate of 190 mm/30 min for water. It is very absorbent, will accept heavy loadings and is recommended for preparative paper chromatography and electrophoresis.

20 Chr. This is a thin paper (0.16 mm) with a slow flow rate of 85 mm/30 min for water. It shows excellent resolution at low loadings. With a white, smooth, normally hard surface it is recommended for the separation of samples of unknown composition.

31 ET Chr. This is a thick (0.53 mm) extremely fast paper with a flow rate of 225 mm/30 min for water. It has a white, soft surface. Its principal use is for the electrophoresis of large molecules.

Available from Whatman Chemical Separation, Inc.

Ion-Exchange Cellulose Chromatography Papers

Chromatography papers manufactured from cellulose and ion-exchange cellulose fibers enable ion-exchange mechanisms to be combined with paper chromatography. These papers retain the convenience and handling characteristics of paper, while costs are lower than for preformed TLC plates or sheets. The advantage of ion-exchange papers lies in the fact that solvents are mainly aqueous in composition and therefore the system is less susceptible to difficulties such as incomplete tank equilibration, temperature changes, and paper humidification. Their pronounced advantage, however, derives from the fact that as solutes are held at fixed ionic sites, the ionic conditions for loading solutions may be chosen so as to allow solutes to have a high affinity for these sites. In this way high volumes may be loaded in a single application to give an origin spot of normal size.

Under conditions of ion-exchange development, the characteristics of the cellulose are largely masked. If, however, the ion-exchange characteristics are suppressed, separations are obtainable similar to those of unmodified cellulose. The conditions necessary for such separations are achieved by working at a pH at which either the exchange sites are no longer ionized or at which the components of the mixture being tested carry the same charge as the exchanger. It thus becomes possible to combine ion-exchange and paper techniques on the same sheet of paper in the form of a two-dimensional chromatogram. Separations on these papers are controlled by pH, salt concentration, and valency of the eluate and by dissociation constants, charge density and distribution, molecular size, and configuration of the substances being separated. Separations thus are subject to a high degree of control.

Ion-exchange cellulose papers are hydrophilic and have an open structure. Their rates of exchange and effective capacities for large ions such as proteins are consequently much greater than are attainable on cross-linked ion-exchange resins of high nominal (small ion) capacity. Anion exchangers should be used when the pH range is above the isoelectric points of the substances to be separated and cation exchangers when working in the pH range below the substances' isoelectric points.

DE 81. DE 81 is a weakly basic anion exchanger with diethylaminoethyl functional groups. It is a thin paper (0.20 mm) with a flow rate of 95 mm/30 min and an ion-exchange capacity of 1.7 μeq/cm^2. DE 81 is used for the separation of nucleotides and related substances, including complex nucleotide "fingerprints" by two-dimensional techniques.

P 81. P 81 is a cellulose phosphate paper. It is a strong cation exchanger with an ion exchange capacity of 18.0 μeq/cm^2. P 81 is a thin paper (0.23 mm) with a flow rate of 125 mm/30 min, which is useful for the separation of biogenic amines, antibiotics, histamines, and certain metals (Fe, Cu) at very low concentrations.

SG 81. This paper combines cellulose and large pore silica gel and can be used for separations in which both partition and adsorption are important mechanisms. It is a moderately thick (0.27 mm) paper with a flow rate of 110 mm/30 min. SG 81 has advantages over TLC for the routine quantification of phospholipids and is very efficient in the separation of polar lipids, steroids, phenols, dyes, natural pigments, and keto-acids.

MACHEREY-NAGEL AND CO.[a]

Chromatography Papers

MN 214	140[b]	0.28[c]	90—100[d]	Smooth[e]
MN 218	180	0.36	90—100	Smooth
MN 260	90	0.20	130—150	Smooth
MN 261	90	0.18	90—100	Smooth
MN 827	270	0.70	130—140[f]	Soft carton
MN 866	650	1.70	150—160[f]	Soft carton
MN 214 ff	140	0.28	90—100	MN 214 defatted
MN 214 AC-10	160	0.30		Acetylated MN 214, 10% acetyl content

[a] Machery-Nagel and Co., Werkstrasse 6-8, D-5160, Düren, Germany.
[b] Weight, g/m^2.
[c] Thickness, mm.
[d] Flow rate, mm/30 min.
[e] Characteristics.
[f] Flow rate, mm/10 min.

REFERENCES

1. **Kovats, E.** *Helv. Chim. Acta,* 42, 2709,1959.
2. **McReynolds, W. O.,** *J. Chromatogr. Sci.,* 8, 685, 1970.
3. **Betts, T. J., Finucane, G. J., and Tweedie, H. A.,** *J. Chromatogr.,* 213, 317, 1981.
4. **Preston, S. T.,** *J. Chromatogr. Sci.,* 8, 18A, 1970.
5. **Leary, J. J., Justice, J. B., Tsuge, S., Lowry, S. R., and Isenhour, T. L.,** *J. Chromatogr. Sci.,* 11, 201, 1973.
6. **Hawkes, S., Grossman, D., Hartkopf, A., Isenhour, T., Leary, J., Parcher, J., Wold, S., and Yancy, J.,** *J. Chromatogr. Sci.,* 13, 115, 1975.

Section VI
Chromatography Book Directory

Section VI

CHROMATOGRAPHY BOOK DIRECTORY

Selected books on chromatography are listed below. Readers will find reviews of new books on chromatography in, among others, *The Journal of Chromatography, The Journal of Chromatographic Science, The Journal of the American Chemical Society,* and *The Journal of Chemical Education.* Lists of recent books are published at regular intervals in the *Analytical Chemistry Lab Guide* and *The Journal of Chromatography.*

Academic Press Inc., 111 Fifth Avenue, New York, N.Y. 10003

1. **Brown, P. R.,** *High Pressure Liquid Chromatography (Biochemical and Biomedical Applications),* 1973.
2. **Dixon, P. F., Gray, C. H., Lim, C. K., and Stoll, M. S.,** *High Pressure Liquid Chromatography in Clinical Chemistry,* 1976.
3. **Hearn, M. T. W., Regnier, F. E., and Wehr, C. T., Eds.,** *High-Performance Liquid Chromatography of Proteins and Peptides. Proceedings of the First International Symposium,* 1983.
4. **Horvath, C., Ed.,** *High Performance Liquid Chromatography: Advances and Perspectives,* Vol. 1, 1980.
5. **Horvath, C., Ed.,** *High Performance Liquid Chromatography: Advances and Perspectives,* Vol. 2, 1980.
6. **Horvath, C., Ed.,** *High-Performance Liquid Chromatography: Advances and Perspectives,* Vol. 3, 1983.
7. **Jennings, W.,** *Gas Chromatography with Glass Capillary Columns,* 2nd ed., 1980.
8. **Lawrence, J. F.,** *Organic Trace Analysis by Liquid Chromatography,* 1981.
9. **Lawson, A. M., Lim, C. K., and Richmond, W., Eds.,** *Current Developments in the Clinical Applications of HPLC, GC and MS,* 1980.
10. **Ma, T. S. and Ladas, A. S.,** *Organic Functional Group Analysis by Gas Chromatography. (The Analysis of Organic Materials,* Belcher, R. and Anderson, D. M. W., Eds., Vol. 10), 1976.
11. **Walker, J. G., Jackson, M. T., Jr., and Maynard, J. B.,** *Chromatographic Systems — Maintenance and Troubleshooting,* 2nd ed., 1977.
12. **Zweig, G. and Sherma, J.,** *Paper Chromatography and Electrophoresis,* Vol. 2, 1971.

Ann Arbor Science Publishers Inc., P. O. Box 1425, Ann Arbor, Mich. 48106

1. **Budde, W. L. and Eichelberger, J. W.,** *Organics Analysis Using Gas Chromatography/Mass Spectrometry: A Techniques and Procedures Manual,* 1979.
2. **Grushka, E., Ed.,** *Bonded Stationary Phases in Chromatography,* 1974.
3. **Niederweiser, A. and Pataki, G., Eds.,** *Progress in Thin-Layer Chromatography and Related Methods,* Vol. 1, 1970.
4. **Niederweiser, A. and Pataki, G., Eds.,** *Progress in Thin-Layer Chromatography and Related Methods,* Vol. 2, 1971.
5. **Niederweiser, A. and Pataki, G., Eds.,** *Progress in Thin-Layer Chromatography and Related Methods,* Vol. 3, 1972.
6. **Neiderweiser, A. and Pataki, G., Eds.,** *New Techniques in Amino Acid, Peptide and Protein Analysis,* 1973.
7. **Scott, R. M. and Lundeen, M.,** *Thin-Layer Chromatography Abstracts,* 1974.

American Chemical Society, 1155 Sixteenth Street N.W., Washington, D.C., 20036

1. *Size Exclusion Chromatography (GPC),* ACS Symposium Series 121, 1980.

American Society for Testing and Materials (ASTM), 1916 Race Street, Philadelphia, Pa. 19103

1. Bibliography on Liquid Exclusion Chromatography (Gel Permeation Chromatography), AMD 40, 1972—1975.
2. Bibliography on Liquid Exclusion Chromatography (Gel Permeation Chromatography), AMD 40-S1, 1977.
3. Gas Chromatographic Data Compilation, AMD 25 A S1, 1971.
4. Liquid Chromatographic Data Compilation, AMD 41, 1975.

Applied Science Publishers Ltd., 22 Rippleside Commercial Estate, Barking Essex, England.

1. **Knapman, C. E. H., Ed.,** *Developments in Chromatography,* I, 1978.
2. **Knapman, C. E. H., Ed.,** *Developments in Chromatography,* II, 1980.

Avondale Division, Hewlett-Packard Co., Avondale, Pa.

1. **Rowland, F. R.,** *The Practice of Gas Chromatography,* 1973.

Barnes and Noble Inc., 105 Fifth Avenue, New York, N.Y. 10003

1. **Simpson, C.,** *Gas Chromatography,* 1970.

Chapman & Hall Ltd., 11 New Fetter Lane, London, E.C.4P 4EE, England

1. **Pryde, A. and Gilbert, M. T.,** *Applications of High Performance Liquid Chromatography,* 1979.
2. **Stock, R. and Rice, C. B. F.,** *Chromatographic Methods,* 3rd ed., 1974.

Marcel Dekker Inc., 270 Madison Avenue, New York, N.Y. 10016

1. **Blackburn, S., Ed.,** *Amino Acid Determination,* 2nd ed., revised and expanded, 1978.
2. **Cazes, J. and Delamare, X.,** *Chromatographic Science Series,* Vol. 13, 1980.
3. **Domsky, I. I. and Perry, J. A.,** *Recent Advances in Gas Chromatography,* 1971.
4. **Fried, B. and Sherma, J.,** *Thin Layer Chromatography, Techniques and Applications,* Vol. 17, Chromatographic Science Series, 1982.
5. **Giddings, J. C. and Keller, R. A., Eds.,** *Advances in Chromatography,* Vol. 10, 1974.
6. **Giddings, J. C. and Keller, R. A., Eds.,** *Advances in Chromatography,* Vol. 11, 1974.
7. **Giddings, J. C., Grushka, E., Keller, R. A., and Cazes, J., Eds.,** *Advances in Chromatography,* Vol. 12, 1975.
8. **Giddings, J. C., Grushka, E., Keller, R. A., and Cazes, J., Eds.,** *Advances in Chromatography,* Vol. 13, 1975.
9. **Giddings, J. C., Grushka, E., Cazes, J., and Brown, P. R., Eds.,** *Advances in Chromatography,* Vol. 14, 1976.
10. **Giddings, J. C., Grushka, E., Cazes, J., and Brown, P. R., Eds.,** *Advances in Chromatography,* Vol. 15, 1977.
11. **Giddings, J. C., Grushka, E., Cazes, J., and Brown, P. R., Eds.,** *Advances in Chromatography,* Vol. 16, 1978.
12. **Giddings, J. C., Grushka, E., Cazes, J., and Brown, P. R., Eds.,** *Advances in Chromatography,* Vol. 17, 1979.
13. **Giddings, J. C., Grushka, E., Cazes, J., and Brown, P. R., Eds.,** *Advances in Chromatography,* Vol. 18, 1980.
14. **Giddings, J. C., Grushka, E., Cazes, J., and Brown, P. R., Eds.,** *Advances in Chromatography,* Vol. 19, 1981.
15. **Giddings, J. C., Grushka, E., Cazes, J., and Brown, P. R., Eds.,** *Advances in Chromatography,* Vol. 20, 1982.
16. **Giddings, J. C., Grushka, E., Cazes, J., and Brown, P. R., Eds.,** *Advances in Chromatography,* Vol. 21, 1983.
17. **Gudzinowicz, B. J., Budzinowicz, M. J., and Martin, H. F.,** *Fundamentals of Integrated GC-MS. Part I. Gas Chromatography,* Vol. 7, Chromatographic Science Series, 1976.
18. **Gudzinowicz, B. J., Budzinowicz, M. J., and Martin, F. M.,** *Fundamentals of Integrated GC-MS. Part II. Mass Spectrometry,* Vol. 7, Chromatographic Science Series, 1976.
19. **Hawk, G. L., Champlin, P. B., Jordi, H. C., and Wenke, D., Eds.,** *Biological/Biomedical Applications of Liquid Chromatography,* Vol. 10, Chromatographic Science Series, 1979.
20. **Hawk, G. L., Ed.,** *Biological/Biomedical Applications of Liquid Chromatography III,* Vol. 18, Chromatographic Science Series, 1981.
21. **Irwin, W. J.,** *Analytical Pyrolysis: A Comprehensive Guide,* Vol. 22, Chromatographic Science Series, 1982.
22. **Jennings, W. G., Ed.,** *Applications of Glass Capillary Gas Chromatography,* 1981.

23. **Kautsky, M. P., Ed.,** *Steroid Analysis by HPLC: Recent Applications,* Vol. 16, Chromatographic Science Series, 1981.
24. **Novak, J.,** *Quantitative Analysis by Gas Chromatography,* Vol. 5, Chromatographic Science Series, 1976.
25. **Perry, J. A.,** *Introduction to Analytical Gas Chromatography. History, Principles and Practice,* Vol. 14, Chromatographic Science Series, 1981.
26. **Rajcsanyi, P. M. and Rajcsanyi, E.,** *High-Speed Liquid Chromatography,* 1975.
27. **Vickery, T. M., Ed.,** *Liquid Chromatography Detectors,* Vol. 23, Chromatographic Science Series, 1983.

Bowden, Hutchinson & Ross Inc., Stroudsburg, Pa.

1. **Walton, H. F., Ed.,** *Ion-Exchange Chromatography,* 1976.

Edinburgh University Press, 22, George Square, Edinburgh EG8 9LF, Scotland

1. **Knox, J. H., Done, J. N., Fell, A. F., Gilbert, M. T., Pryde, A., and Wall, R. A.,** *High-Performance Liquid Chromatography,* 1978.

Elsevier Scientific Publishing Co., P.O. Box 211, Amsterdam, The Netherlands
Elsevier/North-Holland Inc., 52 Vanderbilt Avenue, New York, N.Y. 10017

1. **Deyl, Z. and Kopecky, J., Eds.,** Bibliography of Liquid Column Chromatography, 1971—1973 and Survey of Applications (suppl. to *J. Chromatogr.*), 6, 1976.
2. **Deyl, Z., Macek, K., and Janak, J., Eds.,** *Liquid Column Chromatography,* Journal of Chromatography Library, Vol. 3, 1975.
3. **Drozd, J.,** *Chemical Derivatization in Gas Chromatography,* Journal of Chromatography Library, Vol. 19, 1981.
4. **Fischer, L.,** in *Gel Filtration Chromatography. Laboratory Techniques in Biochemistry and Molecular Biology,* Vol. 1, part 2, Wok, T. S. and Burdon, R. H., Eds., 1980.
5. **Fishbein, L.,** *Drugs of Abuse. Chromatography of Environmental Hazards,* Vol. 4, 1981.
6. **Frigerio, A.,** *Chromatography in Biochemistry, Medicine and Environmental Research,* Vol. 13, Analytical Symposia Chemistry Series, 1983.
7. **Frigerio, A., Ed.,** *Chromatography and Mass Spectrometry in Biomedical Sciences,* 2, Vol. 4: Analytical Chemistry Symposia, 1983.
8. **Frigerio, A. and McCamish, M., Eds.,** *Recent Developments in Chromatography and Electrophoresis,* Vol. 10, *Proc. 10th Int. Symp. Chromatogr. Electrophoresis,* Venice, 1980.
9. **Gribnau, T. C. J., Visser, J., and Nivard, R. J. F., Eds.,** *Affinity Chromatography and Related Techniques, 1981,* Vol. 9, Analytical Chemistry Symposia Series, 1981.
10. **Grob, R. L. and Kaiser, M.,** *Environmental Problem Solving Using Gas and Liquid Chromatography,* Vol. 21, Journal of Chromatography Library, 1982.
11. **Heftmann, E., Ed.,** *Chromatography. Fundamentals and Applications of Chromatographic and Electrophoretic Methods. Part A: Fundamentals and Techniques. Part B: Applications,* Vol. 22A and B, Journal of Chromatography Library, 1983.
12. **Huber, J. F. K., Ed.,** *Instrumentation for High-Performance Liquid Chromatography,* Vol. 13, Journal of Chromatography Library, 1978.
13. **Lawrence, J. F. and Frei, R. W.,** *Chemical Derivatization in Liquid Chromatography,* Vol. 13, Journal of Chromatography Library, 1978.
14. **Macek, K., Hais, I. M., Kopecky, J., Schwarz, V., Gaspario, J., and Churacek, J.,** Bibliography of Paper and Thin-layer Chromatography, 1970—1973 and Survey of Applications (suppl. to *J. Chromatogr.*), 5, 1976.
15. **Parris, N. A.,** *Instrumental Liquid Chromatography. A Practical Manual on High-Performance Liquid Chromatographic Methods,* Vol. 5, Journal of Chromatography Library, 1976.
16. **Roberts, T. R.,** *Radiochromatography. The Chromatography and Electrophoresis of Radio-Labelled Compounds,* Vol. 14, Journal of Chromatography Library, 1978.
17. **Scott, R. P. W.,** *Liquid Chromatography Detectors,* Vol. 11, Journal of Chromatography Library, 1977.
18. **Sevcik, J.,** *Detectors in Gas Chromatography,* Vol. 4, Journal of Chromatography Library, 1976.
19. **Turkova, J.,** *Affinity Chromatography,* Vol. 12, Journal of Chromatography Library, 1978.
20. **Unger, K. K.,** *Porous Silica. Its Properties and Use as Support in Column Liquid Chromatography,* Vol. 16, Journal of Chromatography Library, 1979.
21. **Zlatkis, A., Ed.,** *Advances in Chromatography, 1975, Proc. 10th Int. Symp.,* Munich, 1975.
22. **Zlatkis, A. and Ettre, L. S., Eds.,** *Advances in Chromatography, Proc. 9th Int. Symp.,* 1974 .

23. **Zlatkis, A. and Kaiser, R. E., Eds.**, *HPTLC-High Performance Thin-Layer Chromatography*, Vol. 9, Journal of Chromatography Library, 1977.
24. **Zlatkis, A. and Poole, C. F., Eds.**, *Electron Capture-Theory and Practice in Chromatography*, Vol. 20, Journal of Chromatography Library, 1981.

Walter de Gruyter Inc., 200 Saw Mill River Road, Hawthorne, N.Y. 10532

1. **Lottspeich, F., Henschen, A., and Hupe, K.-P., Eds.**, *High Performance Liquid Chromatography in Protein and Peptide Chemistry*, 1981.

Halsted Press, see John Wiley & Sons, Inc.

Heinemann Educational Books Ltd., London, England

1. **Allsop, R. T. and Healey, J. A. D.**, *Chemical Analysis, Chromatography and Ion Exchange*, 1974.

Heyden and Sons Inc., 247, South, 41st Street, Philadelphia, Pa. 19104
Heyden and Sons Ltd., Spectrum House, Alderton Crescent, London, NW4, England

1. **Blair, K. and King, G., Eds.**, *Handbook of Derivatives for Chromatography*, 1977.
2. **Kolb, B., Ed.**, *Applied Headspace Gas Chromatography*, 1980.
3. **Pattison, J. B.**, *A Programmed Introduction to Gas-Liquid Chromatography*, 1973.
4. **Simpson, C. F., Ed.**, *Practical High Performance Liquid Chromatography*, 1976.

Alfred Hüthig Verlag, Heidelberg, Basel, New York

1. **Bertsch, W., Hara, S., Kaiser, R. E., and Zlatkis, A., Eds.**, *Instrumental HPTLC*, 1980.
2. **Bertsch, W., Jennings, W. G., and Kaiser, R. E., Eds.**, *Recent Advances in Capillary Gas Chromatography*, 1981.
3. **Jennings, W.**, *Comparisons of Fused Silica and Other Glass Columns in Gas Chromatography*, 1981.
4. **Kaiser, R. E. and Oelrich, E.**, *Optimierung in der HPLC*, Chromatographische Methoden Series, 1979.
5. **Schwedt, G.**, *Chemische Reaktionsdetektoren für die Schnelle. Flüssigkeits-Chromatographie. Grundlagen und Anwendungen in der Spurenanalyse*, 1980.

E. Merck, Frankfurter Strasse 250, P. O. Box 4119, 6100 Darmstadt 1, Federal Republic of Germany

1. *Chromatography in Clinical Chemistry. Manual for the Practising Chromatographer.*
2. *Chromatography in Food Anaysis. Ready Reference Manual.*
3. *Chromatography in Pharmaceutical Chemistry. Manual for the Practising Chromatographer.*
4. *Dyeing Reagents for Thin Layer and Paper Chromatography.*
5. *Information on Thin-Layer Chromatography. 1. General Description of Procedure and Materials.*

Pergamon Press Inc., Maxwell House, Fairview Park, Elmsford, N.Y. 10523
Pergamon Press Ltd., Headington Hill Hall, Oxford OX3 OBW, England

1. **Angele, P.**, *Four-Language Technical Dictionary of Chromatography*, 1970.
2. **Jeffery, P. G. and Kipping, P. J.**, *Gas Analysis by Gas Chromatography*, 1972.

The Perkin-Elmer Corporation, Norwalk, Conn. 06856

1. **Ettre, L. S.**, *Introduction to Open Tubular Columns*, 1978.
2. **Rhys Williams, A. T.**, *Fluorescence Detection in Liquid Chromatography*, 1981.

Phase Separations Inc., River View Plaza, 16 River Street, Norwalk, Conn. 06850
Phase Separations Ltd., Deeside Industrial Estate, Queensferry, Clwyd CH5 2LR, Wales

1. *A Guide to the Analysis of Alcohols by Gas Chromatography*, 1976.
2. *A Guide to the Analysis of Amines by Gas Chromatography*, 1973.
3. *A Guide to the Analysis of Fatty Acids and their Esters by Gas Chromatography*, 1971.
4. *A Guide to the Analysis of Hydrocarbons by Gas Chromatography*, 1969.
5. *A Guide to the Analysis of Ketones by Gas Chromatography*, 2nd ed., 1975.
6. *A Guide to the Analysis of Pesticides by Gas Chromatography*, 2nd ed., 1969.
7. *A Guide to the Analysis of Phenols by Gas Chromatography*, 1978.
8. *A Guide to the Analysis of Thioalcohols and Thioethers (Mercaptans and Alkyl Sulfides) by Gas Chromatography*, 1980.

Plenum Publishing Corp., 227 West 17th Street, New York, N.Y. 10011

1. **Frei, R. W. and Lawrence, J. F., Eds.,** *Chemical Derivatization in Analytical Chemistry*, Vol. 1, 1981.
2. **Frei, R. W. and Lawrence, J. F., Eds.,** *Chemical Derivatization in Analytical Chemistry*, Vol. 2, 1982.
3. **Perry, S. G., Amos, R., and Brewer, P. I.,** *Practical Liquid Chromatography*, 1972.
4. **Signeur, A. V.,** *Guide to Gas Chromatography Literature*, Vol. 4, 1979.

Polyscience Corporation, 6366 Gross Point Road, Niles, Ill. 60648

1. **Preston, S. T.,** *A Guide to the Analysis of Amines by Gas Chromatography*, 1973.

Prentice-Hall Inc., Englewood Cliffs, N.J. 07632

1. **Khym, J. X.,** *Analytical Ion-Exchange Procedures in Chemistry and Biology, (Theory, Equipment, Techniques),* 1975.

Preston Publications Inc., P.O. Box 312, 6366 Gross Point Road, Niles, Ill. 60648

1. *Gas Chromatography Literature Abstracts and Index*, published monthly.
2. *Liquid Chromatography Literature, Abstracts and Index*, published bi-monthly.
3. **Kovats, E., Ed.,** *5th Int. Symp. Separation Methods: Column Chromatography*, 1970.
4. **Struppe, H. G., Ed.,** *Aspects in Gas Chromatography*, 1971.
5. **Walker, J. W., Ed.,** *Chromatographic Systems — Problems and Solutions*, 1980.

Regis Chemical Co., 8210 Austin Avenue, Morton Grove, Ill. 60053

1. *A User's Guide to Chromatography, Gas, Liquid, TLC*, 1976.

Supelco Inc., Bellefonte, Pa.

1. **Supina, W. R.,** *The Packed Column in Gas Chromatography*, 1974.

Van Nostrand-Reinhold Co., 450 West 33rd Street, New York, N.Y. 10001

1. **Heftmann, E., Ed.,** *Chromatography: A Laboratory Handbook of Chromatographic and Electrophoretic Methods*, 3rd ed., 1975.

Varian Instrument Group, 220 Humboldt Court, Sunnyvale, Calif. 94086

1. GC Applications Library, 1969—1975 (1977).
2. **Hadden, N., Baumann, F., MacDonald, F., Munk, M., Stevenson, R., Gore, D., Zamaroni, F., and Majors, R.,** *Basic Liquid Chromatography*, 1971.
3. **Johnson, E. L. and Stevenson, R.,** *Basic Liquid Chromatography*, 1978.

Whatman Chemical Separation Inc., 9 Bridewell Place, Clifton, N.J. 07014

1. Advanced Ion Exchange Celluloses Laboratory Manual.
2. **Olivier, R. W. A. and Corrie, M.,** *A Guide to the Literature of Chromatography and Electrophoresis.*

John Wiley & Sons, Inc., 605 Third Avenue, New York, N.Y. 10016
Wiley Interscience — a division of John Wiley & Sons Inc.

1. **David, D. J.,** *Gas Chromatographic Detectors,* 1974.
2. **Denney, R. C.,** *A Dictionary of Chromatography,* 2nd ed., 1982.
3. **Done, J. N., Knox, J. H., and Loheac, J.,** *Applications of High-Speed Liquid Chromatography,* 1974.
4. **Grob, R. L., Ed.,** *Modern Practice of Gas Chromatography,* 1977.
5. **Hamilton, R. J. and Sewell, P. A.,** *Introduction to High Performance Liquid Chromatography,* 1977.
6. **Kabra, P. and Marton, L., Eds.,** *Liquid Chromatography in Clinical Analysis,* 1981.
7. **Kirchner, J. G. and Perry, E. S., Eds.,** *Thin-Layer Chromatography,* 2nd ed., 1978.
8. **Knapp, D. R.,** *Handbook of Analytical Derivatization Reactions,* 1979.
9. **Kremmer, T. and Boross, L.,** *Gel Chromatography: Theory, Methodology, Applications,* 1979.
10. **Krstulovic, A. M. and Brown, P. R.,** *Reversed Phase High Performance Liquid Chromatography —
    Theory, Practice and Biomedical Applications,* 1982.
11. **Mikes, O., Ed.,** *Laboratory Handbook of Chromatography and Allied Methods,* Translated from Czech-
    oslovakian, Chalmers, R. A., transl. Ed., 1979.
12. **Miller, J. M.,** *Separation Methods in Chemical Analysis,* 1975.
13. **Runser, D. J.,** *Maintaining and Troubleshooting HPLC Systems. A User's Guide,* 1981.
14. **Sedlacek, B., Overberger, C. G., and Mark, H. F., Eds.,** *Polymer Catalysts and Affinants-Polymers
    in Chromatography,* 1980.
15. **Scott, R. P. W.,** *Contemporary Liquid Chromatography. Techniques of Chemistry,* Vol. 11, Weissberger,
    A., Ser. Ed., 1976.
16. **Scouten, W. H.,** *Affinity Chromatography. A Bioselective Adsorption on Inert Matrices.* Vol. 59, Chemical
    Analysis Series, Elving, D. J. and Winiforder, J. M., Ser. Eds., 1981.
17. **Simpson, C. F., Ed.,** *Techniques in Liquid Chromatography,* 1983.
18. **Smith, F. C. and Chang, R. C.,** *The Practice of Ion Chromatography,* 1982.
19. **Snyder, L. R. and Kirkland, J. J.,** *Introduction to Modern Liquid Chromatography,* 2nd ed., 1980.
20. **Touchstone, J. C. and Dobbins, M. F.,** *Practice of Thin Layer Chromatography,* 1978.
21. **Touchstone, J. C., Ed.,** *Quantitative Thin-Layer Chromatography,* 1973.
22. **Touchstone, J. V. and Rogers, B., Eds.,** *Thin-Layer Chromatography. Quantitative Environmental and
    Clinical Applications,* 1980.
23. **Touchstone, J. C. and Sherma, J., Eds.,** *Densitometry in Thin-Layer Chromatography: Practice and
    Applications,* 1979.
24. **Yau, W. W., Kirkland, J. J., and Bly, D. D.,** *Modern Size-Exclusion Liquid Chromatography: Practice
    of Gel Permeation and Gel Filtration Chromatography,* 1979.

Section VII
Reviews of Chromatographic Methods and Equipment

Section VII

REVIEWS OF CHROMATOGRAPHIC METHODS AND EQUIPMENT

This selected list of review articles supplements information that may be obtained by consulting appropriate books cited in the Book Directory.

1. **Abbot, S. R.,** Practical aspects of normal-phase chromatography, *J. Chromatogr. Sci.,* 18, 540, 1980.
2. **Amy, J. W.,** GC equipment for university laboratories, *J. Chromatogr. Sci.,* 20, 412, 1982.
3. **Aue, W. A. and Kapila, S.,** The electron capture detector-controversies, comments and chromatograms, *J. Chromatogr. Sci.,* 11, 255, 1973.
4. **Barry, E. F., Li, K. P., and Merritt, C., Jr.,** Behavior of chromatography columns with a varying stationary phase. I. Segmented liquid phase loadings, *J. Chromatogr. Sci.,* 20, 357, 1982.
5. **Bayer, F. L.,** An overview of chromatographic instrumentation: problems and solutions, *J. Chromatogr. Sci.,* 20, 393, 1982.
6. **Blades, A. T.,** The flame ionization detector, *J. Chromatogr. Sci.,* 11, 251, 1973.
7. **Bristow, P. A. and Knox, J. H.,** Standardization of test conditions for high performance liquid chromatography columns, *Chromatographia,* 10, 279, 1977.
8. **Brown, P. R. and Krstulovic, A. M.,** Practical aspects of reversed-phase liquid chromatography applied to biochemical and biomedical research, *Anal. Biochem.,* 99, 1, 1979.
9. **Chandler, C. D. and McNair, H. M.,** High pressure liquid chromatography equipment, *J. Chromatogr. Sci.,* 11, 468, 1973.
10. **Colin, H. and Guichon, G.,** Introduction to reversed-phase high-performance liquid chromatography, *J. Chromatogr.,* 141, 289, 1977.
11. **Cooks, N. H. C. and Olsen, K.,** Some modern concepts in reversed-phase liquid chromatography on chemically bonded alkyl stationary phases, *J. Chromatogr. Sci.,* 18, 512, 1980.
12. **Cooper, A. R. and Van Deerveer, D. S.,** Recent advances in aqueous gel permeation chromatography, *J. Liq. Chromatogr.,* 1, 693, 1978.
13. **Davankov, V. A. and Semechkin, A. V.,** Ligand-exchange chromatography, *J. Chromatogr.,* 141, 313, 1977.
14. **Davis, L. C.,** A simplified calibration procedure for elution chromatography on gel filtration columns, *J. Chromatogr. Sci.,* 21, 214, 1983.
15. **Deyl, Z.,** Advances in separation techniques in sequence analysis of proteins and peptides, *J. Chromatogr.,* 127, 91, 1976.
16. **Deyl, Z. and Kopecky, J., Eds.,** Bibliography of Liquid Column Chromatography 1971—1973 and Survey of Applications (suppl. to *J. Chromatogr.,* 6, 1976).
17. **Drozd, J.,** Chemical derivatization in gas chromatography, *J. Chromatogr.,* 113, 303, 1975.
18. **Drushel, H. V.,** Needs of the chromatographer-detectors, *J. Chromatogr. Sci.,* 21, 375, 1983.
19. **Dupré, G. D. and Schulz, W. W.,** GC and LC: ideals for the future, *J. Chromatogr. Sci.,* 20, 423, 1983.
20. **Farwell, S. O., Gage, D. R., and Kagel, R. A.,** Current status of prominent selective gas chromatographic detectors: a critical assessment, *J. Chromatogr. Sci.,* 19, 358, 1981.
21. **Ford-Holevinski, T. S., Agranoff, B. W., and Radin, N. S.,** An inexpensive, microcomputer-based, video densitometer for quantitating thin-layer chromatographic spots, *Anal. Biochem.,* 132, 132, 1983.
22. **Haken, J. K.,** Polysiloxane stationary phases in gas chromatography, *J. Chromatogr.,* 141, 247, 1977.
23. **Hurtubise, R. J., Lott, P. F., and Dias, J. R.,** Instrumentation of thin layer chromatography, *J. Chromatogr. Sci.,* 11, 476, 1973.
24. **Ishii, D. and Takeuchi, T.,** Open tubular capillary LC, *J. Chromatogr. Sci.,* 18, 462, 1980.
25. **Jennings, W.,** Some aspects of troubleshooting in capillary GC, *J. Chromatogr. Sci.,* 21, 337, 1983.
26. **Johns, T. and Stapp, A. C.,** Modern thermal conductivity detectors for gas chromatographs, *J. Chromatogr. Sci.,* 11, 234, 1973.
27. **Keller, R. A.,** Principles of detection, *J. Chromatogr. Sci.,* 11, 223, 1973.
28. **Kirchner, J. G.,** Modern techniques in TLC, *J. Chromatogr. Sci.,* 13, 558, 1975.
29. **Kipiniak, W.,** A basic problem — the measurement of height and area, *J. Chromatogr. Sci.,* 19, 332, 1981.
30. **Landy, J. S., Ward, J. L., and Dorsey, J. G.,** A critical evaluation of some stainless steel and radially compressed reversed-phase HPLC columns, *J. Chromatogr. Sci.,* 21, 49, 1983.
31. **Little, C. J., Whatley, J. A., and Dale, A. D.,** Detection involving post-chromatographic addition of reagents, *J. Chromatogr.,* 171, 63, 1979.
32. **Lochmüller, C. H. and Souter, R. A.,** Chromatographic resolution of enantiomers. Selective review, *J. Chromatogr.,* 113, 283, 1975.

33. **Lott, P. F., Dias, J. R., and Hurtubise, R. J.,** Instrumentation for thin layer chromatography — an update, *J. Chromatogr. Sci.,* 14, 488, 1976.
34. **Lott, P. F., Dias, J. R., and Slakck, S. C.,** Instrumentation for thin layer chromatography. 1978 update, *J. Chromatogr. Sci.,* 16, 571, 1978.
35. **Lundanes, E., Døhl, J., and Greibrokk, T.,** Guard columns in HPLC. An examination of the effect of MPLC cartridge columns of column efficiencies and some theoretical aspects of the use of guard columns, *J. Chromatogr. Sci.,* 21, 235, 1983.
36. **Lyons, J. W. and Faulkner, L. R.,** Optimization of flow cells for fluorescence detection in liquid chromatography, *Anal. Chem.,* 54, 1960, 1982.
37. **Macek, K., Hais, I. M., Kopecky, J., Schwarz, V., Gasparic, J., and Churacek, J.,** Bibliography of Paper and Thin-layer Chromatography 1970—1973 and Survey of Applications (suppl. *J. Chromatogr.,* 5, 1976).
38. **McNair, H. M. and Chandler, C. D.,** Gas chromatography equipment, *J. Chromatogr. Sci.,* 11, 454, 1973.
39. **McNair, H. M.,** Gas chromatography equipment. II., *J. Chromatogr. Sci.,* 16, 578, 1978.
40. **McNair, H. M. and Chandler, C. D.,** High performance liquid chromatography equipment. II., *J. Chromatogr. Sci.,* 14, 477, 1976.
41. **McNair, H. M. and Chandler, C. D.,** High performance liquid chromatography equipment. III., *J. Chromatogr. Sci.,* 14, 477, 1976.
42. **McNair, H. M.,** High performance liquid chromatography equipment. IV., *J. Chromatogr. Sci.,* 16, 588, 1978.
43. **McNair, H. M.,** Equipment for HPLC. V., *J. Chromatogr. Sci.,* 20, 537, 1982.
44. **McNair, H. M.,** Gas chromatography equipment. III., *J. Chromatogr. Sci.,* 21, 529, 1983.
45. **Majors, R. E.,** High performance liquid chromatography columns and column technology. A state-of-the-art review. I., *J. Chromatogr. Sci.,* 18, 393, 1980; II., *J. Chromatogr. Sci.,* 18, 487, 1980.
46. **Majors, R. E.,** Recent advances in HPLC packings and columns, *J. Chromatogr. Sci.,* 18, 488, 1980.
47. **Novotny, M.,** Capillary HPLC: columns and related instrumentation, *J. Chromatogr. Sci.,* 18, 473, 1980.
48. **Nurok, D., Becker, R. M., and Sassic, K. A.,** Time optimization in thin-layer chromatography, *Anal. Chem.,* 54, 1955, 1982.
49. **Ogan, K. L., Reese, C., and Scott, R. P. W.,** Strong, flexible soft-glass capillary columns; a practical alternative to fused silica, *J. Chromatogr. Sci.,* 20, 425, 1982.
50. **O'Hare, M. J., Capp, M. W., Nice, E. C., Cooke, N. H. C., and Archer, B. G.,** Factors influencing chromatography of proteins on short alkylsilane-bonded large pore-size silicas, *Anal. Biochem.,* 126, 17, 1982.
51. **Parcher, J. F.,** A review of vapor phase chromatography: gas chromatography with vapor carrier gases, *J. Chromatogr. Sci.,* 21, 346, 1983.
52. **Polesuk, J. and Howery, D. G.,** Chromatographic detection, *J. Chromatogr. Sci.,* 11, 226, 1973.
53. **Pohl, C. A. and Johnson, E. L.,** Ion chromatography — the state-of-the-art, *J. Chromatogr. Sci.,* 18, 442, 1980.
54. **Reese, C. E. and Scott, R. P. W.,** Microbore LC column technology, *J. Chromatogr. Sci.,* 18, 479, 1980.
55. **Regnier, F. E.,** High performance ion-exchange chromatography of proteins: the current status, *Anal. Biochem.,* 126, 1, 1982.
56. **Regnier, F. E. and Gooding, K. M.,** High-performance liquid chromatography of proteins, *Anal. Biochem.,* 103, 1, 1980.
57. **Rosie, D. M. and Barry, E. F.,** Quantitation of thermal conductivity detectors, *J. Chromatogr. Sci.,* 11, 237, 1973.
58. **Ross, M. S. F.,** Pre-column derivatisation in high-performance liquid chromatography, *J. Chromatogr.,* 141, 107, 1977.
59. **Rudzinski, W. E., Bennett, D., Gracia, V., and Seymour, M.,** Influence of mobile phase and role of added electrolyte in retention of ionized solutes in HPLC, *J. Chromatogr. Sci.,* 21, 57, 1983.
60. **Seiler, N.,** Chromatography of biogenic amines. I. Generally applicable separation and detection methods, *J. Chromatogr.,* 143, 221, 1977.
61. **Scholten, A. H. M. T., Brinkman, U. A. Th., and Frei, R. W.,** Comparison of liquid segmented with nonsegmented systems in postcolumn reactors for liquid chromatography, *Anal. Chem.,* 54, 1932, 1982.
62. **Scott, C. G.,** Quantitation in chromatography — introduction, *J. Chromatogr. Sci.,* 19, 331, 1981.
63. Subcommittee E — 19.08 Task group on liquid chromatography of the American Society for Testing and Materials (ASTM). An evaluation of quantitative precision in high performance liquid chromatography, *J. Chromatogr. Sci.,* 19, 338, 1981.
64. **Telepchak, M. J.,** The mechanism of reverse phase liquid-solid chromatography, *Chromatographia,* 6, 234, 1973.

65. **Verzele, M. and Geeraert, E.,** Preparative liquid chromatography, *J. Chromatogr. Sci.,* 18, 571, 1980.
66. **Wolf, T., Fritz, G. T., and Palmer, L. R.,** An ASTM standard practice for testing fixed-wavelength photometric detectors used in liquid chromatography, *J. Chromatogr. Sci.,* 19, 387, 1981.
67. **Wood, R., Cummings, L., and Jupille, T.,** Recent developments in ion-exchange chromatography, *J. Chromatogr. Sci.,* 18, 551, 1980.

Section VIII
Indexes

Section VIII

AUTHOR INDEX

A

Abbott, S. R., 67, 96, 213 (ref. 33), 236 (ref. 7), 341 (ref. 1)
Abrahamsson, M., 68
Abramson, F. B., 132
Acharya, A. S., 81, 213 (ref. 26)
Agranoff, B. W., 341 (ref. 21)
Allsop, R. T., 336 (ref. 1)
Amos, R., 337 (ref. 3)
Amy, J. W., 341 (ref. 2)
Angele, P., 336 (ref. 1)
Aoyagi, H., 163
Araujo-Viel, M., 175
Archer, B. G., 102, 342 (ref. 50)
Arendt, A., 190, 248
Arison, B. H., 164
Atassi, M. Z., 272 (ref. 1)
Aue, W. A., 341 (ref. 3)
Auger, G., 181

B

Bairoch, A., 8
Bakkum, J. T. M., 58
Balasubramanian, T. M., 89
Bali, J.-P., 133, 178
Banovsky, J. M., 278 (ref. 1)
Barlow, G. B., 94
Barnes, L. D., 73
Barry, E. F., 341 (ref. 4), 342 (ref. 57)
Battersby, J. E., 49, 213 (ref. 29)
Baumann, F., 337 (ref. 2)
Bayer, F. L., 341 (ref. 5)
Bennett, C. D., 164
Bennett, D., 342 (ref. 59)
Bennett, H. P. J., 212 (ref. 25), 236 (ref. 8, 19)
Berchtold, M., 212 (ref. 3), 236 (ref. 10)
Berger, A., 276 (ref. 1)
Bertsch, W., 336 (ref. 1, 2)
Betto, R., 129
Betts, T. J., 282, 283, 329 (ref. 3)
Beyerman, H. C., 51—54, 58, 184
Bielanski, W., Jr., 224 (ref. 25)
Biemann, K., 7
Biggins, J. A., 71
Bij, K. E., 201 (ref. 11), 212 (ref. 11)
Biral, D., 129
Bishop, C. A., 46—49, 212 (ref. 4, 21), 213 (ref. 28, 29), 229 (ref. 4), 236 (ref. 6)
Blackburn, S., 334 (ref. 1)
Blades, A. T., 341 (ref. 6)
Blair, K., 336 (ref. 1)
Blanot, D., 181
Blevins, D. D., 212 (ref. 8)
Bly, D. D., 338 (ref. 24)
Bodanszky, M., 184

Bohlen, P., 208, 213 (ref. 43), 269 (ref. 1)
Bonora, G. M., 186
Borin, G., 185, 192
Boross, L., 338 (ref. 9)
Bosserhoff, A., 213 (ref. 45), 224 (ref. 18)
Bosshard, H. R., 276 (ref. 1)
Brack, A., 193
Brady, S. F., 164
Braunitzer, G., 277 (ref. 1)
Brewer, P. I., 337 (ref. 3)
Bricas, E., 181
Brinkman, U. A. Th., 342 (ref. 61)
Bristow, P. A., 341 (ref. 7)
Brown, P. R., 333 (ref. 1), 334 (ref. 9—16), 338 (ref. 10), 341 (ref. 8)
Browne, C. A., 212 (ref. 25), 232, 235, 236 (ref. 8, 19)
Brugger, M., 254, 256
Brunati, A. M., 192
Brunschweiler, K., 228 (ref. 1)
Buckley, D. I., 78
Budde, W. L., 333 (ref. 1)
Budzinowicz, M. J., 334 (ref. 17, 18)
Burbach, J. P. H., 67, 125, 270 (ref. 1)
Burke, M. F., 59, 60, 212 (ref. 8)
Bush, C. A., 44

C

Cahill, W. R., Jr., 114—116, 236 (ref. 20)
Caille, A., 193
Capp, M. W., 102, 212 (ref. 9), 236 (ref. 16), 342 (ref. 50)
Caprioli, R. M., 275 (ref. 1)
Carrella, M., 264 (ref. 1)
Castillo, F., 269 (ref. 1)
Cazes, J., 334 (ref. 2, 7—16)
Champlin, P. B., 334 (ref. 19)
Chandler, C. D., 341 (ref. 9), 342 (ref. 38, 40, 41)
Chang, J.-K., 63, 65
Chang, J.-Y., 271 (ref. 1)
Chang, R. C., 338 (ref. 18)
Chang, S. Y., 61
Chang, W.-C., 66, 169
Chan Leuthauser, S. W., 151
Chen-Kiang, S., 212 (ref. 17), 229 (ref. 7)
Chessa, G., 192
Chipens, G. I., 176
Choudhury, A. M., 179
Chretien, M., 224 (ref. 9)
Christen, P., 228 (ref. 1)
Chu, I. C., 40
Churacek, J., 335 (ref. 14), 342 (ref. 37)
Chyad Al-Noaemi, M., 71
Clements, P., 264 (ref. 1)
Cochran, D. W., 164
Codd, E. E., 125

COMPOUND/SUBJECT INDEX

HPLC mapping techniques, 227—228
HPLC packings
  Bondapak, 310
  carbohydrate analysis column, 311
  Chromosorb, 301—302
  energy analysis (NH₂) column, 311—312
  fatty acid analysis column, 311
  Florisil, 299—300
  Hypersil, 308—310
  ion exchangers, 305—306
  LiChrosorb, 298—299
  Partisil, 302—304
  Perisorb, 299
  Porasil, 310
  protein analysis columns, 311
  Protesil, 304—305
  Resolve, 310
  Resolve silica column, 311
  Spherisorb, 307—308
  Sugar-pak 1 column, 311
  triglyceride analysis column, 311
  Ultrapore RPSC column, 307
  Ultrasphere, 306—307
  Zorbax, 300—301
HPLC supports, 312—316
H-Pro-Ser-NH₂·TFA, 171
(Hse¹³)-S-peptide (1-13)lactone, 89
H-Ser(Bzl)-Asn-OBzl·HCl, 193
H-Ser(Bzl)-Gln-Gly-Gly-Ser(Bzl)-Asn-OBzl·HCl, 193
H-Ser-(Bzl)-Gln-Gly-OH·HCl, 193
H-Ser(Bzl)-Gln-OH·HCl, 194
H-Ser(Bzl)-Pro-Phe-ONb.TFA, 176
H-Ser-Gln-Gly-Gly-Ser-Asn-OH, 180
H-Ser-HN₂·HCl, 171
H-Thr-Gly-Trp-Leu-Asp-Phe-NH₂, 178
H-Trp-Leu-Asp-Phe-NH₂, 178
H-Trp-Leu-OH, 256
H-Tyr-D-Ala-Phe-Gly-Tyr-Hyp-Ser-NH₂·HCl, 172
H-Tyr-Ala-Phe-Gly-Tyr-Pro-Ser-NH₂·TFA, 172
H-Tyr-D-Ala-Phe-Gly-Tyr-Pro-Ser-NH₂·TFA, 172
H-Tyr(Bzl)-Hyp-Ser(Bzl)-NH₂·TFA, 171
H-Tyr(Bzl)-Pro-Ser-NH₂·HCl, 171
H-Tyr(Bzl)-Pro-Ser-NH₂·TFA, 171
H-Tyr-Pro-Ser-NH₂·TFA, 171
Human albumin, 101, 222
Human placental lactogen, 101
Human transferrin GP, 221
Hydrazide, S-peptide analogue intermediates, 185
cis-Hydrindan, 283
Hydrocarbonaceous silicas, 203
Hydrocarbon phase, 299
Hydrogel IV, 32, 217
Hydrophilic ion pairing, 227
Hydrophilic supports, 200
Hydrophobic effect, 201
Hydrophobic interactions, 217
Hydrophobic ion pairing, 47—48
Hydrophobicity constants, 231—234
Hydrophobic peptides, 203, 204, 219

Hydroxyl groups, see specific HO peptides and OH peptides
8-Hydroxyquinoline (8-HQ) support, 314
N-Hydroxysuccinimide (NHS) glycophase, 313—314
Hyp, 7
Hypersil
  chiral separation of dansyl peptides, 90
  L,L-dipeptide reversed phase chromatography on ODS-Hypersil, 35—37
  ribosomal protein L 29, 209
Hypersil, Silica, 308—309
Hypersil-ODS, 99—101
Hypersil packing materials, 308—310

I

Igepal Co-880, 284
Ile, 7
Ile-Ala, 41, 152
Ile-Glu, 41, 152
Ile-Gly
  ion exchange chromatography, 41
  TLC, 136, 137, 139, 147, 152
(Ile¹³,Hse²⁰)-S-peptide, 89
(Ile¹³,Hse²⁰)-S-peptide-(Gly-Gly), 89
(Ile¹³,Hse²⁰)-S-peptide lactone, 89
Ile-Leu, 41, 152
Ile-Lys, 41, 152
Ile-Met, 41, 152
(Ile¹³,Met²⁰)-S-peptide-(Gly-Gly), 89
(Ile¹³,Met(O)²⁰)-S-peptide-(Gly-Gly), 89
Ile-Phe, 41, 152
Ile-Pro, 41, 152
Ile-Ser, 41, 152
Ile-Trp, 41, 152
L-Ile-L-Tyr, 23
Ile-Val, 41, 152
Immunoglobulin E peptide III, 257
Immunoglobulin G, 186
Indole, 103
Insulin, 87, 100, 145, 221
  capacity ratios, predicted vs. observed, 94
  gel permeation chromatography, 98, 103, 106, 216
  high performance gel permeation chromatography, 98
  HPLC, 45, 97, 204
  TLC, 182
Insulin A chain, 98, 100
Insulin B chain, 94, 100, 145, 221
Iodine-125 analysis, 103, 105—106
Iodine-125 labeled peptides
  angiotensin, 72
  endorphins, 65—66, 119
  techniques, 270, 273—274
  protein, 218
Iodine vapor detection, 176, 178
Iodoacetamide, 220
Iodo-beads, 273

M

U

V

QP
552
P4B55
1986
V. 1
CHEMISTRY
LIBRARY

RETURN   CHEMISTRY LIBRARY   0150
RETURN   CHEMISTRY LIBRARY
          00 Hildebrand Hall   6443753
          2   3

LIBRARY USE ONLY